Molecular Aspects of the Stress Response:
Chaperones, Membranes and Networks

ADVANCES IN EXPERIMENTAL MEDICINE AND BIOLOGY

A Continuation Order Plan is available for this series. A continuation order will bring delivery of each new volume immediately upon publication. Volumes are billed only upon actual shipment. For further information please contact the publisher.

Molecular Aspects of the Stress Response: Chaperones, Membranes and Networks

Edited by

Peter Csermely

Department of Medical Chemistry, Semmelweis University, Budapest, Hungary

László Vígh

Institute of Biochemistry, Biological Research Center of the Hungarian Academy of Sciences, Szeged, Hungary

Springer Science+Business Media, LLC

Landes Bioscience / Eurekah.com

Springer Science+Business Media, LLC
Landes Bioscience / Eurekah.com

Printed in the U.S.A.

Springer Science+Business Media, LLC, 233 Spring Street, New York, New York 10013, U.S.A.

Please address all inquiries to the publishers:
Landes Bioscience / Eurekah.com, 1002 West Avenue, Second Floor, Austin, TX 78701, U.S.A.
Phone: 512/ 637 6050; FAX: 512/ 637 6079
http://www.eurekah.com
http://www.landesbioscience.com

Molecular Aspects of the Stress Response: Chaperones, Membranes and Networks, edited by Peter
Csermely and László Vígh, Landes Bioscience / Eurekah.com / Springer Science+Business Media,
LLC dual imprint / Springer series: Advances in Experimental Medicine and Biology

ISBN-10: 0-387-39974-7
ISBN-13: 978-0-387-39974-4

Library of Congress Cataloging-in-Publication Data

A C.I.P. Catalogue record for this book is available from the Library of Congress.

PREFACE

We are extremely happy to present the reader this book containing a summary of a well-known research field, the phenomenon of cellular stress defense from two new angles: networks and membranes. The volume starts with an introduction to the concept of molecular chaperones in their original sense: R. John Ellis, the founder of the chaperone concept describes chaperones as mediators of correct assembly and/or misassembly of other macromolecular complexes. This sets the tone of the book, where later chapters give detailed examples of the richness of chaperone action by hundreds of other proteins and membrane structures.

The reader will learn the role of chaperone classes such as Hsp27 or Hsp90, the action of highly organized chaperone networks in various cellular compartments such as the ER or mitochondrial/ER networks as well as the molecular details of the signaling mechanisms leading to chaperone induction during stress. Various special stress defense mechanisms against oxidative stress or dryness will also be covered.

Membranes comprise a surprising mixture of stability and dynamics in the cell. Their role in the regulation of the stress response has been accepted only slowly in the field. Two chapters summarize this important aspect of the stress response showing the importance of membrane hyperstructures, lipid species composition, protein/membrane interactions and cold adaptation.

Protein aggregation is a typical example of protein misassembly leading to devastating consequences. The book gives a summary of both the molecular mechanisms protecting against aggregation as well as the consequences of protein aggregation in neurodegenerative diseases and aging.

Chaperones modulate not only individual protein complexes and their hosting cells, but also have profound effects on complex cellular networks, such as the immune system or on the phenotype of whole organisms regulating their development and evolution.

We believe the reader will be also convinced after studying the chapters of this book and checking some of the vast number of original references that chaperones play a key role in the organization of various molecular, organellar and cellular networks efficiently shaping their emergent properties at even higher levels. This story is just at the beginning, opening a wide range of possibilities for efficient applications in the medical treatment of diseases and aging.

Budapest – Szeged, August 2006
Peter Csermely and László Vígh

ABOUT THE EDITORS...

PETER CSERMELY (48) is a professor at the Semmelweis University (Budapest, Hungary). His major fields of study are molecular chaperones (www.chaperone.sote.hu) and networks (www.weaklink.sote.hu). In 1995 Dr. Csermely launched a highly successful initiative, which provides research opportunities for more than 10,000 gifted high school students (www.nyex.info). He wrote and edited ten books (including *Weak Links* at Springer in 2006) and has published 200 research papers with total citations over 2000. Dr. Csermely is the Vice President of the Hungarian Biochemical Society, the President of Cell Stress Society International, an Ashoka Fellow, was a Fogarty and Howard Hughes Scholar and received several other national and international honors and awards including the 2004 Descartes Award of the European Union for Science Communication.

LÁSZLÓ VÍGH (56) is a member of the Hungarian Academy of Sciences (2004) and of the European Academy of Sciences (2002) at Brussels, in the Biomedical Sciences section. He is an honorary professor at Szeged University (2005). He leads the Molecular Stress Biology research group at the Institute of Biochemistry, Biological Research Centre (BRC) of the Hungarian Academy of Sciences at Szeged. The BRC has been a Centre of Excellence of the European Union since 2000. For his outstanding work in lipid-membrane and stress research, Professor Vígh was awarded the highest State Prize for Science in Hungary, the "Széchenyi Award", in 1998. He was the Director of his Institute from 1994 to 2004. He has over 140 publications with total citations over 2200.

PARTICIPANTS

Julius Anckar
Department of Biology
Åbo Akademi University and Turku
  Centre for Biotechnology
University of Turku and Åbo
  Akademi University
Turku
Finland

André-Patrick Arrigo
Laboratoire Stress Oxydant
Chaperons et Apoptose
CNRS UMR 5534
Centre de Génétique Moléculaire
  et Cellulaire
Université Claude Bernard Lyon-1
Villeurbanne
France

Gábor Balogh
Institute of Biochemistry
Biological Research Center of the
  Hungarian Academy of Sciences
Szeged
Hungary

Gregory L. Blatch
Chaperone Research Group
Department of Biochemistry,
  Microbiology and Biotechnology
Rhodes University
Grahamstown
South Africa

Frank Boellmann
Carolina Cardiovascular
  Biology Center
School of Medicine
University of North Carolina
Chapel Hill, North Carolina
U.S.A.

Heather R. Brignull
Department of Biochemistry,
  Molecular Biology and Cell Biology
Rice Institute for Biomedical Research
Northwestern University
Evanston, Illinois
U.S.A.

Andrew R. Cossins
School of Biological Sciences
Liverpool University
The Biosciences Building
Liverpool
U.K.

John H. Crowe
Section of Molecular
  and Cellular Biology
University of California
Davis, California
U.S.A.

Peter Csermely
Department of Medical Chemistry
Semmelweis University
Budapest
Hungary

R. John Ellis
Department of Biological Sciences
University of Warwick
Coventry CV4 7AL
U.K.

Attila Glatz
Institute of Biochemistry
Biological Research Center of the
    Hungarian Academy of Sciences
Szeged
Hungary

Pierre Goloubinoff
DBMV
Faculty of Biology and Medicine
University of Lausanne
Lausanne
Switzerland

Andrew Y. Gracey
School of Biological Sciences
Liverpool University
The Biosciences Building
Liverpool
U.K.

Scott A.L. Hayward
School of Biological Sciences
Liverpool University
The Biosciences Building
Liverpool
U.K.

Linda M. Hendershot
Department of Genetics and Tumor
    Cell Biology
St. Jude Children's Research Hospital
Memphis, Tennessee
U.S.A.

Marie-Pierre Hinault
DBMV
Faculty of Biology and Medicine
University of Lausanne
Lausanne
Switzerland

Ibolya Horváth
Institute of Biochemistry
Biological Research Center of the
    Hungarian Academy of Sciences
Szeged
Hungary

Walid A. Houry
Department of Biochemistry
University of Toronto
Toronto, Ontario
Canada

Jennifer R. Knapp
Division of Basic Sciences
Fred Hutchinson Cancer
    Research Center
Seattle, Washington
U.S.A.

Jacques Landry
Centre de recherche en cancérologie
    de l'Université Laval
L'Hôtel-Dieu de Québec
Québec
Canada

Richard I. Morimoto
Department of Biochemistry,
   Molecular Biology and Cell Biology
Rice Institute for Biomedical Research
Northwestern University
Evanston, Illinois
U.S.A.

James F. Morley
Department of Biochemistry,
   Molecular Biology and Cell Biology
Rice Institute for Biomedical Research
Northwestern University
Evanston, Illinois
U.S.A.

Patricia A. Murray
School of Biological Sciences
Liverpool University
The Biosciences Building
Liverpool
U.K.

Sébastien Ian Nadeau
Centre de recherche en cancérologie
   de l'Université Laval
L'Hôtel-Dieu de Québec
Québec
Canada

Stefano Piotto
Department of Chemical
   and Food Engineering
University of Salerno
Fisciano-Salerno
Italy

Zoltán Prohászka
IIIrd Department of Internal Medicine
Semmelweis University
Budapest
Hungary

Rosario Rizzuto
Department of Experimental
   and Diagnostic Medicine
University of Ferrara
Ferrara
Italy

Suzannah Rutherford
Division of Basic Sciences
Fred Hutchinson Cancer
   Research Center
Seattle, Washington
U.S.A.

Yuichiro Shimizu
Department of Genetics and Tumor
   Cell Biology
St. Jude Children's Research Hospital
Memphis, Tennessee
U.S.A.

Lea Sistonen
Department of Biology
Åbo Akademi University and Turku
   Centre for Biotechnology
University of Turku and Åbo
   Akademi University
Turku
Finland

Csaba Söti
Department of Medical Chemistry
Semmelweis University
Budapest
Hungary

György Szabadkai
Department of Experimental
   and Diagnostic Medicine
University of Ferrara
Ferrara
Italy

Zsolt Török
Institute of Biochemistry
Biological Research Center of the
    Hungarian Academy of Sciences
Szeged
Hungary

László Vígh
Institute of Biochemistry
Biological Research Center of the
    Hungarian Academy of Sciences
Szeged
Hungary

Richard Voellmy
HSF Pharmaceuticals SA
Pully
Switzerland

Rongmin Zhao
Department of Biochemistry
University of Toronto
Toronto, Ontario
Canada

CONTENTS

8. HEAT SHOCK FACTOR 1 AS A COORDINATOR OF STRESS AND DEVELOPMENTAL PATHWAYS 78

Julius Anckar and Lea Sistonen

9. CHAPERONE REGULATION OF THE HEAT SHOCK PROTEIN RESPONSE 89

Richard Voellmy and Frank Boellmann

10. MECHANISMS OF ACTIVATION AND REGULATION OF THE HEAT SHOCK-SENSITIVE SIGNALING PATHWAYS 100

Sébastien Ian Nadeau and Jacques Landry

14. CHAPERONES AS PART OF IMMUNE NETWORKS 159

Zoltán Prohászka

15. THE STRESS OF MISFOLDED PROTEINS:
C. ELEGANS MODELS FOR NEURODEGENERATIVE
DISEASE AND AGING 167

Heather R. Brignull, James F. Morley and Richard I. Morimoto

CHAPTER 1

Protein Misassembly:
Macromolecular Crowding and Molecular Chaperones

R. John Ellis*

Abstract

The generic tendency of proteins to misassemble into nonfunctional, and sometimes cytotoxic, structures poses a universal problem for all types of cell. This problem is exacerbated by the high total concentration of macromolecules found within most intracellular compartments but it is solved by the actions of molecular chaperones. This review discusses some of the basic evidence and key concepts relating to this conclusion.

Introduction

A recent article in the journal Nature recommended authors to begin their review with a story, a story that is amusing but also sets the scene. My story concerns a professor of biology who decided to test the knowledge of his students at the end of his lecture course. He pointed to a young woman in the front row and said to her "There is an organ of the human body that under appropriate circumstances can increase in size by a factor of six-fold. What is that organ and what are the circumstances?" The young woman blushed and exclaimed that she found this question so offensive she would report the professor to the head of the university. The professor ignored this response and repeated the question to the next student "What is the organ of the body that can increase in size by six-fold?" This student replied that this organ is the iris of the human eye, which expands from a pinpoint in bright light to wide open in dim light. "Correct" said the professor, who then returned to the first student. "I have two things to say to you. Firstly, you clearly have not been paying sufficient attention to my lectures and secondly, I predict that at some point in the future you are going to be greatly disappointed!"

The point of this story is that the first student made an unwarranted assumption, unwarranted in the quantitative sense. This point of this article is to explain why those people who study the properties of isolated macromolecules in uncrowded buffers are similarly making unwarranted assumptions about the quantitative relevance of their measurements to the properties of these molecules inside the cell. They are thereby ignoring the universal cellular feature that explains the existence of molecular chaperones.

Inside the Cell

The term 'crowded' is a quick way of referring to the fact that the total concentration of macromolecules inside cells is very high, a phenomenon called macromolecular crowding. The cartoon in Figure 1A represents the densities and shapes of the macromolecules in part of the cytosol of an animal cell, as drawn by David Goodsell and it is obvious why the term 'crowded'

*R. John Ellis—Department of Biological Sciences, University of Warwick, Coventry CV4 7AL, U.K. Email: jellis@bio.warwick.ac.uk

Molecular Aspects of the Stress Response: Chaperones, Membranes and Networks, edited by Peter Csermely and László Vígh. ©2007 Landes Bioscience and Springer Science+Business Media.

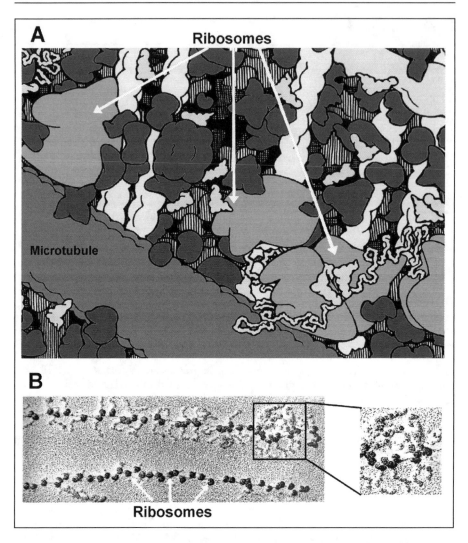

Figure 1. Proteins fold in highly crowded environments. A) representation of part of the cytosol of an animal cell. Reprinted with permission (from ref. 31). B) Electron micrograph of a polysome isolated from the salivary gland of the insect *Chironomus* x 140,000. At the bottom is the start of the polysome, at the top the end, the rest looping out of shot. Reprinted with permission from reference 32, ©1989 Springer Science and Business Media, LLC.

is appropriate. Note that the ribosomes are connected by strands of messenger RNA to form polysomes. Figure 1B shows an actual polysome, isolated from the salivary gland of an insect and spread out on the grid of an electron microscope. You can see the newly synthesized polypeptide chains growing longer as each ribosome slides along the messenger RNA, and you will notice that these chains are close together—within touching distance. You can also see that, as the chains get longer, they show signs of folding into more compact structures whilst still attached to the ribosomes.

This proximity of identical, partly folded chains in the crowded environment of the cytosol creates a danger—a danger that these chains will not fold and assemble correctly into functional proteins but will bind to one another to form nonfunctional misassemblies. It follows that protein misassembly is a universal cellular problem created by the advantage of making several protein chains at the same time from one molecule of messenger RNA. But the misassembly problem is even wider than this, because it affects mature proteins as well, which can partly unfold as a result of environmental stresses. The most distressing consequences of protein misassembly are the human neurodegenerative diseases, characterised by the occurrence of protein aggregates called amyloid plaques in the brain. In these diseases misassembly does not take place at the stage of protein synthesis but when mature proteins partly unfold, but the principle is the same—misassembly occurs when partly folded identical chains are close to one another.

In the remainder of this article I shall discuss the reasons for believing that the dramatic stimulation of protein misassembly by macromolecular crowding accounts for the existence of proteins acting as molecular chaperones to combat this problem. This discussion will focus on the historical origins of the molecular chaperone concept because the fact that most computer databases rarely extend back more than about two decades is leading to ignorance of how this paradigm shift occurred.

The Principle of Protein Self-Assembly: Yesterday and Today

If we ask why protein chains fold as they do, the answer was provided by the classic refolding experiments initiated by Christian Anfinsen around 1956. In this type of experiment a pure native protein is dissolved in dilute buffer and denatured by a high concentration of a chaotrope such as urea. This treatment causes the protein to unfold and thereby lose its biological properties, but if the concentration of chaotrope is lowered by 10 to 50-fold, Anfinsen observed that many of the chains refold into their original functional conformations.[1] This type of experiment has been repeated many times with many proteins, and it is clear that the majority of denatured proteins will refold into their original conformations in the absence of either an energy source or other macromolecules—in other words, proteins can self-assemble spontaneously by a process requiring only their primary structures. Despite the fact that the rates and yields of refolded proteins are often too small to meet biological requirements, as was pointed out by Anfinsen,[2] this observation led to the assumption that spontaneous assembly also happens inside the cell. This assumption was reasonable because a basic principle of scientific enquiry is Occam's razor, which advises us not to complicate hypotheses unnecessarily, but a series of observations made in the 1980s challenged this view. Two reports appeared that newly synthesized polypeptide chains bind transiently to a preexisting protein before they fold.

The first report concerns the synthesis of the photosynthetic enzyme rubisco in higher plants. Rubisco is a chloroplast enzyme, consisting of eight catalytic large subunits made by chloroplast ribosomes and eight structural small subunits made by cytosolic ribosomes. It is highly water-soluble, occurring at around 300g/l in the chloroplast stroma, but it is also one of the few proteins that fails to renature after dilution from a chaotrope. This failure stems from the extreme hydrophobicity of the large subunits, which aggregate with one another to form a water-insoluble precipitate. So we were surprised to find that when we allowed isolated chloroplasts to synthesize large subunits, these accumulate in a water-soluble form, despite the fact that they do not assemble into the holoenzyme until late in the incubation (Fig. 2). This solubility arises because newly synthesized large subunits bind transiently to another water-soluble protein before they assemble with small subunits.[3] Sequencing of this large subunit binding protein revealed that it is 50% identical to the GroEL protein of *Escherichia coli*, a protein required for some bacteriophages to assemble in this bacterium but whose role in uninfected cells was unknown at that time. This identity was interpreted in terms of the molecular chaperone concept, and it was Sean Hemmingsen who coined the name 'chaperonin' for this family of molecular chaperone.[4]

The second report of a newly synthesized polypeptide binding to a preexisting protein before folding was found during experiments with a cultured cell line derived from lymphocyte

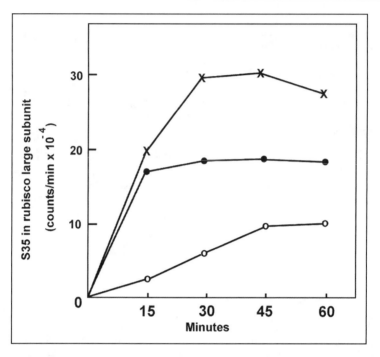

Figure 2. The newly synthesized large subunit of rubisco binds to another protein before assembling into the holoenzyme. Intact chloroplasts were isolated from young pea leaves and incubated with light as the energy source for protein synthesis and S35-methionine. Samples were removed at intervals and stromal extracts prepared by osmotic lysis and centrigugation. Stromal extracts were run on both native and SDS polyacrylamide gels and autoradiographed. Bands corresponding to rubisco large subunit were excised from the gels and their radioactivity measured. Symbols: crosses, total large subunit from SDS gel; filled circles, large subunit attached to binding protein from native gel; open circles, large subunit in holoenzyme from native gel. Reprinted from reference 33.

precursors. These cells synthesize immunoglobin, an oligomeric protein composed of heavy and light chains. Haas and Wabl found that in a cell line of preB lymphocytes that are unable to make light chains, the heavy chains become bound to a preexisting small protein as they enter the lumen of the endoplasmic reticulum. They called this protein BiP and suggested that it is involved in the regulation of immunoglobulin assembly.[5] BiP was later identified by Munro and Pelham as a member of the heat shock protein 70 family.[6] We now know that both the chaperonins and the hsp70 chaperones assist the folding of many newly synthesized proteins in all types of cell.[7]

The Molecular Chaperone Concept

The term 'molecular chaperone' was coined by Ron Laskey in a Nature paper published in 1978 to describe a nuclear protein that solves a misassembly problem during the assembly of nucleosomes in amphibian eggs. Nucleosomes are octamers of basic histone protein bound to DNA by electrostatic interactions. Disruption of these interactions by high salt concentrations enables the histones to be separated from the DNA, but mixing these components together at physiological salt concentrations results in a failure of self-assembly—an insoluble precipitate forms instead of nucleosomes. Laskey and his coworkers discovered that this failure can be prevented by adding an abundant acidic nuclear protein called nucleoplasmin, which results in the correct assembly of nucleosome cores. The detailed mechanism is unclear but in general terms nucleoplasmin is thought to solve the misassembly problem by binding its acidic groups to the positively charged groups on the histones. This binding lowers their overall surface charge and allows the intrinsic self-assembly properties of the histones to predominate over the incorrect interactions favoured by the high densities of opposite charge. In the Discussion of this paper, the term 'molecular chaperone' is used for the first time.[8]

Control experiments show that nucleoplasmin does not provide steric information required for histones to bind correctly to DNA, nor is it a component of assembled nucleosomes. It is these latter two features that laid the basic foundation of our current general concept of the chaperone function, as I shall discuss later, but at this point I wish to emphasise that the role of nucleoplasmin is in the later stages of protein assembly, beyond the folding of newly synthesized histone chains, so the current common perception that chaperones are concerned solely with protein folding is incorrect and has been incorrect since this subject started. I suspect that this erroneous perception is retarding the search for chaperones that deal with the misassembly of folded subunits rather than with the problems posed by the folding of newly synthesized chains. It was only three years ago that a chaperone was found in red blood cells that mops up the excess of alpha subunits of haemoglobin that would otherwise misassemble and damage these cells.[9] I predict that more examples of this type of subunit chaperone will be found in the future.

Ron Laskey did not use the term molecular chaperone to describe any other protein or develop the idea into a more general concept. This is where I enter the story. I read the Laskey paper in 1985 and it struck me that his observations with nucleoplasmin could be thought of in the same functional terms as the observations we had made about the rubisco binding protein, and others such as Hugh Pelham and Jim Rothman had made about hsp70. So I proposed in 1987 the existence of a new type of general cellular function, defined as ensuring that the folding of certain polypeptide chains and their assembly into oligomeric structures occur correctly.[10]

The term molecular chaperone rapidly caught on and the number of papers using this term contains to rise steadily, as you can see in Figure 3. The numbers of families of chaperone, as defined by sequence similarity, is now well over 30. Some of these families, but by no means all,

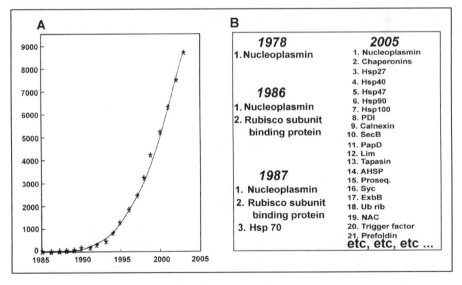

Figure 3. The onward march of molecular chaperones. A) Total number of papers published containing the word 'chaperone' in title or abstract, as derived from the Web of Science database. B) The rising number of chaperone families, as defined by sequence similarity.

are heat shock proteins and this reflects the fact that the need for chaperone function increases after proteins have been denatured by environmental stresses. Many chaperones are involved in protein folding but some are also assist disassembly processes, such as the remodelling of chromatin during fertilization and transcription, and the resolubilisation of insoluble aggregates that have escaped the attention of other chaperones. This accumulation of discoveries supports a new view of protein assembly. In this new view the principle of self-assembly is retained, but before chaperones, this assembly was thought to be spontaneous and require no energy expenditure, whereas now self-assembly is seen in many instances to require assistance by chaperones, some of which hydrolyse ATP. Thus the concept of spontaneous protein self-assembly has been replaced by the concept of assisted protein self-assembly. This new view was first articulated in a TiBS article in 1989 (see ref. 11), and has so far stood the test of time.

There is now a huge literature on the details of chaperone structure and function, but what I wish to concentrate on in this article is why chaperones exist. Why do cells need a chaperone function, given that most denatured proteins know how to fold correctly in the absence of other macromolecules?

The Problem of Protein Misassembly

There are two answers to this question. The fact that proteins are made by polysomes ensures that identical partly folded chains are close together, as I indicated earlier. Moreover, these chains grow vectorially so they cannot fold correctly until a complete folding domain has been made, raising the possibility that incomplete domains may misfold. In addition to these factors, polysomes function within a highly crowded compartment. All these features favour a process that competes with folding, the process of misassembly.

Misassembly is defined as the association of two or more polypeptide chains to form nonfunctional structures; these structures may be as small as dimers or large enough to be insoluble. This emphasis on function serves to distinguish misassembly from the formation of functional oligomers, which is thus termed oligomerization. Many authors use the term 'aggregation' to describe misassembly, and point to the amyloid fibrils found in the brains of people suffering from neurodegenerative disease as typical examples. The problem with this term is that there are increasing reports of amyloid structures that do have biological functions, so the essential distinction between aggregation and oligomerization has disappeared.[12] Misassembly should be distinguished from misfolding, which I define as the formation of a conformation which cannot proceed to the functional conformation on a biologically relevant time scale. Misassemblies are by definition misfolded, but there are very few reports of renaturing proteins that misfold but remain monomeric. It seems likely that the vast majority of primary translation products are capable of folding correctly, provided they can avoid misassembly. It is for this reason that I suggest that the essential problem that chaperones have evolved to combat is misassembly rather than misfolding.

Protein misassembly has not been widely studied but two important aspects are established. It has been known for many years that misassembly competes with folding in an Anfinsen-type refolding experiment (Fig. 4A). Since misassembly is a high order process, it is very sensitive to the concentration of the unfolded chains, so as this concentration rises, misassembly outcompetes folding. The molecular basis for this competition is simply that polypeptide chains do not distinguish between inter- and intra-molecular interactions. Protein chemists traditionally solve this problem by lowering the concentration of the chains, and/or the temperature, but this solution is not open to cells.

A more surprising feature of misassembly is that in many cases it is highly specific - chains misassemble only if they are identical or very similar. This observation is interpreted to mean that the aggregating species have a degree of secondary or tertiary structure - in other words they are partly folded. This point was first established by adding total crude extracts of *E. coli* cells to the 8M urea in a standard Anfinsen refolding experiment applied to tryptophanase (Fig. 4B). The numbers marked with asterisks in the box indicate the enzymic activity of the

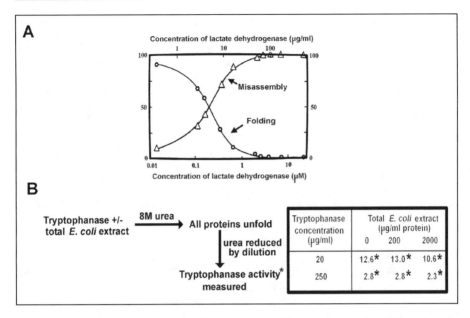

Figure 4. Two properties of misassembly. A) Misassembly competes with refolding. Lactate dehydrogenase was denatured by acid pH and then diluted into buffer at pH 7.0 and the regain of enzymic activity measured. Reprinted with permission from reference 13, ©1979 American Chemical Society. B) Misassembly is highly specific. Tryptophanase was denatured in 8M urea with and without a crude total extract of *E. coli* and diluted into refolding buffer. The numbers in the box are the specific activities of the tryptophanase recovered. Reprinted with permission from reference 34, ©1974 Blackwell.

tryptophanase chains that have refolded after dilution of the urea, and you can see that this activity is unaffected by the addition of 100 times as much total protein from *E. coli* to the urea-containing buffer. Thus the several thousand different unfolded polypeptide chains in the *E. coli* extract do not affect the competition between correct folding and misassembly.[13] What does affect this competition however, is the phenomenon of macromolecular crowding.

Macromolecular Crowding

We use the term 'crowded' rather than 'concentrated' because in general no single macromolecular species occurs at a high concentration, but taken together, macromolecules occupy a significant fraction of the total volume. This fraction is in the range 8-40%. That is, 8-40% of the total volume is physically occupied by macromolecules and therefore is unavailable to other molecules, just as in a football crowd most of the space is occupied by people and is unavailable to other people—other people are sterically excluded. This steric exclusion of part of the volume generates considerable energetic consequences, whose magnitude is not generally appreciated because it is so counter-intuitive—8-40% reduction in available volume does not sound very much. You may think that it simply means that the concentration of macromolecules is 8-40% larger than it appears to be, but in reality this steric exclusion produces large highly nonlinear effects on the effective concentration of macromolecules. This nonlinearity results in an exquisite sensitivity of macromolecular properties to changes in their environment.

There exists a quantitative theory of crowding developed largely by Allen Minton, a biophysicist at the National Institutes of Health in America.[14-16] This theory predicts two major consequences of crowding. The first prediction is that crowding will reduce the diffusion rates of both large and small molecules. There is good experimental evidence that diffusion is reduced in the range three-four fold in eukaryotic cells and eleven-fold in *E. coli* relative to the rate in water; prokaryotic cells appear to be more crowded than eukaryotic cells.[17] Much more important with respect to misassembly however is the large increase in association constants for macromolecules. For example, let us consider the association of two 40 kDa monomers into a dimer. Suppose that the association constant for this oligomerization in water is 1.0. Crowding theory predicts that the value of this constant inside a cell of *E. coli* will be between 8 and 40, depending on what specific volume of the protein is assumed. If we suppose that the dimers can form a homotetramer, the effect is even larger—the association constant is now 10,000 (see ref. 18).

This dramatic consequence arises because crowding increases the effective concentration of macromolecules—in other words, it increases their thermodynamic activity. This increase in activity is produced by the reduction in excluded volume when molecules bind to one another. In fact macromolecular crowding is more precisely called 'the excluded volume effect'. This term emphasises the fact that crowding is a purely physical nonspecific effect based solely on steric exclusion. As the number and the size of molecules in a solution increase, the less randomly they can be distributed. So as the concentration rises, the free energy of the solution also rises, because the entropy of each molecule becomes less. But if these molecules bind to one another, the volume available to them increases, their entropy therefore is greater and the total free energy of the solution decreases. In other words, the most favoured state is the state that excludes the least volume to other macromolecules.

This is a general conclusion—it applies not just to associating macromolecules, but to all processes where a change in excluded volume occurs. So it applies to the folding of newly synthesized polypeptide chains, the folding of nucleic acid chains, the unfolding of mature proteins as a result of environmental stresses such as high temperature, the condensation of DNA in chromosomes, the operation of motile systems such as actin filaments and microtubules, and, of particular relevance here, to the formation of protein misassemblies.

It is important to grasp that the effect of crowding on thermodynamic activity is highly nonlinear with respect to the concentration of the crowding agent. The term 'crowding agent' is used to describe the molecules that cause the crowding. Figure 5A shows how the activity coefficient, that is the ratio of the effective concentration to the actual concentration, varies

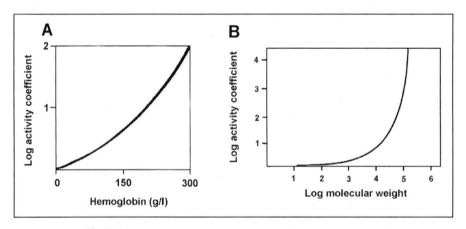

Figure 5. A) Activity coefficients increase nonlinearly with cellular concentrations of macromolecules. The log of the activity coefficient of hemoglobin is plotted against the concentration of hemoglobin. Reprinted from reference 19. B) The effect of crowding on activity is exerted by large molecules on large molecules. The change in activity coefficient of a test molecule introduced into a solution of hemoglobin as crowding agent was measured with respect to the molecular weight of the test molecule. Reprinted with author permission from reference 20.

with the concentration of haemoglobin.[19] Note that this is a log/linear plot and that the activity coefficient rises in a nonlinear fashion with respect to the actual concentration. The concentration of hemoglobin in red blood cells is about 340 g/l, so you can see that the activity of haemoglobin inside the cell is more than two orders of magnitude greater than it is in the dilute buffer where its properties are commonly studied.

Figure 5B plots the activity coefficient against molecular weight for a test molecule placed in a background of hemoglobin at 300 g/l. So in this experiment the crowding agent is haemoglobin and we are asking how the activity of another molecule placed in that solution depends on the molecular weight of that molecule. Note that this is a log/log plot. You can see that the effect of crowding on activity becomes significant only after about 10,000 molecular weight.[20] This is why we use the term 'macromolecular crowding' because the effect on the activity of small molecules such as metabolites and inorganic ions is small by comparison.

I have so far talked about the effect of crowding on thermodynamics. What about effects on kinetics? Figure 6 summarises the effects of crowding on reaction rates for a bimolecular association reaction.[21] The vertical axis of the graph is the log of the association rate constant. We can consider two situations in turn. If the rate-limiting step is the encounter rate of the two components A and B, then the reaction is diffusion-limited. But crowding reduces diffusion so in this case the reaction rate will decrease as the concentration of crowding agent rises. However in the other situation, the rate-limiting step is the conversion of the activated complex (or transition state) to the dimer, so in this case the rate depends on the concentration of this complex. But crowding increases association, so the equilibrium between A + B and the activated complex is displaced to the right. Thus the reaction rate will rise as the concentration of crowding reagent increases. However, if you think about it, the absolute upper limit of any bimolecular reaction must be ultimately set by the encounter rate of the two molecules, so even for transition state-limited reactions, the rate must eventually come down as the concentration of the crowding agent gets high enough. So we have a resultant curve between two opposing effects.

Thus we see that the effect of crowding on reaction rate is complex; it depends on the nature of the reaction and where you are on the concentration axis. The sad fact however, is that although these dramatic effects of crowding have been known for at least 25 years, the vast majority of studies on isolated macromolecules, including those on protein folding, protein

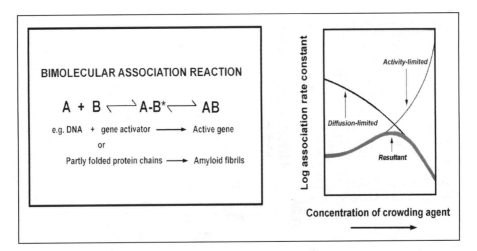

Figure 6. Effects of crowding on reaction rate. The change in association rate constant for the bimolecular reaction A + B = AB* + AB with respect to crowding agent is plotted in two situations: A) where the encounter of A and B is rate-limiting; and B) where the conversion of the activated complex AB* to AB is rate-limiting. Reprinted from reference 21.

misassembly and molecular chaperones, continue to be done in uncrowded buffers. In my view this is a mistake. Until the effect of crowding agents becomes as routine to study as the effects of pH or redox potential, the extrapolation of interpretations made from studies of isolated systems to the intact cell must have a question mark hanging over them.[22]

Stimulation of Misassembly by Crowding Agents

The first demonstration that crowding agents increase the misassembly of refolding chains was obtained with hen lysozyme.[23,24] Oxidised lysozyme refolds impeccably, achieving almost complete recovery of enzymic activity in one or two seconds, but when the two four disulfide bonds are reduced, the chains misassemble in the refolding buffer, the extent depending on their concentration. Misassembly occurs because at least two disulfide bonds have to form by slow air oxidation before the chains collapse and during this time the chains have a chance to misassemble with one another. When several different crowding agents are added to the refolding buffer, all the reduced chains misassemble but the refolding of oxidised chains is unaffected (Fig. 7). Missassembly can be prevented by adding protein disulfide isomerase (PDI) to the refolding buffer. Low concentrations of PDI act by speeding up the rate of disufide formation, which stabilises the correctly folded chains, while higher concentrations act in a chaperone fashion. Further examples of the stimulation of protein misassembly by crowding are discussed in reference 25.

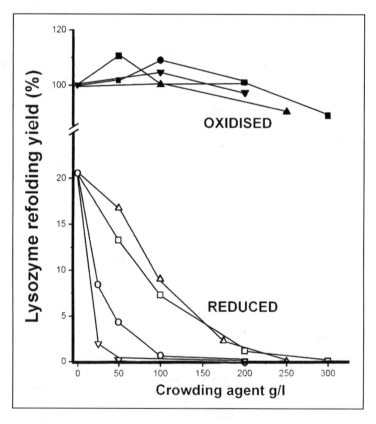

Figure 7. Crowding stimulates the misassembly of reduced lysozyme. Hen lysozyme was denatured under reducing conditions, diluted into refolding buffer containing varying amounts of four crowding agents and the regain of enzymic activity measured. In the absence of crowding agent, the yield is about 20% under the conditions used. Symbols: squares, Ficoll 70; triangles, Dextran 70; circles, bovine serum albumin; inverted triangles, ovalbumin. Reprinted from reference 23.

How do Chaperones Combat Misassembly?

If we survey the literature on the roles of many different chaperones in preventing the misassembly of newly synthesized polypeptides, the picture looks complex, but I suggest that the available data can be encompassed within just two basic principles of chaperone operation.[7]

It is useful to divide chaperones into two classes on the basis of size. Small chaperones are defined as those less than 200 kDa in size and include hsp70, hsp40, PDI and the very recently discovered cases of membrane chaperones i.e., proteins that occur inside membranes and prevent the aggregation of other transmembrane proteins.[26] Small chaperones bind transiently to short hydrophobic sequences as these appear on nascent or just released polypeptide chains, and prevent them from both folding prematurely or misassembling together by binding to these sequences for a time. These chaperones do not appear to affect protein conformation, but function essentially by reducing the time potentially interactive surfaces are exposed by cycling on and off these surfaces until they are buried by folding. Such a simple mechanism can be thought of as analogous to tossing a hot potato from hand-to-hand until it has cooled, an analogy suggested by Ulrich Hartl.

Large chaperones, exemplified by the chaperonins such as GroEL, function by a much more sophisticated mechanism that uses what I call the Anfinsen cage principle.[27] The chaperonins function essentially by providing a molecular cage made of one oligomer of GroEL capped by one oligomer of GroES. Single partly folded chains are encapsulated one at a time inside this cage. The enclosed chain continues to fold in the absence of other folded chains until the hydrophobic surfaces that cause misassembly are buried within the final folded structure. The time of folding inside this cage is set by the slow ATPase activity of the GroEL subunits. Completion of ATP hydrolysis allows ATP to bind to the opposite end of the GroEL/ES complex from the enclosed protein, and this binding triggers via allosteric interactions the release of the folded chain into the cytosol.[7]

But suppose the released chain has not managed to bury its hydrophobic regions during the time it is enclosed? It would be dangerous to release such misassembly-prone chains into the cytosol but this problem is prevented by the crowded state of the cytosol. Experiment with crowding agents added to isolated GroEL/ES complexes by Jorg Martin and Ulrich Hartl show that crowding ensures that any such released, but partly folded chains, bind back to the same GroEL molecule from which they have just been released.[28,29] So we reach the pleasing conclusion that the chaperonins use the crowded state of the cytosol to combat the problem of protein aggregation that has been created by crowding in the first place.

The Molecular Chaperone Function

My current definition of molecular chaperones is that they are a large and diverse group of proteins that share the property of assisting the noncovalent assembly and/or disassembly of other macromolecular structures, but which are not permanent components of these structures when these are performing their normal biological functions.[30] It is important to note that this definition is functional and not structural, but it contains no constraints on the mechanisms by which different chaperones may act; this is the reason for the use of the imprecise term 'assist'. Thus molecular chaperones are not defined by a common mechanism or by sequence similarity. In my view only two criteria need be satisfied to designate a macromolecule a molecular chaperone. Firstly, it must in some sense assist the noncovalent assembly or disassembly of some other macromolecular structure, the mechanism being irrelevant, and secondly, it must not be a component of these structures when they are performing their normal biological function. Note that the term 'assembly' is used in this definition in a very broad sense, and includes the folding of newly synthesized or stress-denatured proteins, the unfolding of proteins during transport across membranes, the association of monomers into oligomers and macromolecular disassembly processes.

All these considerations can be reduced to a simple unifying concept of the chaperone function, defined as the prevention and reversal of incorrect interactions that may occur when potentially interactive surfaces are exposed to the crowded intracellular environment. These surfaces occur on nascent and newly synthesized polypeptide chains, on mature proteins unfolded by environmental stresses, and also on folded proteins in near-native conformations, as in signalling proteins such as the steroid receptor. Thus the term 'molecular chaperone' is not a metaphor, nor an example of academic whimsy, but a precise description. The word 'chaperone' is appropriate because the traditional role of the human chaperone is to prevent incorrect interactions between pairs of people without either providing the steric information required for their correct interaction, or being present during married life - but often reappearing during divorce and remarriage!

Acknowledgements

I thank Robert Freedman for his generous provision of Departmental facilities during my retirement and Allen Minton for correcting my misconceptions about crowding theory.

References

1. Anfinsen CB. Principles that govern the folding of polypeptide chains. Science 1973; 181:223-230.
2. Epstein CJ, Goldberg RF, Anfinsen CB. The genetic control of tertiary protein structure: Studies with model systems. Cold Spring Harbor Symp Quant Biol 1963; 28:439-448.
3. Barraclough R, Ellis RJ. Protein synthesis in chloroplasts. IX. Assembly of newly synthesized large subunits into ribulose bisphosphate carboxylase in isolated intact chloroplasts. Biochim Biophys Acta 1980; 608:19-31.
4. Hemmingsen SM, Woolford C, van der Vies SM et al. Homologous plant and bacterial proteins chaperone oligomeric protein assembly. Nature 1988; 333:330-334.
5. Haas IG, Wabl M. Immunoglobulin heavy chain binding protein. Nature 1983; 306:387-389.
6. Munro S, Pelham SRB. An hsp70-like protein in the ER: Identity with the 78 kd glucose-regulated protein and immunoglobulin heavy chain binding protein. Cell 1986; 46:291-300.
7. Young JC, Agashe VR, Siegers K et al. Pathways of chaperone-mediated folding in the cytosol. Nature Revs Mol Cell Biol 2004; 5:781-791.
8. Laskey RA, Honda BM, Mills AD et al. Nucleosomes are assembled by an acidic protein which binds histones and transfers them to DNA. Nature 1978; 275:416-420.
9. Kihm AJ, Kong YI, Hong W et al. An abundant erythroid protein that stabilizes free alpha-hemoglobin. Nature 2002; 417:758-767.
10. Ellis RJ. Proteins as molecular chaperones. Nature 1987; 328:378-379.
11. Ellis RJ, Hemmingsen SM. Molecular chaperones: Proteins essential for the biogenesis of some macromolecular structures. Trends Biochem Sci 1989; 14:339-342.
12. Fowler DM, Koulov AV, Alory-Jost C. Functional amyloid formation in mammalian tissue. PloS Biol 2006; 4:0001-0008.
13. Zettlmeiss G, Rudolph R, Jaenicke R. Reconstitution of lactic dehydrogenase. Noncovalent aggregation vs. reactivation. 1. Physical properties and kinetics of aggregation. Biochemistry 1979; 18:5567-5571.
14. Minton AP. Molecular crowding: Analysis of effects of high concentrations of inert cosolutes on biochemical equilibria and rates in terms of volume exclusion. Methods Enzym 1988; 295:127-149.
15. Minton AP. Implications of macromolecular crowding for protein assembly. Curr Opin Struct Biol 2000; 10:34-39.
16. Minton AP. The influence of macromolecular crowding and macromolecular confinement on biochemical reactions in physiological media. J Biol Chem 2001; 276:10577-10580.
17. Ellis RJ. Macromolecular crowding: An important but neglected aspect of the intracellular environment. Curr Opin Struct Biol 2001; 11:114-119.
18. Zimmerman SB, Trach SO. Estimation of macromolecular concentrations and excluded volume effects for the cytoplasm of Escherichia coli. J Mol Biol 1991; 222:599-620.
19. Minton AP. The effect of volume occupancy upon the thermodynamic activity of proteins: Some biochemical consequences. Mol Cell Biochem 1983; 55:119-140.
20. Minton AP, Colclasure GC, Parker JC. Model for the role of macromolecular crowding in regulation of cellular volume. Proc Natl Acad Sci USA 1992; 89:10504-10506.
21. Zimmerman SB, Minton AP. Macromolecular crowding: Biochemical, biophysical and physiological consequences. Annu Rev Biophys Biomol Struct 1993; 22:27-65.
22. Ellis RJ. Macromolecular crowding: Obvious but underappreciated. Trends Biochem Sci 2001; 26:597-604.

23. van den Berg B, Ellis RJ, Dobson CM. Effects of macromolecular crowding on protein folding and aggregation. EMBO J 1999; 18:6927-6933.
24. van den Berg B, Wain R, Dobson CM et al. Macromolecular crowding perturbs protein refolding kinetics: Implications for folding inside the cell. EMBO J 2000; 19:3870-3875.
25. Ellis RJ, Minton AP. Protein aggregation in crowded environments. Biol Chem 2006, (in press).
26. Kota J, Ljungdahl PO. Specialized membrane-localized chaperones prevent aggregation of polytopic proteins in the ER. J Cell Biol 2005; 168:79-88.
27. Ellis RJ. Revisiting the Anfinsen cage. Folding and Design 1996; 1:R9-R15.
28. Martin J, Hartl FU. The effect of macromolecular crowding on chaperonin-mediated protein folding. Proc Natl Acad Sci USA 94:1107-1112.
29. Martin J. Chaperonin function - Effects of crowding and confinement. J Mol Recog 2004; 17:465-472.
30. Ellis RJ. Chaperone function: The orthodox view. In: Henderson B, Pockley AG, eds. Molecular Chaperones and Cell Signalling. Cambridge University Press, 2005:3-41.
31. Goodsell DS. The Machinery of Life. Springer-Verlag, 1992:68.
32. Kiseleva EV. Secretory protein synthesis in Chironomus salivary gland cells is not coupled with protein translocation across endoplasmic reticulum membranes. Electron microscopic evidence. FEBS Lett 1989; 257:251-253.
33. Musgrove JE, Ellis RJ. The Rubisco large subunit binding protein. Phil Trans R Soc London B 1986; 313:419-428.
34. London J, Skrzynia C, Goldberg ME. Renaturation of E. coli tryptophanase after exposure to 8 M urea. Evidence for the existence of nucleation centers. Eur J Biochem 1974; 47:409-415.

CHAPTER 2

The Cellular "Networking" of Mammalian Hsp27 and Its Functions in the Control of Protein Folding, Redox State and Apoptosis

André-Patrick Arrigo*

Abstract

Cells possess effective mechanisms to cope with chronic or acute disturbance of homeostasis. Key roles in maintaining or restoring homeostasis are played by the various heat shock or stress proteins (Hsps). Among the Hsps, the group of proteins characterized by low molecular masses (between 20 to 30 kDa) and homology to α-crystallin are called small stress proteins (denoted sHsps). The present chapter summarizes the actual knowledge of the protective mechanisms generated by the expression of mammalian Hsp27 (also denoted HspB1 in human) against the cytotoxicity induced by heat shock and oxidative stress. It also describes the anti-apoptotic properties of Hsp27 and their putative consequences in different pathological conditions.

Introduction

sHsps have been described first in *Drosophila* as being expressed concomitantly with the high molecular masses heat shock proteins in cells exposed to a heat treament.[1] At the molecular level, sHsps are characterized in their C-terminal moeity by a conserved sequence (the alpha-crystallin domain) shared by mammalian alpha-crystallin polypeptide (Fig. 1). sHsps have been described in every eukaryots studied so far[2-4] as well as in prokaryots.[5] For example, the human genome contains 10 genes encoding different small stress proteins[6] (see Table 1). Until recently, the most studied sHsps were the four *Drosophila* sHsps as well as Hsp27 and αB-crystallin from mammals.[7] Hsp27 and αB-crystallin are characterized by a remarkable variety of cellular functions, the first one being to protect the cell against conditions that would otherwise be lethal to the cell. These sHsps increase the cellular resistance to different types of stress, including heat shock,[8] oxidative conditions,[7,9-12] exposure to cytotoxic drugs[13,14] and apoptotic inducers.[15-19] In vitro analysis have demonstrated that Hsp27 and αB-crystallin share an ATP-independent chaperone activity that counteracts protein denaturation and helps in the refolding of denatured polypeptides.[20,21] However, a chaperone activity is not shared by all the members of the human sHsps family (Table 1). Hsp27 is also known to modulate cell growth, differentiation,[10,22] intracellular redox state[11] and tumorigenicity.[15,23] This sHsp also interacts

*André-Patrick Arrigo—Laboratoire Stress Oxydant, Chaperons et Apoptose, CNRS UMR 5534, Centre de Génétique Moléculaire et Cellulaire, Université Claude Bernard, 16 rue Dubois, 69622 Villeurbanne Cedex, France. Email: arrigo@univ-lyon1.fr

Molecular Aspects of the Stress Response: Chaperones, Membranes and Networks, edited by Peter Csermely and László Vígh. ©2007 Landes Bioscience and Springer Science+Business Media.

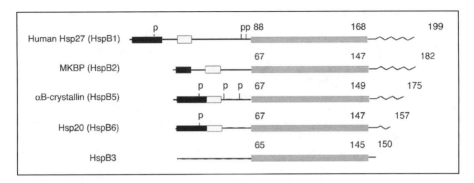

Figure 1. Organization of Hsps polypeptides. Light box: conserved region; black box: WD/EPF domain; gray box: alpha crystallin domain ; ⋀⋀⋀⋀ flexible domain ; P: phosphorylated sites. Amino acids number are indicated.

with the cytoskeleton and membranes and is involved in signaling via MAPKAP kinase 2[24] or estrogen receptors[25] pathways.

Hsp27 in Cells Exposed to Heat Shock

Cells are usually devoid of Hsp27 expression in the absence of heat shock. However, in human, several types of cells, as for example cancer cells of carcinoma origin, constitutively express Hsp27 (HspB1). In both types of cells, sub-lethal heat shock treatments up-regulate or induce the synthesis of Hsp27 as a consequence of the transcriptional activation of hsp27 gene. This transcriptional activation is mediated by HSF-1, a specific transcription factor, that once trimerized binds HSE elements located in the promoter region of the genes encoding Hsps.[26] The heat induced synthesis of the different Hsps is coordinately regulated. This results in the transient accumulation of the Hsps during and after the heat stress. In heat shock conditions, human Hsp27 (HspB1) is phosphorylated at three serine sites by the MAPKAPK2/3 kinase. This kinase is itself activated by the P38 stress kinase.[27] Analysis of the structural organization of the constitutively expressed and therefore preexisting Hsp27 polypeptide present in un-stressed HeLa cells has revealed that this protein is in the form of spherical oligomers of heterogenous native sizes, ranging from 140 to more than 800 kDa.[28,29] No X-ray structure has yet been obtained due to the extreme difficulty to obtain pure crystals of mammalian Hsp27. In heat shock conditions, the oligomerization of preexisting Hsp27 is drastically altered and resembles that of the newly made Hsp27.[29] Indeed, heat shock induces the rapid formation of small sized oligomers, a phenomenon which is followed by the generation of very large Hsp27 containing structures (also called heat shock granules).[29] Phosphorylation is though to be responsible of the dissociation of Hsp27 oligomers during heat shock.[30] The phenomenon is transient and the structural organization of Hsp27 is back to normal several hours after the heat stress.[29] In unstressed cells, the constitutively expressed Hsp27 polypeptide is mainly cytoplasmic with a fraction being associated with membranous structures.[29] In contrast, during heat shock, Hsp27 is mostly associated with detergent resistant structures and redistributes inside the nucleus if the heat stress is drastic enough (45°C, 30 min, HeLa cells).[29] In thermotolerant cells, the structural organization and localization of Hsp27 is not altered by heat shock.[29]

Cells, genetically modified to constitutively express Hsp27 or αB-crystallin, are characterized by an enhanced resistance to the deleterious damages induced by heat shock.[8,31] Similar observations were made in the case of *Drosophila* Hsp27[32] and HspB8, a recently described new human sHsps[33] (see Table 1). Since protein denaturation and aggregation are likely the cause of the noxious effects of heat shock, studies have been performed to test the possibility of an in vivo chaperone activity associated to Hsp27 that could explain its cellular protective

Table 1. The 10 different human sHsps and the proposed new (HspB) names for human sHsps. Note that only four sHsps share a chaperone activity

HspB Name	Other Name	Gene Localization on Chromosomes	Tissue Expression	Function(s)
HspB1	Hsp27, Hsp28	4,7 p12,3	Many tissues including muscles Tumors	Chaperone Protection against stress, apoptosis
HspB2	MKBP	11 q22		Binds myotonic dystrophy protein kinase
HspB3	–	5 q11.2	Muscles	Differentiation
HspB4	αA-crystallin	1 q22.3	Eye lens	Chaperone
HspB5	αB-crystallin	11 q22.3-q23.1	Many tissues including muscles	Chaperone Protection against stress, apoptosis
HspB6	Hsp20	19 q13.13	Smooth muscles	Chaperone Relaxation of vascular smooth muscle
HspB7	cvHsp	1 p36.2-p34.3	Cardiovascular tissues	?
HspB8	H11/Hsp22	12 q24.23	Muscle	Protein Kinase (Thr/ser)
HspB9	–	17 q21.2	Testis	?
HspB10	ODF1	8 q22	Testis	?

Adapted from Kappe et al, Cell Stress and Chaperones 2003; 8:53-61.[6]

activity. Extinction of thermosensitive luciferase light emission was used as an intracellular marker of protein aggregation. No effect of Hsp27 was noticed towards the loss of luciferase activity during heat shock. In contrast, after the heat shock, Hsp27 was found to accelerate the recovery of luciferase activity. This suggests that, in vivo, Hsp27 accelerates the dissociation of the aggregates formed by luciferase.[34] In heat shock treated cells, Hsp27 also play a role in the inhibition of translation of non heat shock induced mRNAs. This phenomenon results probably of the ability of Hsp27 to bind eIF4G initiation factor and to facilitate dissociation of cap-initiation complexes.[35] Following a heat shock treatment, Hsp27 stimulates the recovery of RNA splicing as well as RNA and protein synthesis.[36,37] In addition, phosphorylated Hsp27, as it is observed in heat shock treated cells, helps in maintaining cytoskeletal architecture.[38,39] Taken together, these phenomena provide the cell with a survival advantage.

In vitro analysis have revealed that Hsp27 is a dynamic tetramer of tetramers with a unique ability to refold and reassembles into its active quaternary structure after denaturation. At high temperatures, the structures formed by Hsp27 increase in size but in slow equilibrium with the hexadecameric form.[40] In heat shock conditions, Hsp27 is in the form of an oligomeric active chaperone that binds and stores non native proteins (molten globules) and prevents their aggregation[41-44] before subsequent refolding by ATP-dependent protein chaperones (Hsp70, Hsp40, Hsp90 and co-chaperones) (see Fig. 2). Indeed, Hsp27 ability to refold denatured polypeptides is poor compared to the high efficiency provided by the ATP-dependent Hsp70/Hsp40 chaperone machinery.[45] Further in vitro analysis have demonstrated that phosphorylation significantly reduces Hsp27 ability to form reservoirs containing denatured substrates.[46] It is not known whether phosphorylation is a negative regulator of the formation of reservoirs or is

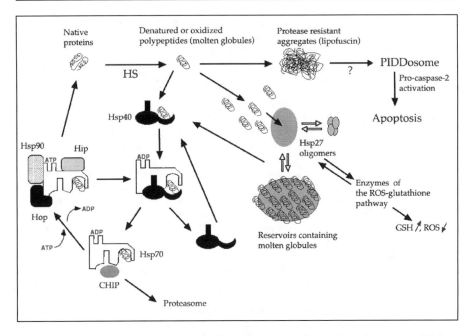

Figure 2. Hsp27 function in cells exposed to agents or conditions that alter protein folding. Denatured or oxidized are either rapidly refolded by ATP-dependent protein chaperones (Hsp70, Hsp40, Hsp90 and co-chaperones Hip, Hop and Hsp90) or recognized by the CHIP co-chaperone and send to degradation by the proteasome machinery. In case of an intense stress, the high loads in denatured or oxidized proteins (molten globules) rapidly saturate the Hsp70 refolding machinery, this results in the formation of aggregates (lipofuscin) that are very deleterious to the cell. These aggregates may activate the PIDDosome machinery (made of the polypeptides PIDD, RIADD and pro-caspase-2) leading to pro-caspase-2 activation and apoptosis through the activation of downstream caspases. To counteract the problem and to down-regulate the formation of aggregates, Hsp27 oligomers bind non native proteins and trap them in very high molecular weight structures. These structures act as storage reservoirs. Denatured proteins present in these reservoirs are less prone to aggregate. A more secure way to process denatured proteins is then created by Hsp27 oligomeric structures. Hsp27 oligomers generate also a pro-reduced state in the cells through the modulation of enzymes involved in gluthatione-ROS pathway. This pro-reduced state counteracts ROS formation and probably also aberrant protein interactions and lipofuscin formation. HS: heat shock.

part of a mechanism that promotes their dynamic recycling. On the other hand, the actin polymerization-inhibiting activity associated to Hsp27 is inhibited when this protein in the form of phosphorylated tetramers.[47] This phenomenon therefore favors F-actin reformation.[48,49] Indeed, in heat treated cells, Hsp27 forms stable complexes with denatured F-actin preventing its aggregation and facilitating its reformation.[50]

Hsp27 in Cells Exposed to Oxidative Stress

Oxidative stress is common in eukaryotic cells since the use of molecular oxygen that leads to respiratory energy production generates side products such as reactive oxygen species (ROS) that become cytotoxic when they are produced at a too high concentration. As mentioned above in the case of heat shock, the type of death that is triggered depends on the intensity of the stress. Apoptosis is usually induced by a moderate oxidative stress, while a drastic exposure to oxidant conditions induces necrosis.[51] In the case of apoptosis, endogenous ROS induce a glutathione depletion which triggers the mitochondrial apoptotic pathway[52] (see below). In contrast, high levels of ROS result in necrosis probably because of irreversible oxidations that inhibit caspases and decrease ATP intracellular levels.[53,54]

Sophisticated defenses (i.e., detoxifiant enzymes, vitamines C and E and thiol-containing molecules, such as glutathione) are induced by the cells to protect themselves in the face of an extracellular oxidizing environment.[55] To the list of protectors one can add the anti-apoptotic protein Bcl2[56,57] and several sHsps, such as mammalian Hsp27 and αB-crystallin and *Drosophila* sHsps.[11] Indeed, these sHsps have a cytoprotector effect in cells exposed to oxidative stress induced by agents or conditions such as hydrogen peroxide, menadione, doxorubicin and tumor necrosis factor (TNFα).[13,38,58-63] In unstressed cells, devoid of constitutively expressed sHsps, such as murine L929,[46,60,61] murine NIH 3T3-ras[64] and human colorectal cancer HT-29[65] cells, sHsps expression significantly decreases the basal level of intracellular reactive oxygen species (ROS) as well as the burst of ROS generated by oxidative stress. This effect is correlated with decreased lipid peroxidation and protein oxidation[59] and F-actin architecture disruption.[38,66,67] Glutathione[68] is an important factor of the protective activity of Hsp27 against oxidative stress.[59] Consequently, Hsp27 does not counteract the oxidative stress generated by either buthionine sulfoximine (BSO) (a specific and essentially irreversible inhibitor of γ-glutamyl-cysteine synthetase), or diethyl maleate (DEM) which binds the free sulfhydryl groups of glutathione. In unstressed murine cells that are devoid of endogenous sHsps expression, Hsp27 expression upregulates glutathione level and upholds this redox modulator in its reduced form.[59,60,69,70] However, in cells that already constitutively express high levels of Hsp27 (i.e HeLa cells), an expression vector mediated increase in Hsp27 level does not necessarily upregulates glutathione level (probably because the maximum intracellular level of glutathione is already reached in normal HeLa cells).[59] The expression and/or activity of enzymes involved in the ROS-glutathione pathway, such as glucose-6-phosphate dehydrogenase, glutathione reductase and glutathione transferase is often up-regulated in cells (particularly murine fibroblasts) expressing Hsp27. This phenomenon could be at the origin of the glutathione mediated antioxidant power generated by the expression of Hsp27.[60] Recently, we have observed that the expression of Hsp27 or αB-crystallin downregulates iron intracellular level[71] (Virot S and Arrigo A.-P., manuscript in preparation). This down-regulation could represent a key factor in the protective mechanism of Hsp27 since free iron, through Fenton reaction, catalyzes the formation of hydroxyl radicals that oxidize proteins.

In oxidative conditions, Hsp27 mediated rise in reduced glutathione and/or the upholding of this redox modulator in its reduced form appears directly responsible of the protection observed at the level of the cell morphology, cytoskeletal architecture[60,67] and mitochondrial membrane potential.[60,70] In addition, the cellular locale, oligomerization and phosphorylation of Hsp27 are also drastically modified in cells undergoing oxdative stress. Formation of dimers is also stimulated in cells exposed to oxidative stress. This occurs through the unique cysteine residue present in Hsp27 which is essential for the protective activity of the protein.[72] In this respect, it is intriguing to note that only medium sized oligomers contain cysteine-dependent dimers. In L929 murine fibroblasts genetically modified to express human Hsp27, a one hour exposure to TNFα, shifted, in an ROS-dependent manner, Hsp27 oligomers from an heterogenous native size distribution to large structures (up to 800 kDa). Thereafter, TNFα decreased the size of Hsp27 oligomers to about 200 kDa.[73,74] Hsp27 phosphorylated isoforms are recovered as small or medium-sized oligomers (<300 kDa). These studies, together with the analysis of Hsp27 mutants, led to the conclusion that the large unphosphorylated oligomers of Hsp27 represent the active form of the protein which modulates ROS and glutathione levels,[64] probably as a consequence of their in vitro chaperone activity.[46] Moreover, Hsp27 large oligomers may bind irreversibly oxidized polypeptides and stimulate their presentation to the ubiquitin-independent 20S proteasome, a protein degradation machine with a high affinity for oxidized proteins.[75] This activity may then counteract the accumulation of lipofuscin made of proteolysis resistant large aggregates of oxidized proteins.

Hsp27 was also reported to sensitize some cells to oxidative stress in spite of its ability to confer heat shock resistance.[31,76,77] Hence, despite the proeminent role played by the chaperone activity associated to its large oligomers, the mechanism of action of Hsp27 is probably stress- and cell-specific.

Hsp27 in Cells Committed to Apoptosis

Depending on their intensity, heat, oxidative or other types of stress induce at least two major two types of cell death processes. In the case of cells exposed to mild stress, cells have the opportunity to decide either to repair the damages or to commit suicide through an apoptotic process. In contrast, drastic stress usually trigger cell necrosis. The execution phase of apoptosis occurs through the proteolytic activation of specific proteases called caspases.[78,79] Several different pathways lead to the activation of caspases, such as those triggered by death receptors[80] and those which depend on the mitochondria-apoptosome machinery.[81,82] Apoptosis induced by either heat shock or oxidative stress usually occurs through the mitochondria pathway of activation.[83]

Depending on its intensity heat shock induces first an apoptotic death while necrosis is observed when the stress is drastic and impairs Hsps expression. Heat induced apoptotic process requires cytochrome c release from mitochondria and caspase 3 activation[83] and is blocked by the broad caspase inhibitor z-VAD-fmk.[84,85] It has recently been described that apoptosis induced by heat shock is dependent on the PIDDosome complex and pro-caspase-2 activation.[86] For example, caspase-2 (-/-) and RIADD (-/-) cells that are devoid of functional PIDDosome are resistant to heat-induced apoptosis. Whether accumulation of denatured proteins, as it occurs during heat shock, triggers the activation of the PIDDosome complex is an interesting hypothesis to be tested (Fig. 2). Moreover, downstream of pro-caspase-2, the multidomain proapoptotic molecules Bax and Bak have been described to be directly activated by heat.[87] It is not yet known if Hsp27 or the other Hsps interfere specifically with these pathways.

We and others have reported that mammalian Hsp27, *Drosophila* Hsp27 as well as α-B crystallin counteract apoptosis induced by the kinase C inhibitor staurosporine. Hsp27 from human also protects against apoptosis induced by actinomycin D, camptothecin, Fas/APO-1 receptor,[88] cisplatin,[65] the topoisomerase II inhibitor etoposide,[85,89] and doxorubicin.[65,90,91] Hsp27 is less efficient in the protection against T antigen/p53[92] and stimulated CTL cells mediated apoptotic cell deaths.[93]

Concerning death ligands, we and others have shown that Hsp27 or alphaAB-crystallin expression are negative regulators of TNFα and Fas ligand mediated cell death.[9,88,94] Only the phosphorylated dimeric form of Hsp27 has been shown to interact with DAXX, a mediator of Fas-induced apoptosis, preventing the interaction of DAXX with both Fas and apoptotic signaling kinase 1 (Ask1) and therefore blocking DAXX-mediated apoptosis.[95] In the case of the ROS dependent necrotic cell death generated by TNFα, the formation of the large unphosphorylated oligomers of Hsp27 which bear chaperone activity is observed. However, since Hsp27 protects also against the apoptotic death induced by TNFα in the presence of actinomycine D, an inhibitory action of this Hsp should also occur at the level of caspases.

We have observed that Hsp27 is a negative regulator of several different pathways whih converge to mitochondria and trigger the release in the cytosol of apoptogenic factors, such as cytochrome c[96] and Smac/Diablo.[97] This activity of Hsp27 could be related, at least in part, to the ability of Hsp27 to protect the integrity of F-actin microfilaments against the damages induced by cytochalasin D or other apoptosis inducers (staurosporine, ROS, oxidative stress, heat shock...). This study also led to the discovery of an apoptotic-signaling pathway linking cytoskeleton damages to mitochondria.[96] Hsp27 also acts downstream of mitochondria. For example, Hsp27 was shown to interact with cytochrome c once this apoptogenic polypeptide is released from mitochondria. Once bound to Hsp27, cytochrome c looses its ability to trigger apoptosome formation.[89,98] Cytochrome c probably interacts with the N-terminal part of Hsp27 polypeptide, between amino acids 51 and 88;[98] a domain which is not related with the chaperone activity of Hsp27.[99] In contrast, Hsp27 does not interfere with the caspase independent apoptosis inducing factor (AIF) which, once released from mitochondria, triggers chromatin condensation and large-scale DNA fragmentation. Hsp27 has also been reported to bind pro-caspase 3 and to interfere with its activation ; however, the structural organization of Hsp27 responsible of this event is not known.[100]

Studies performed with genetically manipulated HeLa cells which express reduced levels of Hsp27 have revealed that the constitutive expression of this protein provides apoptotic resistance to staurosporine, etoposide, Fas, or cytochalasin D.[96] In HeLa cells, inducers-specific changes in the structural organization and phosphorylation of Hsp27 are observed that correlate with the decreased activation of caspases (Paul C, Arrigo A.-P. in preparation). For example, etoposide induces the transient formation of large Hsp27 oligomers suggesting that only the chaperone-like structure of the protein protects against this apoptotic agent. In contrast, staurosporine and cytochalasin D induce the transient formation of two populations of oligomeric Hsp27 structures characterized by large and small molecular masses. As already mentioned above, inhibition of DAXX-mediated cell death requires only the unphosphorylated dimers of Hsp27.[95] Hence, future studies will have to determine which oligomeric form of Hsp27 is active at the different steps of the apoptotic pathways and whether activities other than the chaperone function (formation of reservoirs) are also involved in the protection (see Fig. 3).

Hsp27 is transiently expressed during the early differentiation of different cell types.[7,101,102] Inhibition of Hsp27 accumulation results in differentiation failure, a phenomenon often accompanied by massive apoptosis induction.[22,103-105] The mechanism by which Hsp27 controls

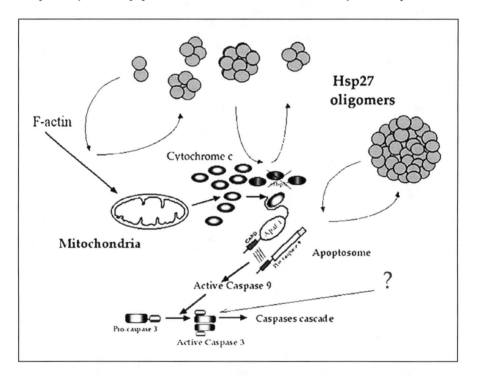

Figure 3. Hsp27 function in the mitochondrial apoptotic pathway. In vitro caspase assay performed with cytosolic post-mitochondrial fraction treated with cytochrome c and dATP revealed that Hsp27, once in the form of large unphosphorylated oligomers, interferes with the activation of pro-caspase-9 and pro-caspase-3 and copurifies with apoptosome containing fractions. Hsp27 in the form of medium sized or small oligomers may bind cytochrome c and only small oligomers appear to be active upstream of mitochondria at the level of F-actin. The structural organization of Hsp27 which is active in the case of the interaction with pro-caspase-3 is unknown.

the differentiation program and negatively modulates apoptosis is unknown but correlates with the formation of chaperone-like large Hsp27 oligomers. These large oligomers may avoid the accumulation of junk protein structures that trigger abberrant differentiation processes and apoptosis.

Conclusions and in Vivo Perspectives

It is now well established that the expression of Hsp27 and alphaB-crystallin protect, through their chaperone activity, cells against the deleterious damages induced by heat or oxidative injuries. These proteins also reduce and/or delay different apoptotic processes. However, more work is needed to better understand what is the exact role played by the « classical chaperone function » of sHsps (such as the formation of reservoirs containing folding intermediates) in their anti-apoptotic activity. Indeed, it is conceivable that, in addition to their chaperone function, other activities associated to these proteins may be required to fight against apoptotic cell death. This hypothesis is strenghtened by the fact that some Hsp27 mutants which protect against cytochrome c induced apoptosis are not efficient against oxidative stress induced cell death.[99]

In spite of the fact that the knock out of Hsp27 encoding gene is lethal, recent reports now provide the evidences that, in vivo, Hsp27 also acts as a negative regulator of programmed cell death,[106] and therefore may play a crucial role in the development and integrity of organisms. It is also well known that several human tumor cells constitutively express high levels of Hsp27.[107-110] Moreover, Hsp27 has been described to have a potent tumorigenic activity,[23] probably because of its ability to counteracts apoptosis mediated by death ligands, hence allowing cancer cells to escape the surveillance mediated by the immune system. In this respect, modification of Hsp27 by the methylglyoxal-arginine adduct argpyrimidine (due to the high level of non enzymatic glycation in glycolitic active cancer cells) stimulates its anti-apoptotic activity[111] and may be part of the mechanism triggered by some cancer cells for evasion of apoptosis.[112] It is also intriguing to note that elevated titers of serum antibodies to Hsp27 accompany several human diseases such as cancer and glaucoma. Hsp27 auto-immune antibodies are associated with improved survival in patients with breast cancer[113] probably because of their ability to induce apoptosis by inactivating or attenuating the anti-apoptotic activity of native Hsp27.[114] These observations have led to the conclusion that Hsp27 is a potential therapeutic target for future clinical studies.[110,115-119]

There are also several other pathological situations where high levels of Hsp27 are detected, such as in the case of chronic liver damage,[120] neurodegenerescence pathologies and asthma.[121,122] In these pathologies, the protective mechanism provided by Hsp27 expression is supposed to be beneficial as it fights against the pathological state. This assumption is supported by the fact that mutations in several human sHsps, such as Hsp27 (HspB1), are associated with inherited peripheral neuropathologies.[122] Hence, more work is needed before future strategies aimed at targetting Hsp27 function(s) in a specific disease could be planned.

Acknowledgements

Work in the author laboratory was supported by the Région Rhône-Alpes and the Association pour la Recherche sur le Cancer (ARC).

References

1. Tissieres A, Mitchell H, Tracy U. Protein synthesis in salivary glands of Drosophila melanogaster: Relation to chromosome puffs. J Mol Biol 1974; 84:389-398.
2. de Jong W, Leunissen J, Voorter C. Evolution of the alpha-crystallin/small heat-shock protein family. Mol Biol Evol 1993; 10:103-126.
3. de Jong WW, Caspers GJ, Leunissen JA. Genealogy of the alpha-crystallin—small heat-shock protein superfamily. Int J Biol Macromol 1998; 22:151-162.
4. Kappe G, Verschuure P, Philipsen RL et al. Characterization of two novel human small heat shock proteins: Protein kinase-related HspB8 and testis-specific HspB9. Biochim Biophys Acta 2001; 1520:1-6.

5. Kappe G, Leunissen JA, de Jong WW. Evolution and diversity of prokaryotic small heat shock proteins. Prog Mol Subcell Biol 2002; 28:1-17.
6. Kappe G, Franck E, Verschuure P et al. The human genome encodes 10 alpha-crystallin-related small heat shock proteins: HspB1-10. Cell Stress Chaperones 2003; 8:53-61.
7. Arrigo AP, Landry J. In: Morimoto RI, Tissieres A, Georgopoulos C, eds. The Biology of Heat Shock Proteins and Molecular Chaperones. Cold Spring Harbor, NY: Cold Spring Harbor Laboratory Press, 1994:335-373.
8. Landry J, Chretien P, Lambert H et al. Heat shock resistance confered by expression of the human HSP 27 gene in rodent cells. J Cell Biol 1989; 109:7-15.
9. Arrigo AP. Small stress proteins: Chaperones that act as regulators of intracellular redox state and programmed cell death. Biol Chem 1998; 379:19-26.
10. Arrigo AP, Préville X. In: Latchman DS, ed. Stress Proteins, Vol. 136, Handbook of Experimental Pharmacology. Springer, 1999:101-132.
11. Arrigo AP. Hsp27: Novel regulator of intracellular redox state. IUBMB Life 2001; 52:303-307.
12. Arrigo AP, Paul C, Ducasse C et al. Small stress proteins: Modulation of intracellular redox state and protection against oxidative stress. Prog Mol Subcell Biol 2002; 28:171-184.
13. Huot J, Roy G, Lambert H et al. Increased survival after treatments with anticancer agents of Chinese hamster cells expressing the human Mr 27,000 heat shock protein. Cancer Res 1991; 51:5245-5252.
14. Fuqua SA, Oesterreich S, Hilsenbeck SG et al. Heat shock proteins and drug resistance. Breast Cancer Res Treat 1994; 32:67-71.
15. Arrigo AP. sHsp as novel regulators of programmed cell death and tumorigenicity. Pathol Biol (Paris) 2000; 48:280-288.
16. Garrido C, Gurbuxani S, Ravagnan L et al. Heat shock proteins: Endogenous modulators of apoptotic cell death. Biochem Biophys Res Commun 2001; 286:433-442.
17. Arrigo AP, Paul C, Ducasse C et al. Small stress proteins: Novel negative modulators of apoptosis induced independently of reactive oxygen species. Prog Mol Subcell Biol 2002; 28:185-204.
18. Concannon CG, Gorman AM, Samali A. On the role of Hsp27 in regulating apoptosis. Apoptosis 2003; 8:61-70.
19. Latchman DS. HSP27 and cell survival in neurones. Int J Hyperthermia 2005; 21:393-402.
20. Jakob U, Gaestel M, Engels K et al. Small heat shock proteins are molecular chaperones. J Biol Chem 1993; 268:1517-1520.
21. Jakob U, Buchner J. Assisting spontaneity: The role of Hsp90 and small Hsps as molecular chaperones. Trends Biochem Sci 1994; 19:205-211.
22. Mehlen P, Mehlen A, Godet J et al. hsp27 as a switch between differentiation and apoptosis in murine embryonic stem cells. J Biol Chem 1997; 272:31657-31665.
23. Garrido C, Fromentin A, Bonnotte B et al. Heat shock protein 27 enhances the tumorigenicity of immunogenic rat colon carcinoma cell clones. Cancer Res 1998; 58:5495-5499.
24. Stokoe D, Engel K, Campbell D et al. Identification of MAPKAP kinase 2 as a major enzyme responsible for the phosphorylation of the small mammalian heat shock proteins. FEBS Lett 1992; 313:307-313.
25. Ciocca DR, Luque EH. Immunological evidence for the identity between the hsp27 estrogen-regulated heat shock protein and the p29 estrogen receptor-associated protein in breast and endometrial cancer. Breast Cancer Res Treat 1991; 20:33-42.
26. Morimoto RI. Regulation of the heat shock transcriptional response: Cross talk between a family of heat shock factors, molecular chaperones, and negative regulators. Genes Dev 1998; 12:3788-3796.
27. Rouse J et al. A novel kinase cascade triggered by stress and heat shock that stimulates MAPKAP kinase-2 and phosphorylation of the small heat shock proteins. Cell 1994; 78:1027-1037.
28. Arrigo AP, Welch W. Characterization and purification of the small 28,000-dalton mammalian heat shock protein. J Biol Chem 1987; 262:15359-15369.
29. Arrigo AP, Suhan JP, Welch WJ. Dynamic changes in the structure and intracellular locale of the mammalian low-molecular-weight heat shock protein. Mol Cell Biol 1988; 8:5059-5071.
30. Kato K, Hasegawa K, Goto S et al. Dissociation as a result of phosphorylation of an aggregated form of the small stress protein, hsp27. J Biol Chem 1994; 269:11274-11278.
31. Trautinger F, Kokesch C, Herbacek I et al. Overexpression of the small heat shock protein, hsp27, confers resistance to hyperthermia, but not to oxidative stress and UV-induced cell death, in a stably transfected squamous cell carcinoma cell line. J Photochem Photobiol 1997; 39:90-95.
32. Rollet E, Lavoie J, Landry J et al. Expression of Drosophila's 27 kDa heat shock protein into rodent cells confers thermal resistance. Biochem Biophys Res Commun 1992; 185:116-120.
33. Carra S, Sivilotti M, Chavez Zobel AT et al. HspB8, a small heat shock protein mutated in human neuromuscular disorders, has in vivo chaperone activity in cultured cells. Hum Mol Genet 2005; 14:1659-1669.

34. Kampinga HH, Brunsting JF, Stege GJ et al. Cells overexpressing Hsp27 show accelerated recovery from heat-induced nuclear protein aggregation. Biochem Biophys Res Commun 1994; 204:1170-1177.

35. Cuesta R, Laroia G, Schneider RJ. Chaperone Hsp27 inhibits translation during heat shock by binding eIF4G and facilitating dissociation of cap-initiation complexes. Genes Dev 2000; 14:1460-1470.

36. Carper SW, Rocheleau TA, Cimino D et al. Heat shock protein 27 stimulates recovery of RNA and protein synthesis following a heat shock. J Cell Biochem 1997; 66:153-164.

37. Marin-Vinader L, Shin C, Onnekink C et al. Hsp27 enhances recovery of splicing as well as rephosphorylation of SRp38 after heat shock. Mol Biol Cell 2005; 7:7-17.

38. Huot J, Houle F, Spitz DR et al. HSP27 phosphorylation-mediated resistance against actin fragmentation and cell death induced by oxidative stress. Cancer Res 1996; 56:273-279.

39. Lavoie JN, Lambert H, Hickey E et al. Modulation of cellular thermoresistance and actin filament stability accompanies phosphorylation-induced changes in the oligomeric structure of heat shock protein 27. Mol Cell Biol 1995; 15:505-516.

40. Ehrnsperger M, Lilie H, Gaestel M et al. The dynamics of hsp25 quaternary structure. Structure and function of different oligomeric species. J Biol Chem 1999; 274:14867-14874.

41. Ehrnsperger M, Graber S, Gaestel M et al. Binding of nonnative protein to Hsp25 during heat shock creates a reservoir of folding intermediates for reactivation. EMBO J 1997; 16:221-229.

42. Lee GJ, Roseman AM, Saibil HR et al. Small heat shock protein stably binds heat-denatured model substrates and can maintain a substrate in a folding-competent state. EMBO J 1997; 16:659-671.

43. Stromer T, Ehrnsperger M, Gaestel M et al. Analysis of the interaction of small heat shock proteins with unfolding proteins. J Biol Chem 2003; 278:18015-18021.

44. Haslbeck M, Franzmann T, Weinfurtner D et al. Some like it hot: The structure and function of small heat-shock proteins. Nat Struct Mol Biol 2005; 12:842-846.

45. Ehrnsperger M, Gaestel M, Buchner J. Analysis of chaperone properties of small Hsp's. Methods Mol Biol 2000; 99:421-429.

46. Rogalla T, Ehrnsperger M, Preville X et al. Regulation of Hsp27 oligomerization, chaperone function, and protective activity against oxidative stress/tumor necrosis factor alpha by phosphorylation. J Biol Chem 1999; 274:18947-18956.

47. Benndorf R, Hayess K, Ryazantsev S et al. Phosphorylation and supramolecular organization of murine small heat shock protein HSP25 abolish its actin polymerization-inhibiting activity. J Biol Chem 1994; 269:20780-20784.

48. Guay J, Lambert H, GingrasBreton G et al. Regulation of actin filament dynamics by p38 map kinase-mediated phosphorylation of heat shock protein 27. J Cell Sci 1997; 110:357-368.

49. Mounier N, Arrigo AP. Actin cytoskeleton and small heat shock proteins: How do they interact? Cell Stress Chaperones 2002; 7:167-176.

50. Pivovarova AV, Mikhailova VV, Chernik IS et al. Effects of small heat shock proteins on the thermal denaturation and aggregation of F-actin. Biochem Biophys Res Commun 2005; 331:1548-1553.

51. Vayssier M, Banzet N, Francois D et al. Tobacco smoke induces both apoptosis and necrosis in mammalian cells: Differential effects of HSP70. Am J Physiol 1998; 275:771-779.

52. Zucker B, Hanusch J, Bauer G. Glutathione depletion in fibroblasts is the basis for apoptosis-induction by endogenous reactive oxygen species. Cell Death and Diff 1997; 4:388-395.

53. Samali A, Nordgren H, Zhivotovsky B et al. A comparative study of apoptosis and necrosis in HepG2 cells: Oxidant-induced caspase inactivation leads to necrosis. Biochem Biophys Res Commun 1999; 255:6-11.

54. Jacobson MD. Reactive oxygen species and programmed cell death. Trends Biochem Sci 1996; 21:83-86.

55. Powis G, Briehl M, Oblong J. Redox signalling and the control of cell growth and death. Pharmacol Ther 1995; 68:149-173.

56. Hockenbery DM, Oltvai ZN, Yin XM et al. Bcl-2 functions in an antioxidant pathway to prevent apoptosis. Cell 1993; 75:241-251.

57. Kane DJ, Sarafian TA, Anton R et al. Bcl-2 inhibition of neural death: Decreased generation of reactive oxygen species. Science 1993; 262:1274-1277.

58. Mehlen P, Briolay J, Smith L et al. Analysis of the resistance to heat and hydrogen peroxide stresses in COS cells transiently expressing wid type or deletion mutants of the Drosophila 27-kDa heat-shock protein. Eur J Biochem 1993; 215:277-284.

59. Mehlen P, Préville X, Kretz-Remy C et al. Human hsp27, Drosophila hsp27 and human aB-crystallin expression-mediated increase in glutathione is essential for the protective activity of these protein against TNFα-induced cell death. EMBO J 1996; 15:2695-2706.

60. Preville X, Salvemin F, Giraud S et al. Mammalian small stress proteins protect against oxidative stress through their ability to increase glucose-6-phosphate dehydrogenase activity and by maintaining optimal cellular detoxifying machinery. Exp Cell Res 1999; 247:61-78.

61. Mehlen P, Préville X, Chareyron P et al. Constitutive expression of human hsp27, Drosophila hsp27, or human alpha B-crystallin confers resistance to TNF- and oxidative stress-induced cytotoxicity in stably transfected murine L929 fibroblasts. J Immunol 1995; 154:363-374.

62. Wang G, Klostergaard J, Khodadadian M et al. Murine cells transfected with human Hsp27 cDNA resist TNF-induced cytotoxicity. J Immunother Emphasis Tumor Immunol 1996; 19:9-20.

63. Park YM, Han MY, Blackburn RV et al. Overexpression of HSP25 reduces the level of TNF alpha-induced oxidative DNA damage biomarker, 8-hydroxy-2'-deoxyguanosine, in L929 cells. J Cell Physiol 1998; 174:27-34.

64. Mehlen P, Hickey E, Weber L et al. Large unphosphorylated aggregates as the active form of hsp27 which controls intracellular reactive oxygen species and glutathione levels and generates a protection against TNFα in NIH-3T3-ras cells. Biochem Biophys Res Comm 1997; 241:187-192.

65. Garrido C, Ottavi P, Fromentin A et al. HSP27 as a mediator of confluence-dependent resistance to cell death induced by anticancer drugs. Cancer Res 1997; 57:2661-2667.

66. Huot J, Houle F, Marceau F et al. Oxidative stress-induced actin reorganization mediated by the p38 mitogen-activated protein kinase/heat shock protein 27 pathway in vascular endothelial cells. Circ Res 1997; 80:383-392.

67. Préville X, Gaestel M, Arrigo AP. Phosphorylation is not essential for protection of L929 cells by Hsp25 against H2O2-mediated disruption actin cytoskeleton, a protection which appears related to the redox change mediated by Hsp25. Cell Stress Chaperones 1998; 3:177-187.

68. Meister A, Anderson ME. Glutathione. Annu Rev Biochem 1983; 52:711-760.

69. Baek SH, Min JN, Park EM et al. Role of small heat shock protein HSP25 in radioresistance and glutathione-redox cycle. J Cell Physiol 2000; 183:100-107.

70. Paul C, Arrigo AP. Comparison of the protective activities generated by two survival proteins: Bcl-2 and Hsp27 in L929 murine fibroblasts exposed to menadione or staurosporine. Exp Gerontol 2000; 35:757-766.

71. Arrigo AP, Firdaus WJ, Mellier G et al. Cytotoxic effects induced by oxidative stress in cultured mammalian cells and protection provided by Hsp27 expression. Methods 2005; 35:126-138.

72. Diaz-Latoud C, Buache E, Javouhey E et al. Substitution of the unique cysteine residue of murine hsp25 interferes with the protective activity of this stress protein through inhibition of dimer formation. Antioxid Redox Signal 2005; 7:436-445.

73. Mehlen P, Mehlen A, Guillet D et al. Tumor necrosis factor-a induces changes in the phosphorylation, cellular localization, and oligomerization of human hsp27, a stress protein that confers cellular resistance to this cytokine. J Cell Biochem 1995; 58:248-259.

74. Mehlen P, Kretz-Remy C, Briolay J et al. Intracellular reactive oxygen species as apparent modulators of heat-shock protein 27 (hsp27) structural organization and phosphorylation in basal and tumour necrosis factor alpha-treated T47D human carcinoma cells. Biochem J 1995; 312:367-375.

75. Sitte N, Merker K, Grune T. Proteasome-dependent degradation of oxidized proteins in MRC-5 fibroblasts. FEBS Lett 1998; 440:399-402.

76. Arata S, Hamaguchi S, Nose K. Effects of the overexpression of the small heat shock protein, Hsp27, on the sensitivity of human fibroblast cells exposed to oxidative stress. J Cell Physiol 1995; 163:458-465.

77. Mairesse N, Bernaert D, Del Bino G et al. Expression of HSP27 results in increased sensitivity to tumor necrosis factor, etoposide, and H2O2 in an oxidative stress-resistant cell line. J Cell Physiol 1998; 177:606-617.

78. Nicholson DW, Thornberry NA. Caspases: Killer proteases. Trends Biochem Sci 1997; 22:299-306.

79. Thornberry NA, Lazebnik Y. Caspases: Enemies within. Science 1998; 281:1312-1316.

80. Scaffidi C, Fulda S, Srinivasan A et al. Two CD95 (APO-1/Fas) signaling pathways. Embo J 1998; 17:1675-1687.

81. Reed JC. Cytochrome c: Can't live with it—can't live without it. Cell 1997; 91:559-562.

82. Green DR, Reed JC. Mitochondria and apoptosis. Science 1998; 281:1309-1312.

83. Samali A, Robertson JD, Peterson E et al. Hsp27 protects mitochondria of thermotolerant cells against apoptotic stimuli. Cell Stress Chaperones 2001; 6:49-58.

84. Samali A, Cotter TG. Heat shock proteins increase resistance to apoptosis. Exp Cell Res 1996; 223:163-170.

85. Samali A, Orrenius S. Heat shock proteins: Regulators of stress response and apoptosis. Cell Stress Chaperones 1998; 3:228-236.

86. Tu S, McStay GP, Boucher LM et al. In situ trapping of activated initiator caspases reveals a role for caspase-2 in heat shock-induced apoptosis. Nat Cell Biol 2006; 8:72-77.

87. Pagliari LJ, Kuwana T, Bonzon C et al. The multidomain proapoptotic molecules Bax and Bak are directly activated by heat. Proc Natl Acad Sci USA 2005; 102:17975-17980.

88. Mehlen P, Schulze-Osthoff K, Arrigo AP. Small stress proteins as novel regulators of apoptosis. Heat shock protein 27 blocks Fas/APO-1- and staurosporine-induced cell death. The J Biol Chem 1996; 271:16510-16514.

89. Garrido C, Bruey JM, Fromentin A et al. HSP27 inhibits cytochrome c-dependent activation of procaspase-9. Faseb J 1999; 13:2061-2070.

90. Garrido C, Mehlen P, Fromentin A et al. Inconstant association between 27-kDa heat-shock protein (Hsp27) content and doxorubicin resistance in human colon cancer cells. The doxorubicin-protecting effect of Hsp27. Eur J Biochem 1996; 237:653-659.

91. Hansen RK, Parra I, Lemieux P et al. Hsp27 overexpression inhibits doxorubicin-induced apoptosis in human breast cancer cells. Breast Cancer Res Treat 1999; 56:187-196.

92. Guenal I, Sidoti-de Fraisse C, Gaumer S et al. Bcl-2 and Hsp27 act at different levels to suppress programmed cell death. Oncogene 1997; 15:347-360.

93. Beresford PJ, Jaju M, Friedman RS et al. A role for heat shock protein 27 in CTL-mediated cell death. J Immunol 1998; 161:161-167.

94. Andley UP, Song Z, Wawrousek EF et al. Differential protective activity of αA- and {alpha}B-crystallin in lens epithelial cells. J Biol Chem 2000; 30:30-40.

95. Charette SJ, Lavoie JN, Lambert H et al. Inhibition of daxx-mediated apoptosis by heat shock protein 27. Mol Cell Biol 2000; 20:7602-7612.

96. Paul C, Manero F, Gonin S et al. Hsp27 as a negative regulator of cytochrome C release. Mol Cell Biol 2002; 22:816-834.

97. Chauhan D, Li G, Hideshima T et al. Hsp27 inhibits release of mitochondrial protein Smac in multiple myeloma cells and confers dexamethasone resistance. Blood 2003; 102:3379-3386.

98. Bruey JM, Ducasse C, Bonniaud P et al. Hsp27 negatively regulates cell death by interacting with cytochrome c. Nat Cell Biol 2000; 2:645-652.

99. Arrigo AP, Virot S, Chaufour S et al. Hsp27 consolidates intracellular redox homeostasis by up-holding glutathione in its reduced form and by decreasing iron intracellular levels. Antioxid Redox Signal 2005; 7:414-422.

100. Pandey P, Farber R, Nakazawa A et al. Hsp27 functions as a negative regulator of cytochrome c-dependent activation of procaspase-3. Oncogene 2000; 19:1975-1981.

101. Arrigo AP. Expression of stress genes during development. Neuropathol and Applied Neurobiol 1995; 21:488-491.

102. Arrigo AP. In search of the molecular mechanism by which small stress proteins counteract apoptosis during cellular differentiation. J Cell Biochem 2005; 94:241-246.

103. Chaufour S, Mehlen P, Arrigo AP. Transient accumulation, phosphorylation and changes in the oligomerization of Hsp27 during retinoic acid-induced differentiation of HL-60 cells: Possible role in the control of cellular growth and differentiation. Cell Stress Chaperones 1996; 1:225-235.

104. Davidson SM, Loones MT, Duverger O et al. The developmental expression of small HSP. Prog Mol Subcell Biol 2002; 28:103-128.

105. Mehlen P, Coronas V, Ljubic-Thibal V et al. Small stress protein Hsp27 accumulation during dopamine-mediated differentiation of rat olfactory neurons counteracts apoptosis. Cell Death Differ 1999; 6:227-233.

106. Brar BK, Stephanou A, Wagstaff MJ et al. Heat shock proteins delivered with a virus vector can protect cardiac cells against apoptosis as well as against thermal or hypoxic stress. J Mol Cell Cardiol 1999; 31:135-146.

107. Têtu B, Têtu BB, Lacasse HL et al. Prognostic influence of HSP-27 expression in malignant fibrous histiocytoma: A clinicopathological and immunohistochemical study. Cancer Res 1992; 52:2325-2328.

108. Ciocca DR, Oesterreich S, Chamnes GC et al. Biological and clinical implications of heat shock proteins 27000 (Hsp27): A review. J Natl Cancer Inst 1993; 85:1558-1570.

109. Ciocca DR, Calderwood SK. Heat shock proteins in cancer: Diagnostic, prognostic, predictive, and treatment implications. Cell Stress Chaperones 2005; 10:86-103.

110. Rocchi P, Beraldi E, Ettinger S et al. Increased Hsp27 after androgen ablation facilitates androgen-independent progression in prostate cancer via signal transducers and activators of transcription 3-mediated suppression of apoptosis. Cancer Res 2005; 65:11083-11093.

111. Sakamoto H, Mashima T, Yamamoto K et al. Modulation of heat-shock protein 27 (Hsp27) anti-apoptotic activity by methylglyoxal modification. J Biol Chem 2002; 277:45770-45775.

112. Heijst JW, Niessen HW, Musters RJ et al. Argpyrimidine-modified Heat Shock Protein 27 in human nonsmall cell lung cancer: A possible mechanism for evasion of apoptosis. Cancer Lett 2005; 5:5-15.
113. Conroy SE, Sasieni PD, Amin V et al. Antibodies to heat-shock protein 27 are associated with improved survival in patients with breast cancer. Br J Cancer 1998; 77:1875-1879.
114. Tezel G, Wax MB. The mechanisms of hsp27 antibody-mediated apoptosis in retinal neuronal cells. J Neurosci 2000; 20:3552-3562.
115. Berrieman HK, Cawkwell L, O'Kane SL et al. Hsp27 may allow prediction of the response to single-agent vinorelbine chemotherapy in nonsmall cell lung cancer. Oncol Rep 2006; 15:283-286.
116. Feng JT, Liu YK, Song HY et al. Heat-shock protein 27: A potential biomarker for hepatocellular carcinoma identified by serum proteome analysis. Proteomics 2005; 5:4581-4588.
117. Shin KD, Lee MY, Shin DS et al. Blocking tumor cell migration and invasion with biphenyl isoxazole derivative KRIBB3, a synthetic molecule that inhibits Hsp27 phosphorylation. J Biol Chem 2005; 280:41439-41448.
118. Mineva I, Gartne W, Hauser P et al. Differential expression of alphaB-crystallin and Hsp27-1 in anaplastic thyroid carcinomas because of tumor-specific alphaB-crystallin gene (CRYAB) silencing. Cell Stress Chaperones 2005; 10:171-184.
119. Thanner F, Sutterlin MW, Kapp M et al. Heat shock protein 27 is associated with decreased survival in node-negative breast cancer patients. Anticancer Res 2005; 25:1649-1653.
120. Federico A, Tuccillo C, Terracciano F et al. Heat shock protein 27 expression in patients with chronic liver damage. Immunobiology 2005; 209:729-735.
121. Merendino AM, Paul C, Costa MA et al. Heat shock protein-27 protects human bronchial epithelial cells against oxidative stress-mediated apoptosis: Possible implication in asthma. Cell Stress Chaperones 2002; 7:269-280.
122. Dierick I, Irobi J, De Jonghe P et al. Small heat shock proteins in inherited peripheral neuropathies. Ann Med 2005; 37:413-422.

CHAPTER 3

Molecular Interaction Network of the Hsp90 Chaperone System

Rongmin Zhao and Walid A. Houry*

Abstract

Hsp90 is an essential and ubiquitous molecular chaperone that is required for the proper folding of a set of client proteins at a late stage in their folding process. In eukaryotes, cytoplasmic Hsp90 is absolutely essential for cell viability under all growth conditions. The functional cycle of the Hsp90 system requires a cohort of cochaperones and cofactors that regulate the activity of this chaperone. Hence, Hsp90 function is highly complex; in order to understand that complexity, several groups have attempted to map out the interaction network of this chaperone in yeast and mammalian systems using the latest available proteomic and genomic tools. Interaction networks emerging from these large scale efforts clearly demonstrate that Hsp90 plays a central role effecting multiple pathways and cellular processes. In yeast *Saccharomyces cerevisiae*, Hsp90 was shown to interact directly or indirectly with at least 10% of the yeast ORFs. The systematic application of large scale approaches to map out the Hsp90 chaperone network should allow the determination of the mechanisms employed by this chaperone system to maintain protein homeostasis in the cell.

Introduction

Hsp90 is an abundant and essential molecular chaperone present in all eukaryotic cells. Hsp90 has been implicated in mediating the folding of a specific set of substrate proteins, which are typically called 'clients'. Hsp90 clients are thought to interact with the chaperone in a nearly native state at the late stage of de novo folding;[1] alternatively, Hsp90 can also enhance the rate of reactivation of heat denatured proteins under stress conditions.[2] Although no common sequence or structural features have yet been identified for Hsp90 clients, however, many of the early clients that were identified fell into two functional classes: transcription factors[3] and protein kinases.[4] These early observations suggested the involvement of Hsp90 in modulating signaling pathways. Hsp90 is, therefore, considered to be particularly important for cell cycle control, organism development, and response to environmental stresses.[5,6] Hsp90 has become a novel anti-cancer drug target, and there are several drugs that specifically bind to Hsp90 and are in clinical trials.[7,8]

Hsp90 is a dimeric protein consisting of three domains: an N-terminal ATP-binding domain, a middle region, and a C-terminal domain involved in homodimerization. The N-terminal domain has a unique fold termed the Bergerat fold.[9] Although the ATPase activity of Hsp90 is weak, ATP binding and hydrolysis is crucial to the chaperone function.[10] Several drugs that

*Corresponding Author: Walid A. Houry—Department of Biochemistry, University of Toronto, Toronto, Ontario M5S 1A8, Canada. Email: walid.houry@utoronto.ca

Molecular Aspects of the Stress Response: Chaperones, Membranes and Networks, edited by Peter Csermely and László Vígh. ©2007 Landes Bioscience and Springer Science+Business Media.

have been designed to target Hsp90, such as geldanamycin, bind in the ATP pocket at the N-terminal domain. In eukaryotes, the C-terminal domain of cytosolic Hsp90 ends with a conserved EEVD motif. This motif has been shown to be essential for the interaction of cytosolic Hsp90 with cofactors that contain α-helical tetratricopeptide repeat domains (TPR).

The Hsp90 functional cycle has been intensively investigated, however, due to the complexity of this chaperone system and the diversity of the clients, a full understanding of the molecular aspects of the cycle is still lacking. It is known that the activity of Hsp90 requires the involvement of other chaperone systems such as Hsp70 and Hsp40 as well as several cofactors (Fig. 1). The Hsp90 functional cycle has been well-studied for the case of steroid hormone receptors (Fig. 1).[3,11] It is thought that the monomeric inactive receptor is initially recognized by the Hsp70/40 system and, subsequently, transferred to Hsp90. The formation of a complex between Hsp70 and Hsp90 is mediated by the scaffold protein Hop/p60 (Sti1 in yeast). In yeast, Sti1 has been shown to be an activator of the Hsp70 ATPase and an inhibitor of Hsp90 ATPase, although mammalian Hop does not seem to have such an effect.[12,13] Subsequently, other cyclophilins (e.g., yeast Cpr6 and Cpr7) and immunophilins (e.g., yeast Cns1) as well as the cochaperone p23 (Sba1 in yeast) interact with the Hsp70/40-Hop-Hsp90-substrate complex

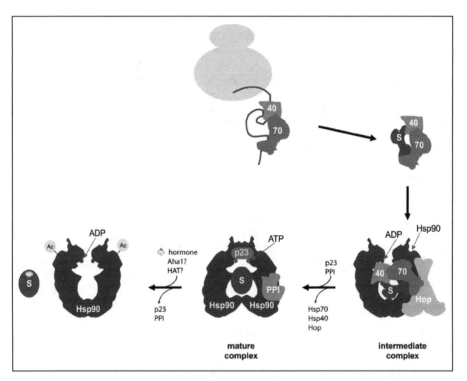

Figure 1. The Hsp90 functional cycle. A nascent polypeptide chain of a steroid hormone receptor, for example, is initially recognized by the Hsp70/Hsp40 system. Hop (yeast Sti1) binds both Hsp70 and Hsp90 to form the intermediate complex allowing the transfer of the substrate protein to Hsp90. Subsequently, p23 (yeast Sba1) and other cyclophilins and immunophilins (indicated by PPI) bind Hsp90, releasing Hsp70/Hsp40 and Hop. In the resulting mature complex, Hsp90 is stabilized in the ATP state by p23. If the substrate is a protein kinase, Cdc37 is also involved in stabilizing this complex (not shown). Upon entry of the hormone into the cell, the receptor binds the hormone and is released from Hsp90. The release is promoted by Aha1 which enhances the ATPase activity of the chaperone and, in mammalian cells, is also possibly promoted by the acetylation of the chaperone.

resulting in the release of Hsp70/40 and Hop. p23 stabilizes the Hsp90-receptor complex in the ATP state. Upon arrival of the hormone into the cell, the receptor binds the hormone and is released from Hsp90 upon hydrolysis of the chaperone-bound ATP to ADP; the receptor is then active as a dimer and translocates into the nucleus. The ATPase activity of Hsp90 is enhanced by the cofactor Aha1/Hch1.[14,15] It has also been demonstrated that Hsp90 activity can be regulated by phosphorylation by the phosphatase Ppt1[16] in yeast and by acetylation/deacetylation by an unknown HAT and by HDAC6 in mammalian cells, respectively.[17,18] Several other Hsp90 cofactors are also known, such as Cdc37 and CHIP, and they affect the chaperone activity in different manners. It is generally thought that different Hsp90 cofactors target the chaperone to different sets of substrates.

Recently, several systematic genomic and proteomic approaches were employed in order to map out the Hsp90 molecular interaction network. These approaches provide a global view of the influence of the Hsp90 system on multiple processes and complexes in the cell. They also shed more light on the mechanism of Hsp90 function. The description of these approaches follows below.

Mapping the Hsp90 Physical Interaction Network

The Hsp90 physical interactors represent a class of proteins that either directly interact with Hsp90 or that are part of complexes that interact with Hsp90. Most of these Hsp90 physical interactors are thought to mainly include Hsp90 cofactors and cochaperones as well as Hsp90 clients. Two different methods have been employed to map out such interactors: (1) pull-down or immunoprecipitation of the chaperone to isolate Hsp90-containing complexes followed by mass spectrometry to identify proteins in those complexes, and (2) two-hybrid method to identify proteins that, predominantly, directly bind to Hsp90.

Isolation of Hsp90 Complexes Followed by Mass Spectrometry

Isolation of Hsp90 complexes followed by mass spectrometry to identify components of the complexes is an effective and straightforward approach to explore the Hsp90 interactors. Matrix-assisted laser desorption ionization time of flight (MALDI-ToF), tandem electrospray mass spectrometry (MS/MS), or liquid chromatography coupled with tandem mass spectrometry (LC-MS/MS) have been extensively used to identify single proteins or protein mixtures[19] in gels or in solution. While protein identification might be robust, this approach greatly depends on the quality of purified complexes.

This method has been used to isolate Hsp90 complexes either by affinity pull-down of tagged Hsp90 in yeast or by coimmunoprecipitation of Hsp90 from mammalian cells. Zhao et al[20] introduced a TAP-tag to endogenous Hsp90 in yeast, while keeping the gene under the control of its own promoter. The TAP-tag is a tandem affinity purification tag containing two IgG binding domains from Protein A and a calmodulin binding peptide motif separated by a tobacco etch virus (TEV) protease cut site.[21] Yeast has two virtually identical *HSP90* genes, one termed *HSP82*, which is heat shock induced, while the other is termed *HSC82*, which is constitutively expressed. The proteins are localized in the cytoplasm. Strains were constructed in which endogenous *HSC82* or *HSP82* were C-terminally TAP-tagged, in addition, a third strain was constructed in which *HSP82* was deleted, while *HSC82* was N-terminally TAP-tagged. Hsp90-containing complexes were then isolated on IgG beads, released using TEV protease, further purified on calmodulin beads in the presence of calcium, and then released by addition of EGTA to chelate the calcium. This approach allowed the isolation of native protein complexes that were purified to more than 10[6] fold to virtual homogeneity. Since the EEVD motif at the C-terminus of Hsp90 is essential for the interaction of Hsp90 cofactors with the chaperone, most of the useful complexes were obtained from the third strain in which *HSC82* was N-terminally TAP-tagged and *HSP82* was deleted. It should be noted that, in this approach, the addition of the tag might affect Hsp90 activity. Indeed, the yeast strain with N-terminally TAP-tagged Hsp90 had slight temperature sensitivity. Isolated proteins were identified by mass spectrometry and some 'hits' were then verified using reverse pull-downs in which the identified proteins were TAP-tagged and pulled-down.

To identify Hsp90 interactors in mammalian systems, Falsone et al[22] used a commercially available anti-Hsp90 antibody to immunoprecipitate Hsp90 complexes from human embryonic kidney (HEK293) cells. One of the hits was verified by determining that its levels were sensitive to the inhibition of Hsp90 by the drug geldanamycin. It should be noted that in the human genome there are seventeen genes which have been detected to belong to the Hsp90 family.[23] Six of these genes were recognized as functional, while the other eleven genes were classified as pseudogenes. These Hsp90s are present in different compartments of the eukaryotic cell further complicating the studies on mammalian Hsp90.

Two Hybrid Screens

The yeast two-hybrid (2H) system is one of the most widely used methods to screen or confirm protein-protein interactions.[24] The basic principle of this system is that a protein X, the bait, is fused to a DNA binding domain (DBD), which is commonly derived from the Gal4 or LexA transcription factors. Protein Y, the prey, is fused to a transcription activation domain (AD). If proteins X and Y interact, then AD and DBD are brought close to each other, triggering the activation of a downstream reporter gene.

Despite the popularity of this method, there are several drawbacks that have to be considered. The interactions have to occur in the nucleus and a nuclear targeting sequence is present in the constructs used. In this regard, yeast Hsp90 is predominantly cytoplasmic and it has been reported that the chaperone has a cytoplasmic localization signal in its sequence.[25] The nuclear localization signal in the DBD domain may not be strong enough to bring the large Hsp90 to the nucleus and this is indeed observed for full-length yeast Hsp90 (R. Zhao and W.A. Houry, unpublished results). Also, the DBD-Hsp90 fusion might not be fully active. It has been reported that a C-terminal Hsp90-Gal4 fusion can support yeast growth better than an N-terminal Gal4-Hsp90 fusion.[26] Also, the interaction between Hsp90 and its client proteins is probably transient and the interaction might not be strong enough to activate the downstream reporter gene. Nevertheless, this approach has been successfully used by the Piper and Houry groups[20,27,28] to map out the Hsp90 2H interaction network using an ordered array of about 6000 yeast ORFs fused to Gal4 activation domain that was created by Uetz et al.[29]

Mapping the Hsp90 Genetic Interaction Network

Hsp90 genetic interactors represent a class of proteins that are involved in a pathway where Hsp90 also plays key functions in parallel. Hsp90 genetic interactors may or may not directly interact with the chaperone, but both, the chaperone and its genetic interactor, are required for the successful function of a particular essential cellular process or pathway, and consequently, for cell viability. The large scale identification of Hsp90 genetic interactors has been successfully used for yeast Hsp90.[20]

Synthetic Genetic Array

Synthetic genetic array (SGA) analysis is based on the principle of synthetic lethality developed for the budding yeast *Saccharomyces cerevisiae*.[30] For this organism, about 80% of the open reading frames are nonessential.[31] Tong et al[30] developed automated methods to screen for synthetic lethality of a given knockout strain against the systematic gene deletion array[32,33] consisting of about 4700 haploid strains each of which has a marked deletion of one nonessential gene. A knockout strain is mated with the deletion array to eventually obtain strains with double deletions. If these strains are sick or not viable, then the two deleted genes are said to be genetic interactors. Putative hits can be verified by tetrad analysis or random spore analysis. The detailed method and its application has recently been reviewed in reference 34.

This screen is ideally designed for a query of nonessential gene; however, it is also suitable for an essential gene query if a conditional knockout, such a temperature sensitive mutant of

the essential gene, is available. Hsp90 is essential in all eukaryotic cells. A number of temperature sensitive yeast Hsp82 mutants such as Hsp82(G170D), Hsp82(G313S), and
Hsp82(E381K) have already been identified and characterized.[35] An SGA analysis of Hsp90
interactors has recently been carried out using a strain in which *hsc82* is deleted and *HSP82* is
rendered temperature sensitive by introducing the G170D mutation.[20] The screen was carried
out at the semi-permissive temperature of 35°C.

Chemical Genetic Screen/Drug Screen

The chemical genetic screen or drug screen is based on the principle of synthetic lethality
and takes advantage of the unique bar codes introduced to mark each deletion mutant in the
systematic deletion array.[32] These bar codes can be amplified using common primers. In this
method, the bar-coded haploid deletion mutants of the nonessential genes[31] are pooled and
grown in liquid culture containing an Hsp90 specific drug inhibitor, geldanamycin, to challenge Hsp90 function. Strains containing deleted genes that are synthetic lethal or synthetic
sick with Hsp90 will die or grow slower compared to other strains. The same experiment is
repeated in the absence of the drug. The common primers are then used to amplify genomic
DNA from the pooled strains and the resultant products are hybridized onto specific DNA
oligo array chips. Strains that do not grow or that grow slowly in the presence of the drug will
not give a signal or will give a lower signal when compared to the signal obtained in the absence
of the drug. The hits obtained from applying this technique should closely overlap with the hits
obtained from the SGA analysis as both methods depend on the same principle of synthetic
lethality. Zhao et al[20] employed this method to determine yeast Hsp90 genetic interactors
using the drug geldanamycin (geldanamycin screen or GS).

The Hsp90 Interactome

Yeast is currently the ideal system for large-scale genetic and proteomic studies due to the
ready availability of many tools for such studies in this organism. Figure 2A gives an overview
of the number of hits obtained using the four different methods described above for yeast
Hsp90 as reported by Zhao et al.[20] The Hsp90 interactors (over 600 ORFs) represent about
10% of the yeast proteome and are distributed among multiple cellular functional categories
(Fig. 3) suggesting that Hsp90 is involved in many processes. One feature that stands out in
Figure 2A is that there is little overlap among the different datasets obtained using the different
methods. This reflects the experimental differences between the methods used as each method
favors certain types of interactions. The best overlap was obtained between the SGA and the
geldanamycin screen (GS). A total of 451 Hsp90 genetic interactors were identified from the
two screens with an overlap of 49 hits. These 49 hits represent a high fidelity dataset that could
then be pursued further for more detailed biochemical studies. The overlap is considered to be
reasonable.

On the other hand, the physical interaction screens gave a total of 208 hits with an overlap
of only 10 hits between the TAP and 2H screens (Fig. 2A), which is less than expected. The
differences might arise because the 2H screen predominantly detects binary interactions, while
the TAP method detects interactions between the bait and a complex of polypeptides that need
not directly bind to the protein bait. Hence, the 2H screen might reveal weaker interactions
than the TAP screen.

It is interesting to note that no protein was simultaneously detected by the four methods
used and that only Cpr6, Cpr7, Sti1, and Mon1 were detected by three of the four screens.
Cpr6, Cpr7, and Sti1 are all well-established Hsp90 cofactors, while the interaction of Hsp90
with Mon1 has not been reported before; therefore, Mon1 is likely an Hsp90 substrate. Mon1
is poorly studied; however, it is known to form a complex with another protein termed Ccz1
that binds the vacuolar membrane and is involved in homotypic vacuole fusion.[36,37] This might
implicate Hsp90 in vacuolar trafficking.

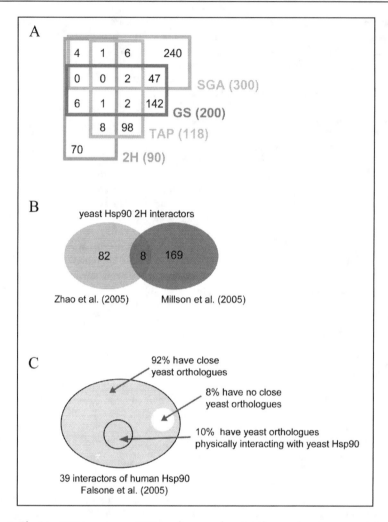

Figure 2. The Hsp90 interactome. A) Venn diagram showing the overlap among proteins identified by Zhao et al[20] to interact physically or genetically with yeast Hsp90. Reproduced wtih permission from reference 20, ©2005 Elsevier. B) Overlap in the 2H yeast Hsp90 interactors identified by Zhao et al[20] and Millson et al.[28] C) Features of the mammalian Hsp90 interactors identified by Falsone et al[22] as compared to yeast orthologues.

There were 22 proteins that were found to have the unique property of interacting with Hsp90 physically and genetically (Fig. 2A).[20] These proteins are either important cofactors of the Hsp90 chaperone system that become essential when the chaperone's activity is reduced. Alternatively, these proteins are substrates of Hsp90 that physically interact with the chaperone. In this scenario, Hsp90 might affect directly or indirectly the folding of two proteins X and Y and is found to physically interact with X (one of the 22 proteins). X and Y have to be genetic interactors such that strains deleted of both the *X* and *Y* genes are not viable. Hence, when Hsp90 function is compromised, X and Y will not fold properly and synthetic lethality results when X is deleted in a strain background with reduced Hsp90 activity. Identifying two proteins to be physically as well as genetically interacting is not expected to be common and might be a peculiar property of chaperones.

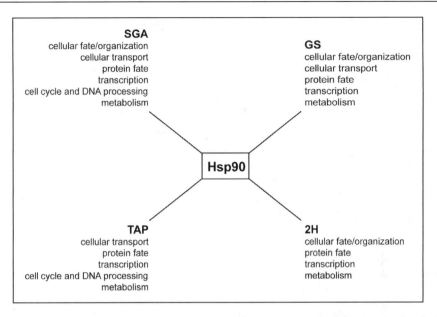

Figure 3. Hsp90 influences multiple cellular processes. The yeast Hsp90 interactors identified by Zhao et al.[20] are distributed among a variety of functional categories. Functional categories which contain more than 10% of interactors for each screening method are listed.

2H screens for yeast Hsp90 have also been carried out by Millson et al.[27,28] Initially, the group used full length Hsc82 as bait; however, even with DBD domain fused to the C-terminus of Hsc82, which is reported to have similar activity as the wild type chaperone, only four interactors were identified including the known Hsp90 cofactors Aha1 and Hch1.[27] To detect Hsp90 clients that probably only weakly interact with Hsp90, Millson et al repeated the 2H screen using a mutant Hsp82(E33A). This form of Hsp90 is barely functional and is thought not to proceed to the final maturation step in the chaperone functional cycle due to failed ATP hydrolysis, which in turn is thought to hold the client proteins tighter rendering them easier to detect by 2H. A total of 177 genes including some known Hsp90 cochaperones and substrates were identified using this approach.[28] In this regard, it is surprising to find that one-third (59 out of 177) of the ORFs identified by Millson et al[28] are putative membrane proteins, putting into question the validity of this screen.

In comparison, Zhao et al[20] used full length yeast Hsp82 as well as the different domains of Hsp82 as baits. This approach might be better in identifying physiological substrates and co-factors of Hsp90 as compared to the use of an inactive mutant of the chaperone.[28] A total of 90 Hsp90 interactors were identified[20] mostly by binding to Hsp90 domains as only 8 interactors were found to bind to full length Hsp82. In this screen, only 6 proteins out of the 90 are putative membrane proteins.

Eight proteins overlap between the 2H datasets of Zhao et al[20] and Millson et al[28] (Fig. 2B). These proteins are: Aha1, Arr3, Bsd2, Cns1, She3, Slt2, Sor2, and Tah1. This low overlap might reflect differences in experimental conditions and baits used. The 8 proteins are probably robust interactors of Hsp90. In fact, it is already known that Aha1[14,15] and Cns1[38] are Hsp90 cofactors that interact with Hsp90 in vivo and in vitro. Furthermore, Zhao et al[20] and Millson et al[28] verified the interaction of Hsp90 with Tah1 and Slt2, respectively, using in vitro assays.

The high throughput analysis of mammalian Hsp90 interactors is limited. Physical interactors of mammalian Hsp90 in HEK293 cells were identified using a simple procedure of immuno-precipitation of Hsp90 complexes and the identification of proteins in those complexes using nano-LC-MS/MS.[22] Falsone et al[22] identified 39 proteins as Hsp90 interactors (Fig. 2C). These proteins include metabolic enzymes, ribosomal subunits, and components of the cytoskeleton. However, many of the known Hsp90 interactors were not detected. The identified proteins are mostly housekeeping proteins and are abundant in the cell. This reflects the nature of the coimmunoprecipitation method; that is, most of the weak and transient interactions, especially with low abundance proteins, are missed. However, even for the short list of human Hsp90 physical interactors obtained, it is noticed that 36 out of 39 of the interactors (except for P04075, Q9NR45 and Q96KM8) have close yeast orthologues (Fig. 2C). Some of these orthologues (yeast Act1, Mdh3, Tdh3, and Yer156c which correspond to human P63261, P00338, P00354, and Q9HB07, respectively) were also found to physically interact with the yeast Hsp90 by TAP[20] or 2H[28] screens (Fig. 2C). This may suggest that not only the Hsp90 chaperone machinery itself, but also the core Hsp90 chaperone network may be somewhat conserved across organisms.

Perspectives and Future Directions

The integrated approach used by Zhao et al[20] could also be used for other chaperone systems in yeast. The ultimate goal of mapping the chaperone interaction networks in the cell is to determine the in vivo mechanisms that govern protein homeostasis. In other words, if these screens are repeated often enough and if high fidelity data sets are obtained using multiple methods under different conditions, can a set of rules be eventually derived that would allow us to a priori predict how a given protein will fold inside the cell: what chaperone systems are recruited to help in the folding process and what types of interactions are important for proper de novo folding?

None of the current reported large-scale interaction studies on Hsp90 have been able to find a consensus sequence or structural motif that is preferentially recognized by this chaperone. This is in contrast to other simpler systems such as Hsp60 in *E. coli* (GroEL) in which preferential binding of GroEL to $\alpha\beta$ folds or TIM barrels has been suggested.[39,40] This seems to point to the functional complexity of the Hsp90 system and the fact that substrate specificity is probably dictated, to a certain extent, by the cofactors and cochaperones of Hsp90. Therefore, mapping the chaperone interaction networks must be carried out using multiple genetic and proteomic screens and varying experimental conditions in order to obtain a comprehensive view of the cellular protein folding landscape and in order to minimize false negative and false positive hits. For example, even with the integrated proteomic and genomic approach employed by Zhao et al,[20] some well-established Hsp90 interactors were missed such as Sba1/p23 and Hap1. The interaction between Hsp90 and Sba1/p23 is stable only in the presence of ATP. In another example, physical interactions with membrane proteins are usually difficult to detect. The application of a recently developed split-ubiquitin membrane yeast two-hybrid system might allow the detection of such interactors.[41,42]

With the development of new techniques, the integrated approach used by Zhao et al[20] could also be applied for mammalian Hsp90: genetic interactors can be revealed using RNA interference technology, while physical interactors can be studied using 2H and pull-down/mass spectrometry methods. Furthermore, investigating Hsp90 interactors in a particular tissue or organ, such as human liver or kidney, might be more meaningful than a whole genome-wide screen of Hsp90 interactors. The interactions identified in these organs or tissues could reveal site-specific roles of Hsp90. Finally, quantitative proteomics has been used as a sensitive method to monitor the dynamics of a proteome.[43] Future studies should concentrate on examining the change of the Hsp90 interactome under different stress conditions. Hence, the hope is that future chaperone networks will evolve from static pictures to dynamic entities.

Acknowledgements

Work in the author's laboratory is supported by research grants from the Canadian Institutes of Health research, Natural Sciences and Engineering Research Council of Canada, and the National Cancer Institute of Canada.

References

1. Buchner J. Hsp90 & Co. - A holding for folding. Trends Biochem Sci 1999; 24(4):136-141.
2. Nathan DF, Vos MH, Lindquist S. In vivo functions of the Saccharomyces cerevisiae Hsp90 chaperone. Proc Natl Acad Sci USA 1997; 94(24):12949-12956.
3. Pratt WB, Galigniana MD, Morishima Y et al. Role of molecular chaperones in steroid receptor action. Essays Biochem 2004; 40:41-58.
4. Sreedhar AS, Soti C, Csermely P. Inhibition of Hsp90: A new strategy for inhibiting protein kinases. Biochim Biophys Acta 2004; 1697(1-2):233-242.
5. Young JC, Moarefi I, Hartl FU. Hsp90: A specialized but essential protein-folding tool. J Cell Biol 2001; 154(2):267-273.
6. Terasawa K, Minami M, Minami Y. Constantly updated knowledge of Hsp90. J Biochem (Tokyo) 2005; 137(4):443-447.
7. Beliakoff J, Whitesell L. Hsp90: An emerging target for breast cancer therapy. Anticancer Drugs 2004; 15(7):651-662.
8. Workman P. Altered states: Selectively drugging the Hsp90 cancer chaperone. Trends Mol Med 2004; 10(2):47-51.
9. Dutta R, Inouye M. GHKL, an emergent ATPase/kinase superfamily. Trends Biochem Sci 2000; 25(1):24-28.
10. Prodromou C, Panaretou B, Chohan S et al. The ATPase cycle of Hsp90 drives a molecular 'clamp' via transient dimerization of the N-terminal domains. EMBO J 2000; 19(16):4383-4392.
11. Zhao R, Houry WA. Hsp90: A chaperone for protein folding and gene regulation. Biochem Cell Biol 2005; 83(6):703-710.
12. Richter K, Muschler P, Hainzl O et al. Sti1 is a noncompetitive inhibitor of the Hsp90 ATPase. Binding prevents the N-terminal dimerization reaction during the atpase cycle. J Biol Chem 2003; 278(12):10328-10333.
13. Wegele H, Haslbeck M, Reinstein J et al. Sti1 is a novel activator of the Ssa proteins. J Biol Chem 2003; 278(28):25970-25976.
14. Panaretou B, Siligardi G, Meyer P et al. Activation of the ATPase activity of hsp90 by the stress-regulated cochaperone aha1. Mol Cell 2002; 10(6):1307-1318.
15. Lotz GP, Lin H, Harst A et al. Aha1 binds to the middle domain of Hsp90, contributes to client protein activation, and stimulates the ATPase activity of the molecular chaperone. J Biol Chem 2003; 278(19):17228-17235.
16. Wandinger SK, Suhre MH, Wegele H et al. The phosphatase Ppt1 is a dedicated regulator of the molecular chaperone Hsp90. EMBO J 2006; 25(2):367-376.
17. Bali P, Pranpat M, Bradner J et al. Inhibition of histone deacetylase 6 acetylates and disrupts the chaperone function of heat shock protein 90: A novel basis for antileukemia activity of histone deacetylase inhibitors. J Biol Chem 2005; 280(29):26729-26734.
18. Kovacs JJ, Murphy PJ, Gaillard S et al. HDAC6 regulates Hsp90 acetylation and chaperone-dependent activation of glucocorticoid receptor. Mol Cell 2005; 18(5):601-607.
19. Lane CS. Mass spectrometry-based proteomics in the life sciences. Cell Mol Life Sci 2005; 62(7-8):848-869.
20. Zhao R, Davey M, Hsu YC et al. Navigating the chaperone network: An integrative map of physical and genetic interactions mediated by the hsp90 chaperone. Cell 2005; 120(5):715-727.
21. Puig O, Caspary F, Rigaut G et al. The tandem affinity purification (TAP) method: A general procedure of protein complex purification. Methods 2001; 24(3):218-229.
22. Falsone SF, Gesslbauer B, Tirk F et al. A proteomic snapshot of the human heat shock protein 90 interactome. FEBS Lett 2005; 579(28):6350-6354.
23. Chen B, Piel WH, Gui L et al. The HSP90 family of genes in the human genome: Insights into their divergence and evolution. Genomics 2005; 86(6):627-637.
24. Fields S, Song O. A novel genetic system to detect protein-protein interactions. Nature 1989; 340(6230):245-246.
25. Passinen S, Valkila J, Manninen T et al. The C-terminal half of Hsp90 is responsible for its cytoplasmic localization. Eur J Biochem 2001; 268(20):5337-5342.
26. Millson SH, Truman AW, Piper PW. Vectors for N- or C-terminal positioning of the yeast Gal4p DNA binding or activator domains. Biotechniques 2003; 35:60-64.
27. Millson SH, Truman AW, Wolfram F et al. Investigating the protein-protein interactions of the yeast Hsp90 chaperone system by two-hybrid analysis: Potential uses and limitations of this approach. Cell Stress Chaperones (Winter) 2004; 9(4):359-368.
28. Millson SH, Truman AW, King V et al. A two-hybrid screen of the yeast proteome for Hsp90 interactors uncovers a novel Hsp90 chaperone requirement in the activity of a stress-activated mitogen-activated protein kinase, Slt2p (Mpk1p). Eukaryot Cell 2005; 4(5):849-860.

29. Uetz P, Giot L, Cagney G et al. A comprehensive analysis of protein-protein interactions in Saccharomyces cerevisiae. Nature 2000; 403(6770):623-627.
30. Tong AH, Evangelista M, Parsons AB et al. Systematic genetic analysis with ordered arrays of yeast deletion mutants. Science 2001; 294(5550):2364-2368.
31. Giaever G, Chu AM, Ni L et al. Functional profiling of the Saccharomyces cerevisiae genome. Nature 2002; 418(6896):387-391.
32. Shoemaker DD, Lashkari DA, Morris D et al. Quantitative phenotypic analysis of yeast deletion mutants using a highly parallel molecular bar-coding strategy. Nat Genet 1996; 14(4):450-456.
33. Winzeler EA, Shoemaker DD, Astromoff A et al. Functional characterization of the S. cerevisiae genome by gene deletion and parallel analysis. Science 1999; 285(5429):901-906.
34. Tong AH, Boone C. Synthetic genetic array analysis in Saccharomyces cerevisiae. Methods Mol Biol 2006; 313:171-192.
35. Nathan DF, Lindquist S. Mutational analysis of Hsp90 function: Interactions with a steroid receptor and a protein kinase. Mol Cell Biol 1995; 15(7):3917-3925.
36. Wang CW, Stromhaug PE, Shima J et al. The Ccz1-Mon1 protein complex is required for the late step of multiple vacuole delivery pathways. J Biol Chem 2002; 277(49):47917-47927.
37. Wang CW, Stromhaug PE, Kauffman EJ et al. Yeast homotypic vacuole fusion requires the Ccz1-Mon1 complex during the tethering/docking stage. J Cell Biol 2003; 163(5):973-985.
38. Dolinski KJ, Cardenas ME, Heitman J. CNS1 encodes an essential p60/Sti1 homolog in Saccharomyces cerevisiae that suppresses cyclophilin 40 mutations and interacts with Hsp90. Mol Cell Biol 1998; 18(12):7344-7352.
39. Houry WA, Frishman D, Eckerskorn C et al. Identification of in vivo substrates of the chaperonin GroEL. Nature 1999; 402(6758):147-154.
40. Kerner MJ, Naylor DJ, Ishihama Y et al. Proteome-wide analysis of chaperonin-dependent protein folding in Escherichia coli. Cell 2005; 122(2):209-220.
41. Iyer K, Burkle L, Auerbach D et al. Utilizing the split-ubiquitin membrane yeast two-hybrid system to identify protein-protein interactions of integral membrane proteins. Sci STKE 2005; 2005(275):pl3.
42. Miller JP, Lo RS, Ben-Hur A et al. Large-scale identification of yeast integral membrane protein interactions. Proc Natl Acad Sci USA 2005; 102(34):12123-12128.
43. Yan W, Chen SS. Mass spectrometry-based quantitative proteomic profiling. Brief Funct Genomic Proteomic 2005; 4(1):27-38.

CHAPTER 4

Organization of the Functions and Components of the Endoplasmic Reticulum

Yuichiro Shimizu and Linda M. Hendershot*

Abstract

The endoplasmic reticulum is the site of entry into the secretory pathway and represents a major and particularly crowded site of protein biosynthesis. In addition to the complexity of protein folding in any organelle, the ER environment poses further dangers and constraints to the process. A quality control apparatus exists to monitor the maturation of proteins in the ER. Nascent polypeptide chains are specifically prevented from traveling further along the secretory pathway until they have completed their folding or assembly. Proteins that cannot achieve a proper conformation are recognized and removed from the ER for degradation by the 26S proteasome. Finally, the homeostasis of the ER is vigilantly monitored and changes that impinge upon the proper maturation of proteins in this organelle lead to the activation of a signal transduction cascade that serves to restore balance to the ER. Recent studies suggest that some of these diverse functions may be achieved due to the organization of the ER into functional and perhaps even physical sub-domains.

Introduction

The endoplasmic reticulum (ER) is a command center of the cell that is second only to the nucleus in terms to the breath of its influence on other organelles and activities. It is a major site of protein synthesis, contains the cellular calcium stores that are an essential component of many signaling pathways, and is the proximal site of a signal transduction cascade that responds to cellular stress conditions and serves to maintain homeostasis of the cell. All eucaryotic cells possess an ER, which can comprise nearly 50% of the membranes of a cell. Its functions can be divided into those that occur on the cytosolic side of the membrane (where protein translation and degradation occur) and the lumenal space (where most other ER functions take place).

Overview of Protein Biosynthesis in the ER

Proteins that will populate single membrane bound organelles of the cell, as well as those destined for expression on the cell surface or secretion, are synthesized on membrane bound polyribosomes or rough ER, where, in the case of mammalian cells, they will be cotranslationally translocated into the lumen of the ER. The nascent polypeptide chain is transported through the transolocon, which is composed of Sec61 α, β, and γ and translocation associated protein (TRAM).[1-3] Once inside the ER, the nascent chain begins to fold cotranslationally, and disulfide bonds are formed to stabilize protein folding and assembly of subunits. Most secretory

*Corresponding Author: Linda M. Hendershot—Department of Genetics and Tumor Cell Biology, St. Jude Children's Research Hospital, 332 N. Lauderdale Memphis, Tennessee 38105, U.S.A. Email: linda.hendershot@stjude.org

Molecular Aspects of the Stress Response: Chaperones, Membranes and Networks, edited by Peter Csermely and László Vígh. ©2007 Landes Bioscience and Springer Science+Business Media.

pathway proteins are modified cotranslationally by the addition of a complex core of sugars that are added in block to asparagine residues within a N-X-S/T sequence as it emerges from the translocon.[4] The N-linked glycans immediately become the target of enzymes that continuously add and remove sugars from the core. The composition of the glycan moiety is the basis of systems that recognize the folding state of the protein (see below). This highly charged moiety attached to an unfolded protein also limits the folding pathways that are available to the nascent protein. Mutations or drugs that prevent N-linked glycosylation dramatically affect protein folding in the ER, suggesting that proteins in this organelle may have evolved to place glycosylation sites at critical regions of the protein to ensure they do not fold improperly.

When protein folding and assembly are completed, the protein can be transferred out of the ER through ER exit sites on the smooth face of the ER and transported to the Golgi for further modification and sorting along the secretory pathway.[5] It is well-accepted that the ER is divided into spatially distinct regions for translocation and transport and in fact these regions can be separated physically by density centrifugation of ER that has been disrupted into vesicles. If however, the nascent polypeptide does not fold properly, it will be detected by an ER quality control system and retrotranslocated to the cytosol for degradation by the 26S proteasome. It is not clear whether this function also exists in a separate region of the ER or how the ER maintains these three distinct functions that would be expected to have different requirements in terms of components and ER conditions (Fig. 1).

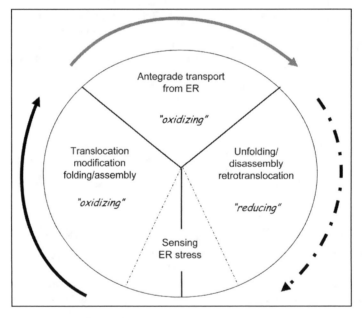

Figure 1. The ER houses numerous diverse functions. The ER is a major site of protein synthesis. Proteins enter the ER through a region termed the rough ER (black arrow), which is an oxidizing environment that promotes protein folding and assembly. Properly matured proteins exit the ER for the Golgi from the smooth ER (grey arrow), and it is believed that this region of the ER would also be oxidizing to maintain the disulfide bonds that stabilize protein folding. Proteins that fail to mature properly are identified and translocated back to the cytosol. It is not known which region of the ER supports this activity (dashed arrow) nor is the oxidizing status of this "region" known, but it is likely to be somewhat more reducing than the other two regions. Finally, the folding environment is monitored and corrected by activating a signal transduction pathway that serves to restore ER homeostasis. The "locations" of the signal transducers are not known, but it is anticipated that they would be required in both the folding and the retrotranslocation sites.

The ER Possesses a Unique Environment for Protein Folding

The environment of the ER introduces additional dangers for the nascent protein beyond those encountered by proteins folding in the cytosol, but in some cases it can offer advantages. The concentration of unfolded or partially folded proteins is particularly high in the ER, which would be expected to dramatically increase the potential for aggregation or inappropriate inter- and intra-molecular interactions between nascent proteins. This is prevented by a correspondingly high concentration of molecular chaperones and folding enzymes in this organelle. In addition, the ER lumen has a much higher oxidizing potential.[6] This promotes disulfide bond formation between cysteine residues, while few disulfide bonds are formed in proteins synthesized in the cytosol due to the low redox state.[7,8] Although disulfide bonds serve to stabilize a properly folded protein or subunit interaction, the fact that the correct disulfide bond often must form between cysteine residues that are far apart on the linear sequence or between cysteines in other subunits increases the possibility that transient, inappropriate interactions could be stabilized by the formation of a disulfide bond. This must be prevented or resolved if it occurs. The maintenance of an oxidizing environment in the ER is achieved by Ero1, which is oxidized by molecular oxygen. Oxidized Ero1 can then oxidize proteins like protein disulfide isomerase,[9,10] which possess oxy-reductase activity. The oxidized forms of these proteins are then able to facilitate oxidation of the unfolded proteins associated with them, whereas the reduced forms of these proteins are thought to reduce substrate proteins. Thus, the balance between formation and disruption of disulfide bonds is regulated by the oxidation status of the ER, which in turn contributes to stabilizing or destabilizing the folding of proteins in this organelle. The latter is likely to be important for the dislocation of proteins back to the cytosol for degradation, although it is unclear how both systems might work in the same organelle and raises the possibility of separate domains or subcompartments within the ER (Fig. 1).

In addition, the ER has high concentrations of Ca^{2+} and ATP, which contribute to the folding efficiency of the chaperones.[11,12] High concentrations of ER calcium are achieved through the action of the ER calcium ATPase and ryanodine receptors and are stored via weak associations with several abundant ER chaperones including BiP, GRP94, calnexin and calreticulin. The binding of calcium to acidic residues on nascent proteins can affect side chain interactions that are a fundamental basis of protein folding. Thus, ER proteins have evolved to fold in a calcium rich environment and decreases in ER calcium levels adversely affect folding in this organelle.[13] Although it would appear that the environment of the ER could interfere with protein folding, some dedicated secretory cells (i.e., hepatocytes, pancreatic β islet cells, and plasma cells) are able to successfully fold and export very large quantities of proteins. For example, a single plasma cell is able to produce and secrete thousands of immunoglobulin (Ig) molecules per second, which requires the formation of ~100,000 disulfide bonds and addition of ~50,000 N-linked glycans.[14]

The ER Quality Control System

A critical checkpoint in the protein maturation process in the ER is the identification of proteins that have not completed folding, have folded improperly, or have not acquired the proper subunit complement. An elaborate quality control system exists, which is able to distinguish between these intermediates based on the identification of hydrophobic residues that have not been properly buried, N-linked sugars that are not properly modified, and unpaired or mispaired cysteines that result in exposed thiols or improper disulfide bond formations, respectively. For the most part, these systems are not specific for individual proteins but instead identify features that are common to unfolded proteins, although protein-specific quality control mechanisms do exist. The proper conformational maturation of most nascent secretory pathway proteins is both aided and monitored by a large number of ER chaperones, their cofactors, and folding enzymes.[15] The components and mechanisms of action of two major chaperone systems have been best studied. The first ER chaperone to be identified in any organism was BiP,[16,17] the ER Hsp70 family member, which recognizes hydrophobic side chains that are exposed on unfolded regions of a nascent protein.[18] BiP can bind to nascent chains cotranslationally and the binding is usually transient, except in the case of unpartnered subunits or mutant proteins. Like all Hsp70 proteins, BiP binds both ADP and ATP, which serve to regulate its binding and release from nascent chains.[19,20] The ATPase cycle of Hsp70 proteins is both positively and negatively regulated by a number of chaperones and cofactors. To date five DnaJ homologues have been identified (ERdj1-5) that stimulate the ATPase activity of BiP[15] and at least one nucleotide releasing factor, BAP/Sls1/Sil1p.[21] ERdj3 binds directly to some secretory pathway proteins and is thought to recruit BiP and stabilize its binding. By releasing ADP and allowing ATP to bind, BAP accelerates the release of BiP from the unfolded protein and allows folding to occur.

The processing of N-linked glycans is closely linked to the folding status of the protein through the action of glucosidases I and II, UDP-glucosyltransferase (UGT), and ER-localized α-mannosidase.[22,23] Glucosidases I and II sequentially trim the outer two glucoses from a precursor high mannose glycan ($Glc_3Man_9GlcNAc_2$) that is added en bloc to the translocating polypeptide chain reducing it to a mono-glucosylated state ($Glc_1Man_9GlcNAc_2$). Calnexin (CNX) and calreticulin (CRT), which constitute the second major chaperone system in the ER, bind specifically to this mono-glucosylated glycan until glucosidase II removes the final glucose from the core resulting in a $Man_9GlcNAc_2$ structure.[24,25] UGT recognizes both unfolded regions on a protein and the $Man_9GlcNAc_2$ glycan,[26] and as long as unfolded regions remain on the protein, UGT can reglucosylate the glycan moiety, thus recreating the CNX/CRT binding motif. If however, the substrate has completed folding and is ready for further transport to the Golgi complex; one mannose is trimmed before the protein exits the ER. This abolishes the ability of UGT to recognize the protein and stops the CNX/CRT cycle.

Finally, oxy-reductases, like protein disulfide isomerase (PDI) and ERp57, catalyze disulfide bond formation between unpaired cysteine residues in proteins as they fold and in some cases break these bonds to either achieve proper folding or to prepare a partially folded protein for retrotranslocation.[27,28] The activity of these enzymes are supported by specific interaction with chaperones (i.e., PDI-BiP or ERp57-CNX/CRT), which bind and consequently stabilize the substrate so that the reductases can access it efficiently.

Once folding and disulfide bond formation are complete and the features recognized by these systems are no longer exposed, the protein in many cases is free to move along the secretory pathway to the Golgi. However, proteins that fail ER quality control inspection are recognized, translocated back through the ER membrane and targeted for degradation by the 26S proteasome (Fig. 2). The components of this "retrotranslocon" are not completely understood but recent studies suggest that it is composed of Derlin, a multi-pass transmembrane protein, p97 a member of the AAA ATPase family that provides energy for extracting the protein, Vimp, Hrd1, an E3 ubiquitin ligase, and Herp.[29,30] Proteins that are in the process of folding must somehow be distinguished from proteins that are unable to fold completely as they share the same common features. In the case of glycoproteins, this involves recognition of a mannose trimmed form of N-linked glycans ($Glc_{0-3}Man_8GlcNAc_2$) by EDEM, and this interaction removes the protein from a "folding" pathway and puts it on a "degradation" pathway.[31,32] However, the Man_8 form is commonly produced during the normal maturation process before transport of the folded protein to the Golgi complex by the action of a resident mannosidase. Therefore it is not entirely clear how the substrate's native and nonnative states are distinguished. This could be explained if the site for retrotranslocation was separate from the sites of exit for Golgi transport. The mechanism for targeting nonglycosylated proteins for degradation is even less well understood.

A balance exists between protein folding and degradation in cells under normal physiological conditions that in most cases favors protein folding. However, changes in the cellular environment can often impact on protein folding in the ER by either altering the redox potential, the ability to glycosylate proteins or the storage of calcium in this organelle. These changes in ER homeostasis can shift the balance and lead to the accumulation of unfolded proteins in the ER.[33,34]

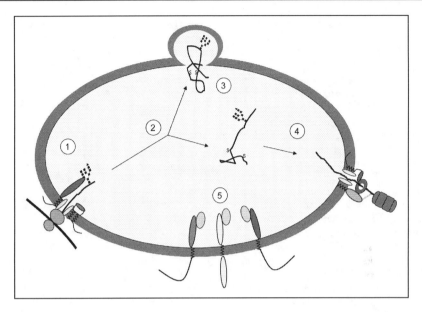

Figure 2. ER functions require a distinct complement of proteins. Secretory pathway proteins enter the ER through a membrane channel known as the translocon (1). This proteinaceous pore is comprised of the Sec61 complex (light blue), a DnaJ family member (red) that serves to recruit BiP and to pull the nascent polypeptide into the ER and oligosaccharyl transferase (purple), which adds complex N-linked sugars to the extended chain. The protein begins to fold (2), a process that is both aided and monitored through the action of a large number of resident ER chaperones and folding enzymes. If a protein achieves the proper tertiary structure, it is no longer a substrate for molecular chaperones and will leave the ER for the Golgi through ER exit sites (3). However, proteins that fail to fold properly are identified, unfolded, and translocated back to the cytosol for degradation by the 26S proteasome (dark pink) (4). The nature of the retrotranslocon is not well understood, but appears to be formed by several transmembrane proteins including Derlin (yellow), which is thought to form the channel, VIMP (purple), which binds to both Derlin and p97 (blue) an ATPase that provides energy for extracting the protein, and Hrd1, an E3 ligase that adds ubiquitin to the substrate. Finally, in order to maintain ER homeostasis in the face of increased demands on the ER or adverse conditions in the ER, 3 transmembrane proteins, Ire1 (green), PERK (dark blue), and ATF6 (yellow) exist that sense ER stress via interactions with BiP (turquoise) and upon activation signal the UPR (5). A color version of this figure is available online at http://www.Eurekah.com.

The cell has the ability to recognize this change via three resident ER transmembrane proteins, Ire1, PERK and ATF6 that activate a signal transduction cascade termed the "unfolded protein response".[35] Together these transducers up-regulate ER chaperones and folding enzymes to prevent aggregation of unfolded proteins, transiently inhibit translation to limit the accumulation of proteins, and increase the degradative capacity of the cell to reduce the load of unfolded proteins. These responses serve to protect the cell and maintain ER homeostasis, but in extreme or prolonged conditions of ER stress, proapoptotic responses, including caspase 12/4, CHOP, and JNK, are activated to eliminate the cell and protect the organism.[36,37]

Chaperone Selection during Protein Maturation in the ER

The nascent polypeptide moves through the ribosome channel and translocon in a primarily unfolded state that does not appear to progress beyond α-helix formation.[38,39] It is largely protected against degradation while in these two channels.[40,41] As the growing polypeptide chain enters inside the ER a choice is usually made between the two major chaperone systems depending on certain features of the nascent polypeptide chain. If an N-linked glycan is added within the first ~50 residues of the protein, the calnexin system will be engaged, whereas if no glycan is present, BiP will bind.[42-44] In addition to protecting the nascent unfolded protein from aggregation, it is thought that the binding of chaperones very early can further serve to pull the substrate into the lumen or prevent it from slipping back to the cytosol. However, the largely cotranslational nature of translocation in higher eukaryotes is likely to have less of a requirement for the latter function.

Proteins that remain in the ER for prolonged times in an unfolded state become substrates for ER associated degradation (ERAD).[45-47] This requires first distinguishing a protein that will not fold from those that are in the process of folding. In the case of substrates using the CNX/CRT system, part of the identification of terminally misfolded proteins involves EDEM, which can recognize glycan structures that are formed due to a competition between mannosidases and UGT.[48] In many cases the protein is likely to have partially folded and must be unfolded for efficient extraction from the ER. In the case of BACE457 and a temperature-sensitive mutant of the VSV protein G, ts045, it appears these CNX substrates are transferred to BiP/PDI prior to degradation for further unfolding.[49,50] Based on the fact that glucosidase I and II can remove glycans irrespective of substrate's conformational state,[4] while UGT only reglucosylates nearly native glycoprotein folding intermediates,[51] Molinari et al (ref. 50.) have hypothesized that if glucosidase II acted on the monoglucosylated glycan of a more unfolded glycoprotein, UGT would no longer bind and reglucosylate the glycan. This would cause the protein to exit the CNX/CRT cycle and allow BiP to bind. The fact that UGT is a component of the BiP chaperone complex[52] (see below) makes this a particularly appealing idea, since competition between neighboring BiP and UGT could possibly determine the fate of the substrate. In addition, the presence of PDI in the BiP complex could serve to break disulfide bonds if it was in a partially oxidized state.

Organization of a Subset of Chaperones into Large Preformed Complexes

In a number of studies, multiple ER chaperones were shown to associate with a given nascent protein. For example, both thyroglobulin[53] and HCGβ[54] can be cross-linked to BiP, GRP94, and ERp72 during their maturation. In addition, multiple chaperones including calnexin, calreticulin, BiP and GRP94 were found to interact with newly synthesized influenza hemagglutinin protein,[55] while unoxidized Ig LC formed transient disulfide bonds with both PDI and ERp72, probably in order to trap the immature (unoxidized) protein in the ER until it is properly processed.[56] However, it was not clear from these studies if the different chaperones were binding to distinct regions of the unfolded protein, if each bound to somewhat different folding states of a protein, or if the chaperones reacted as a complex to the same exposed hydrophobic region on the nascent protein.

In the case of Ig heavy chains, a preformed complex including BiP, GRP94, CaBP1, PDI, ERdj3, Hsp40, cyclophilin B, ERp72, GRP170, UGT, and SDF2-L1 is formed and binds together to the CH1 domain of the heavy chain.[52] Factors involved in the CNX/CRT system were largely absent from the complex, with the exception of UGT, which might serve as a link between the two chaperone systems. The complex existed in multiple cell types where it had a similar composition suggesting that this may represent a common organization of the chaperones in the ER. This could serve to concentrate a number of folding enzymes and chaperones onto the nascent protein making them available if needed. The absence of members of the CNX/CRT system from the complex suggests that there is a spatial separation to the two chaperone systems in addition to the functional and temporal separation that has been reported for many substrates. However, the fact that several proteins have been found to associate with both the BiP and the CNX/CRT system means that either the two systems are in close proximity or that distinct pools of the protein bind to each of the chaperones. Further support for a separation between these two chaperone systems comes from studies on what has been termed the ER quality control compartment. Using fluorescence microscopy, the precursor of human asialoglycoprotein receptor, H2a, and the free heavy chains of MHC class I molecules were shown to accumulate in a compartment containing calnexin and calreticulin, but not BiP, PDI, or UDP-GT, when proteasomal degradation was inhibited.[57,58]

Although it seems reasonable that the organization of chaperones and folding enzymes into a complex would allow them to work together more efficiently, only limited data are available to support this idea. Argon and coworkers showed the sequential interaction of BiP and GRP94 with Ig LC.[59] In a separate study GRP94 remained bound to Ig HC in vitro even after BiP was released with ATP.[60] In the first case the results suggest that BiP delivers the substrate to GRP94 and then leaves, whereas in the second case, they suggest that both BiP and GRP94 bind simultaneously, although independently to the substrate. Data showing that BiP and GRP94 exist as a complex that binds together to the CH1 domain of Ig HC appears inconsistent with these results at first glance.[52] However, it should be noted that neither of these two studies used cross-linkers to stabilize binding of the various components of the complex. Thus, it is possible in the first study that the chaperone complex can associate with the nascent LC through BiP and that as the protein begins to fold that the complex "rolls over" to allow GRP94 to bind directly. In the absence of crosslinker this would appear as BiP binding alone first and GRP94 second. In addition to evidence suggesting a collaboration between BiP and GRP94, a synergistic action between BiP and PDI were also suggested in a yeast expression system,[61] and in an in vitro system.[62] These studies suggested that BiP binds to unfolded or denatured antibody fragments and keeps them in a conformation in which the cysteine residues are more readily accessible for PDI to catalyze disulfide bond formation. The association of both BiP and PDI with an aggregated pancreatic isoform of human β-secretase (BACE457), which had formed nonnative disulfide bonds in vivo, further suggests a cooperation between BiP and PDI in vivo.[49]

Components of the Calnexin/Calreticulin System and Their Organization

CNX and CRT are highly homologous (membrane bound and soluble respectively) lectin-like proteins that bind to the monoglucosylated N-linked glycans of nascent polypeptide chains translocated into the ER.[22,63] Although there are redundancies in substrate specificities between CNX and CRT as expected by their similarities,[64,65] various studies demonstrated that CNX and CRT have different properties in substrate binding,[66,67] including kinetics and duration of binding during substrate maturation.[68,69] These findings led to a number of studies trying to explain the observed differences. Given that both CNX and CRT possess identical lectin binding specificities and affinities,[70] one possible explanation could be distance constraints that might arise for the membrane anchored CNX that could develop as the polypeptide elongates in the ER lumen. This idea was examined by expressing either membrane bound CRT or transmembrane deleted CNX in vivo compared with the activity of wild-type CNX or CRT, respectively.[71,72] The study revealed that the functions were largely interchangeable, which confirmed the importance of their spatial distribution. Also, by expressing translocation intermediates of hemagglutinin, Hebert and coworkers showed that CNX binding could be detected very soon after the polypeptide entered the ER, while the binding of CRT required the synthesis of at least 30 additional amino acids.[68]

Although the association of CXN/CRT with unfolded substrates is functionally dependent on the action of UGT, there are no data to demonstrate that they are in a physical complex with each other. Thus, it is unclear if their concentrations and affinities for substrates in the ER are high enough to allow them to perform their functions efficiently without forming a complex or if they do enjoy an association that has not been revealed. Like BiP, which is associated with oxy-reductases (PDI and ERp72), the CXN/CRT system has its own dedicated oxy-reductase, ERp57, which binds directly to the chaperones at their P domain.[73,74]

Possible Advantages and Constraints That an Organization of ER Chaperones Might Impose

Both the BiP and CXN/CRT systems are associated with a "cofactor" that can either form or reduce disulfide bonds in the substrate protein depending on its oxidation status. This would provide the chaperone systems with the ability to aid in both the folding of nascent proteins as well as the disposal of misfolded ones. How the necessary function is specifically called upon is not known, but could be achieved if there were sub-regions of the ER that varied in terms of their oxidizing state. This could be achieved by regulating the concentration of molecular oxygen, Ero1, or even as yet to be identified reductases within subregions of the ER. However, there are currently no data to support this idea or to clarify how the ER is able to separate its essential functions of folding and retaining nascent proteins in the ER, releasing proteins that have matured properly for transport to the Golgi, and identifying proteins that fail to fold and targeting them for degradation. Recent studies have begun to reveal an organization to the ER that was not previously appreciated. It is likely that studies in the near future will further illuminate how the diverse functions of this complicated organelle are balanced to allow a secretory cell to produce large quantities of properly matured proteins efficiently and with a high degree of fidelity. It is equally likely that the organization of the ER will be at the heart of this.

References

1. Nunnari J, Walter P. Protein targeting to and translocation across the membrane of the endoplasmic reticulum. Curr Opin Cell Biol 1992; 4:573-580.
2. Johnson AE, van Waes MA. The translocon: A dynamic gateway at the ER membrane. Annu Rev Cell Dev Biol 1999; 15:799-842.
3. Meacock SL, Greenfield JJ, High S. Protein targeting and translocation at the endoplasmic reticulum membrane — Through the eye of a needle? Essays Biochem 2000; 36:1-13.
4. Kornfeld R, Kornfeld S. Assembly of asparagine-linked oligosaccharides. Annu Rev Biochem 1985; 54:631-664.
5. LaPointe P, Gurkan C, Balch WE. Mise en place-this bud's for the Golgi. Mol Cell 2004; 14:413-414.
6. Hwang C, Sinskey AJ, Lodish HF. Oxidized redox state of glutathione in the endoplasmic reticulum. Science 1992; 257:1496-1502.
7. Fahey RC, Hunt JS, Windham GC. On the cysteine and cystine content of proteins. Differences between intracellular and extracellular proteins. J Mol Evol 1977; 10:155-160.
8. Freedman R. Native disulphide bond formation in protein biosynthesis: Evidence for the role of protein disulfide isomerase. Trends Biochem Sci 1984; 9:438-441.
9. Tu BP, Weissman JS. The FAD- and O(2)-dependent reaction cycle of Ero1-mediated oxidative protein folding in the endoplasmic reticulum. Mol Cell 2002; 10:983-994.
10. Mezghrani A, Fassio A, Benham A et al. Manipulation of oxidative protein folding and PDI redox state in mammalian cells. EMBO J 2001; 20:6288-6296.
11. Clairmont CA, De Maio A, Hirschberg CB. Translocation of ATP into the lumen of rough endoplasmic reticulum-derived vesicles and its binding to luminal proteins including BiP (GRP 78) and GRP 94. J Biol Chem 1992; 267:3983-3990.
12. Michalak M, Robert Parker JM, Opas M. Ca(2+) signaling and calcium binding chaperones of the endoplasmic reticulum. Cell Calcium 2002; 32:269-278.
13. Lee AS. Coordinated regulation of a set of genes by glucose and calcium ionophores in mammalian cells. Trends Biochem Sci 1987; 12:20-23.
14. Hendershot LM, Sitia R. Antibody synthesis and assembly. In: Alt FW, Honjo T, Neuberger MS, eds. Molecular Biology of B Cells. Elsevier Science, 2004:261-73.
15. Ma Y, Hendershot LM. ER chaperone functions during normal and stress conditions. J Chem Neuro 2004; 28:51-65.
16. Haas IG, Wabl M. Immunoglobulin heavy chain binding protein. Nature 1983; 306:387-389.
17. Bole DG, Hendershot LM, Kearney JF. Posttranslational association of immunoglobulin heavy chain binding protein with nascent heavy chains in nonsecreting and secreting hybridomas. J Cell Biol 1986; 102:1558-1566.
18. Blond-Elguindi S, Cwirla SE, Dower WJ et al. Affinity panning of a library of peptides displayed on bacteriophages reveals the binding specificity of BiP. Cell 1993; 75:717-728.
19. Kassenbrock CK, Kelly RB. Interaction of heavy chain binding protein (BiP/GRP78) with adenine nucleotides. EMBO J 1989; 8:1461-1467.

20. Wei JY, Gaut JR, Hendershot LM. In vitro dissociation of BiP: Peptide complexes requires a conformational change in BiP after ATP binding but does not require ATP hydrolysis. J Biol Chem 1995; 270:26677-26682.

21. Chung KT, Shen Y, Hendershot LM. BAP, a mammalian BiP associated protein, is a nucleotide exchange factor that regulates the ATPase activity of BiP. J Biol Chem 2002; 277:47557-47563.

22. Helenius A, Aebi M. Intracellular functions of N-linked glycans. Science 2001; 291:2364-2369.

23. Chevet E, Cameron PH, Pelletier MF et al. The endoplasmic reticulum: Integration of protein folding, quality control, signaling and degradation. Curr Opin Struct Biol 2001; 11:120-124.

24. Ware FE, Vassilakos A, Peterson PA et al. The molecular chaperone calnexin binds Glc1Man9GlcNAc2 oligosaccharide as an initial step in recognizing unfolded glycoproteins. J Biol Chem 1995; 270:4697-4704.

25. Hammond C, Braakman I, Helenius A. Role of N-linked oligosaccharide recognition, glucose trimming, and calnexin in glycoprotein folding and quality control. Proc Natl Acad Sci USA 1994; 91:913-917.

26. Sousa M, Parodi AJ. The molecular basis for the recognition of misfolded glycoproteins by the UDP-Glc: Glycoprotein glucosyltransferase. EMBO J 1995; 14:4196-4203.

27. Rietsch A, Beckwith J. The genetics of disulfide bond metabolism. Annu Rev Genet 1998; 32:163-184.

28. Frand AR, Cuozzo JW, Kaiser CA. Pathways for protein disulphide bond formation. Trends Cell Biol 2000; 10:203-210.

29. Ye Y, Shibata Y, Yun C et al. A membrane protein complex mediates retro-translocation from the ER lumen into the cytosol. Nature 2004; 429:841-847.

30. Schuberth C, Buchberger A. Membrane-bound Ubx2 recruits Cdc48 to ubiquitin ligases and their substrates to ensure efficient ER-associated protein degradation. Nat Cell Biol 2005; 7:999-1006.

31. Oda Y, Hosokawa N, Wada I et al. EDEM as an acceptor of terminally misfolded glycoproteins released from calnexin. Science 2003; 299:1394-1397.

32. Molinari M, Calanca V, Galli C et al. Role of EDEM in the release of misfolded glycoproteins from the calnexin cycle. Science 2003; 299:1397-1400.

33. Kozutsumi Y, Segal M, Normington K et al. The presence of malfolded proteins in the endoplasmic reticulum signals the induction of glucose-regulated proteins. Nature 1988; 332:462-464.

34. Lee AS. Mammalian stress response: Induction of the glucose-regulated protein family. Curr Opin Cell Biol 1992; 4:267-273.

35. Kaufman RJ. Stress signaling from the lumen of the endoplasmic reticulum: Coordination of gene transcriptional and translational controls. Genes and Development 1999; 13:1211-1233.

36. Ma Y, Hendershot LM. The role of the unfolded protein response in tumour development: Friend or foe? Nat Rev Cancer 2004; 4:966-977.

37. Rao RV, Ellerby HM, Bredesen DE. Coupling endoplasmic reticulum stress to the cell death program. Cell Death Differ 2004; 11:372-380.

38. Mingarro I, Nilsson I, Whitley P et al. Different conformations of nascent polypeptides during translocation across the ER membrane. BMC Cell Biol 2000; 1:3.

39. Johnson AE. Functional ramifications of FRET-detected nascent chain folding far inside the membrane-bound ribosome. Biochem Soc Trans 2004; 32:668-672.

40. Chen W, Helenius A. Role of ribosome and translocon complex during folding of influenza hemagglutinin in the endoplasmic reticulum of living cells. Mol Biol Cell 2000; 11:765-772.

41. Kowarik M, Kung S, Martoglio B et al. Protein folding during cotranslational translocation in the endoplasmic reticulum. Mol Cell 2002; 10:769-778.

42. Hammond C, Helenius A. Folding of VSV G protein: Sequential interaction with BiP and calnexin. Science 1994; 266:456-458.

43. Molinari M, Helenius A. Chaperone selection during glycoprotein translocation into the endoplasmic reticulum. Science 2000; 288:331-333.

44. Wang N, Daniels R, Hebert DN. The cotranslational maturation of the type I membrane glycoprotein tyrosinase: The heat shock protein 70 system hands off to the lectin-based chaperone system. Mol Biol Cell 2005; 16:3740-3752.

45. Werner ED, Brodsky JL, McCracken AA. Proteasome-dependent endoplasmic reticulum-associated protein degradation: An unconventional route to a familiar fate. Proc Natl Acad Sci USA 1996; 93:13797-13801.

46. Meusser B, Hirsch C, Jarosch E et al. ERAD: The long road to destruction. Nat Cell Biol 2005; 7:766-772.

47. Hampton RY. ER-associated degradation in protein quality control and cellular regulation. Curr Opin Cell Biol 2002; 14:476-482.

48. Hosokawa N, Wada I, Hasegawa K et al. A novel ER alpha-mannosidase-like protein accelerates ER-associated degradation. EMBO Rep 2001; 2:415-422.

49. Molinari M, Galli C, Piccaluga V et al. Sequential assistance of molecular chaperones and transient formation of covalent complexes during protein degradation from the ER. J Cell Biol 2002; 158:247-257.
50. Molinari M, Galli C, Vanoni O et al. Persistent glycoprotein misfolding activates the glucosidase II/UGT1-driven calnexin cycle to delay aggregation and loss of folding competence. Mol Cell 2005; 20:503-512.
51. Trombetta ES, Helenius A. Conformational requirements for glycoprotein reglucosylation in the endoplasmic reticulum. J Cell Biol 2000; 148:1123-1129.
52. Meunier L, Usherwood YK, Chung KT et al. A subset of chaperones and folding enzymes form multiprotein complexes in endoplasmic reticulum to bind nascent proteins. Mol Biol Cell 2002; 13:4456-4469.
53. Kuznetsov G, Chen LB, Nigam SK. Multiple molecular chaperones complex with misfolded large oligomeric glycoproteins in the endoplasmic reticulum. J Biol Chem 1997; 272:3057-3063.
54. Feng W, Matzuk MM, Mountjoy K et al. The asparagine-linked oligosaccharides of the human chorionic gonadotropin beta subunit facilitate correct disulfide bond pairing. J Biol Chem 1995; 270:11851-11859.
55. Tatu U, Helenius A. Interactions between newly synthesized glycoproteins, calnexin and a network of resident chaperones in the endoplasmic reticulum. J Cell Biol 1997; 136:555-565.
56. Reddy P, Sparvoli A, Fagioli C et al. Formation of reversible disulfide bonds with the protein matrix of the endoplasmic reticulum correlates with the retention of unassembled Ig light chains. EMBO J 1996; 15:2077-2085.
57. Kamhi-Nesher S, Shenkman M, Tolchinsky S et al. A novel quality control compartment derived from the endoplasmic reticulum. Mol Biol Cell 2001; 12:1711-1723.
58. Frenkel Z, Shenkman M, Kondratyev M et al. Separate roles and different routing of calnexin and ERp57 in endoplasmic reticulum quality control revealed by interactions with asialoglycoprotein receptor chains. Mol Biol Cell 2004; 15:2133-2142.
59. Melnick J, Dul JL, Argon Y. Sequential interaction of the chaperones BiP and Grp94 with immunoglobulin chains in the endoplasmic reticulum. Nature 1994; 370:373-375.
60. Rosser MF, Trotta BM, Marshall MR et al. Adenosine nucleotides and the regulation of GRP94-client protein interactions. Biochemistry 2004; 43:8835-8845.
61. Shusta EV, Raines RT, Pluckthun A et al. Increasing the secretory capacity of Saccharomyces cerevisiae for production of single-chain antibody fragments. Nat Biotechnol 1998; 16:773-777.
62. Mayer M, Kies U, Kammermeier R et al. BiP and PDI cooperate in the oxidative folding of antibodies in vitro. J Biol Chem 2000; 275:29421-29425.
63. Bedard K, Szabo E, Michalak M et al. Cellular functions of endoplasmic reticulum chaperones calreticulin, calnexin, and ERp57. Int Rev Cytol 2005; 245:91-121.
64. Otteken A, Moss B. Calreticulin interacts with newly synthesized human immunodeficiency virus type 1 envelope glycoprotein, suggesting a chaperone function similar to that of calnexin. J Biol Chem 1996; 271:97-103.
65. Peterson JR, Ora A, Van PN et al. Transient, lectin-like association of calreticulin with folding intermediates of cellular and viral glycoproteins. Mol Biol Cell 1995; 6:1173-1184.
66. Halaban R, Cheng E, Zhang Y et al. Aberrant retention of tyrosinase in the endoplasmic reticulum mediates accelerated degradation of the enzyme and contributes to the dedifferentiated phenotype of amelanotic melanoma cells. Proc Natl Acad Sci USA 1997; 94:6210-6215.
67. Pipe SW, Morris JA, Shah J et al. Differential interaction of coagulation factor VIII and factor V with protein chaperones calnexin and calreticulin. J Biol Chem 1998; 273:8537-8544.
68. Daniels R, Kurowski B, Johnson AE et al. N-linked glycans direct the cotranslational folding pathway of influenza hemagglutinin. Mol Cell 2003; 11:79-90.
69. Van Leeuwen JE, Kearse KP. The related molecular chaperones calnexin and calreticulin differentially associate with nascent T cell antigen receptor proteins within the endoplasmic reticulum. J Biol Chem 1996; 271:25345-25349.
70. Vassilakos A, Michalak M, Lehrman MA et al. Oligosaccharide binding characteristics of the molecular chaperones calnexin and calreticulin. Biochemistry 1998; 37:3480-3490.
71. Danilczyk UG, Cohen-Doyle MF, Williams DB. Functional relationship between calreticulin, calnexin, and the endoplasmic reticulum luminal domain of calnexin. J Biol Chem 2000; 275:13089-13097.
72. Wada I, Imai S, Kai M et al. Chaperone function of calreticulin when expressed in the endoplasmic reticulum as the membrane-anchored and soluble forms. J Biol Chem 1995; 270:20298-20304.
73. Frickel EM, Riek R, Jelesarov I et al. TROSY-NMR reveals interaction between ERp57 and the tip of the calreticulin P-domain. Proc Natl Acad Sci USA 2002; 99:1954-1959.
74. Pollock S, Kozlov G, Pelletier MF et al. Specific interaction of ERp57 and calnexin determined by NMR spectroscopy and an ER two-hybrid system. EMBO J 2004; 23:1020-1029.

CHAPTER 5

Molecular Crime and Cellular Punishment:
Active Detoxification of Misfolded and Aggregated Proteins in the Cell by the Chaperone and Protease Networks

Marie-Pierre Hinault and Pierre Goloubinoff*

Abstract

Labile or mutation-sensitised proteins may spontaneously convert into aggregation-prone conformations that may be toxic and infectious. This hazardous behavior, which can be described as a form of "molecular criminality", can be actively counteracted in the cell by a network of molecular chaperone and proteases. Similar to law enforcement agents, molecular chaperones and proteases can specifically identify, apprehend, unfold and thus neutralize "criminal" protein conformers, allowing them to subsequently refold into harmless functional proteins. Irreversibly damaged polypeptides that have lost the ability to natively refold are preferentially degraded by highly controlled ATP-consuming proteases. Damaged proteins that escape proteasomal degradation can also be "incarcerated" into dense amyloids, "evicted" from the cell, or internally "exiled" to the lysosome to be hydrolysed and recycled. Thus, remarkable parallels exist between molecular and human forms of criminality, as well as in the cellular and social responses to various forms of crime. Yet, differences also exist: whereas programmed death is the preferred solution chosen by aged and aggregation-stressed cells, collective suicide is seldom chosen by lawless societies. Significantly, there is no cellular equivalent for the role of familial care and of education in general, which is so crucial to the proper shaping of functional persons in the society. Unlike in the cell, humanism introduces a bias against radical solutions such as capital punishment, favouring crime prevention, reeducation and social reinsertion of criminals.

The Criminal Nature of Protein Aggregation in the Cell

Because of their structural and functional complexity, proteins are particularly fragile macromolecules, especially under environmental stress. Heat-shock, light, oxidation, dehydration or pathogen attacks may cause labile proteins to partially unfold in the cell, resulting in their loss of function, misfolding and aggregation. Anfinsen has shown that the primary sequence of all proteins should, in principle, contain all the information needed to spontaneously fold, or refold, into natively functional proteins.[1] However, when nascent polypeptides exit the ribosome in the unfolded state, or when labile proteins are partially unfolded by stress, they may fault and instead of folding properly, achieve misfolded conformations that abnormally expose hydrophobic surfaces and are enriched in β-strands.[2] Seeking stabilization, newly exposed hydrophobic

*Corresponding Author: Pierre Goloubinoff—DBMV, Faculty of Biology and Medicine, University of Lausanne, CH-1015 Lausanne, Switzerland. Email: Pierre.Goloubinoff@unil.ch

Molecular Aspects of the Stress Response: Chaperones, Membranes and Networks, edited by Peter Csermely and László Vígh. ©2007 Landes Bioscience and Springer Science+Business Media.

surfaces tend to spontaneously associate to form stable protein aggregates,[3] or to combine with cellular membranes, thereby disturbing vital membranal functions (for a review, see ref. 4).

The toxicity of aggregates from proteins such as SOD-1, α-synuclein and PrP[sc], respectively causing familial amyotrophic lateral sclerosis, Parkinson's and prion diseases, may result from several factors:[5] (1) loss of function, (2) filling of precious cytoplasm space, (3) unlawful interactions with native and misfolded proteins and (4) disruption of membranes. Noticeably, the aggregated form of α-synuclein can spontaneously insert into the plasma membranes of neurones and form pores[6] and thus induce programmed cell death and Parkinsonian symptoms (for a review, see ref. 7). Some protein aggregates may induce the aggregation of other proteins, an intra- and inter-cellular self-feeding process known in its most extreme form as prion propagation.[8]

One could thus describe the misfolding and aggregation events that may take place within a small minority of stress- or mutation-labile proteins in the cell, as being a form of molecular criminality. Whereas after a stress, some labile proteins may still choose the native refolding pathway, others may choose a "wrong" or "criminal" path of refolding, leading to confrontational interactions with the surrounding proteins and membranes, with strong negative effects on cell fitness and survival.

Each protein in the cell has its own intrinsic propensity to unfold and misfold spontaneously, a tendency, which increases with mutations and variations of environmental conditions. Thus, a continuous flux of toxic misfolded proteins, which because of cumulative damages may increase with age, is expected to spontaneously form during the lifetime of a cell. Depending on their cellular concentration, misfolded species tend to assemble into stable protein aggregates in the cytoplasm, which in itself is extremely dense and viscous, with 350 mg native proteins per mL.[9] Thus, the cytoplasm resembles the centre of a large city, in which densely crowded proteins, each with a different complementing function, need to move randomly in order to meet and timely interact with rare specific partners. Most native proteins expose repulsive negative charges on their surfaces and refrain from exposing hydrophobic segments. Thus native proteins can optimally move in the highly promiscuous environment of the cytoplasm while avoiding each other.

In this context, the spontaneous conversion of a native protein into a "criminal" misfolded one, exposing positive charges and new hydrophobic surfaces, will greatly increase the friction between the macromolecules and thus the viscosity of the cytoplasm. Increased cytoplasmic viscosity is expected to reduce the freedom of movements and consequently generally impair the function of many cytoplasmic proteins, in addition to the other above-mentioned cytotoxic effects of aggregates.[6,10,11] Similarly, near native protein subunits, on their way to assembly into native oligomers, may also present problematic surfaces to the surrounding, which would need to be neutralised, for example by transient association with binding chaperones (see below).

Defence Mechanisms against Protein Aggregation in the Cell

During their lifetime, cells have to maintain a battery of defences that can reduce the concentration of the toxic misfolded protein species as they form, and maintain them below critical toxic concentrations. Very early in evolution, bacteria and eukaryotes have developed defence mechanisms against "criminal" protein aggregation, in the form of two main classes of proteins: the molecular chaperones (typically Hsp90, Hsp70, Hsp60, Hsp27) and the ATP-dependant proteases (typically Lon, ClpC/X/P, FtsH, HslU/V, and the 26S proteasome),[12] which can be paralleled in the human society to the police and judicial systems, respectively (Fig. 1).

Without stress, the molecular chaperones and the proteases are expressed in the cytoplasm at low concentrations, which are sufficient to carry the physiological and housekeeping functions, and to remove sporadically-forming misfolded protein species. In contrast, during a stress such as heat shock, many molecular chaperones and proteases are massively synthetized

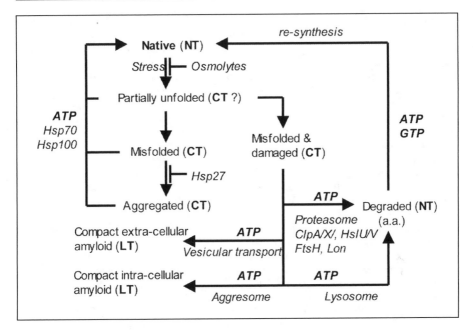

Figure 1. Consecutive lines of defence against the formation of toxic protein conformers in the cell. NT: Non Toxic conformers, LT: Less Toxic conformers, CT: Cytotoxic conformers.

by the cell.[13] The stress-inducible nature of many molecular chaperones has led to their early classification among the heat shock proteins (HSPs), according to their apparent molecular weights in gels: Hsp100, Hsp90, Hsp70 (Hsp40, Hsp20), Hsp60 (Hsp10) and Hsp22/27 in eukaryotes (co-chaperones in brackets), corresponding in bacteria to ClpB, HtpG, DnaK (DnaJ, GrpE), GroEL (GroES), IbpA/B, respectively. Different chaperones display mutually nonexclusive properties. Some "binding" chaperones, such as Hsp90, Hsp70, Hsp60, Hsp40 and Hsp22/27 can provide adhesive surfaces, which upon interaction with partially denatured polypeptides or oligomerizing subunits, can passively reduce the extent of aggregation.[14,15] Unfolding chaperones, such as Hsp100, Hsp70 and Hsp60 (possibly also Hsp90), are involved in the ATP-dependent unfolding (followed by the spontaneous native refolding) of denatured polypeptides[16,17] (for a review, see ref. 18). Misfolded polypeptides may be transferred from "binding" to "unfolding" chaperones, thereby allowing optimal cooperation in the recovery of native proteins by the various chaperone systems.

The common denominator to all classes of molecular chaperones is their ability to identify rare misfolded (delinquent) proteins within a large crowd of native functional ones. Like efficient police officers, molecular chaperones can discriminate between potentially toxic misfolded proteins and apprehend them without disturbing the "law abiding" surrounding functional proteins. Hence, Hsp70, Hsp90, Hsp100 and the ATPases guarding the entrance to the proteasome can interact with abnormally exposed hydrophobic structures at the surface of misfolded proteins.[19,20] Molecular chaperones have a high affinity for misfolded proteins -typical of an enzyme-to-substrate relationship-, and a low affinity for natively folded proteins -typical to an enzyme-to-product relationship.[18] We therefore suggest to define the molecular chaperones as enzymes that can accelerate the unfolding of their misfolded protein substrates and assist the later to reach a state from which they can spontaneously convert into low affinity, natively folded products. To overcome the energetic barrier between the stable misfolded aggregated state of the substrate and the unstable transiently unfolded intermediate leading to the low affinity chaperone product, the energy of ATP hydrolysis is used by some ATPase chaperones to apply an unfolding force on the misfolded substrate.

The capture of an unfolding protein by a molecular chaperone is only the beginning of an evaluation process between a possible subsequent native refolding and a costly degradation, followed by an even costlier resynthesis of the irreversibly damaged molecules. Thus, the unfolding ATPase chaperones Hsp100, Hsp70 and GroEL use the energy of ATP hydrolysis to forcefully unfold transiently bound (apprehended) misfolded and aggregated polypeptides, allowing them, once released, to refold into native functional proteins.[17] This approach resembles that of social reeducation programs offered to minor criminals, leading to their possible social reinsertion.

Laskey has first proposed the term "molecular chaperone" for nucleoplasmin.[21] John Ellis thereafter suggested extending the definition of a social chaperone to describe the molecular mechanism of GroEL-related proteins,[22] whereby a molecular chaperone would become a (mature) protein escorting (young) nascent proteins to (social) gatherings, to prevent improper associations (aggregation). Nearly two decades later, we find that this definition, implying a rather passive, energetically inexpensive mechanism, is still applicable to some simple binding chaperones, such as the small HSPs. However, it has since been demonstrated that ATPase chaperones, such as GroEL,[23] Hsp70 and Hsp100 carry much more active functions:[18] they can convert part of the energy of ATP hydrolysis to repair structural damages in stably misfolded and already aggregated proteins (see Fig. 1). Unfolding chaperones can forcefully disentangle stably associated partners of a protein aggregate and, moreover, unfold ("re-educate") formerly stable "criminal" misfolded polypeptides into "born again" native, law-abiding functional members of the society of proteins in the cell. Unless plagued by recidivist criminal proteins, this solution is apparently less costly, in terms of ATP-consumption, than the alternative of having to degrade and resynthesize indiscriminately all the misfolded proteins that form during the lifetime of a cell. Molecular recidivism is possible when a misfolded protein becomes irreversibly damaged, for example by oxidation or breakage and has lost its intrinsic ability to regain its native functional structure. As in the case of social reeducation and reinsertion services, the problem with recidivism is that "criminal" molecules may block the unfolding chaperone systems and prevent potentially recoverable proteins from being reactivated.

To this aim, the cells have placed the repairing system carried by the chaperones in competition with more radical systems carried by proteases, leading to the elimination and recycling of the toxic protein conformers. Thus, misfolded polypeptides that remain misfolded too long, despite the recurrent futile unfolding attempts by ATPase chaperones, may sooner or later interact with a protease and be degraded.

There are extreme situations in nature where the chaperone and protease systems become overloaded by toxic protein forms. This is the case when mutant proteins are overexpressed, or when the molecular chaperones and proteases are insufficiently produced by aging mammalian cell, which poorly react to inflammation and environmental stresses (for a review, see ref. 24). Then, an excess of toxic misfolded proteins in the cell can be compacted into less toxic inclusion bodies by the aggresome, an active ATP-consuming mechanism involving molecular motors and the microtubule cytoskeleton[25] (Fig. 1). The social equivalent to the aggresome would be incarceration, and to the ATP needed for the process, the salary of the jail keepers (Table 1). Others toxic aggregates can be actively secreted from the cell, socially equivalent to banishment. This is the case of PrP^{Sc} prions and the Aβ-peptides that accumulate as compact protein deposits outside the cell.[26,27] Active secretion of toxic aggregates to the lysosome for degradation by nonspecific proteolytic enzymes could be compared to active deportation to death camps.

By far, the most frequent strategy used by eukaryotic cells to eliminate "recidivist" toxic protein conformers from the cytoplasm is controlled (ATP-consuming) degradation by the proteasome, whose social analogy would be that of a tribunal ruling for capital punishments. The decision to destroy proteins in the cell depends on a very sophisticated molecular system that, in a first step, must discriminate nonrecoverable delinquent molecules from the recoverable ones and from the majority of "law abiding" native proteins in the cytoplasm, and in a second

Table 1. *Equivalent terms between molecular and social organization levels of life*

Protein in the Cell	Human in the Society
Various problems leading to criminality	
Natively folded functional protein	Law abiding mature functional citizen
Folding of a simple nascent protein	Maturing adolescent without problems
Folding of a multi-domain nascent protein	Maturing adolescent with potential problems
Folding of a mutant protein	Maturing individual with an inherited propensity to criminality
Folding of a damaged protein	Maturing individual who has been abused in his childhood
Negative repulsive charges on native proteins	Politeness, courtesy, civility
Stress, environmental pressures on proteins	Stress, environmental pressures on functional individuals
Natively folded stable protein under stress	Functional (employed) individual under pressure
Osmolytes that stabilize native proteins	Social security, welfare that stabilize labile proteins
Protein unfolding and inactivation (stress)	Dismissal of a functional employee
Proteins misfolding	Dismissed employee turning into a criminal
Protein with exposed hydrophobic surfaces	Rude, aggressive beggar
Protein aggregation	Gang formation, association of criminals, mobs
Recoverable toxic protein aggregate,	Redeemable criminal causing short-term damages
Nonrecoverable toxic protein aggregate	Nonredeemable criminal causing long-term damages
Aggregation prone mutant protein	Multi-recidivist criminal
Propagation of protein aggregates	Corruption of functional (key) members of the society
No obvious equivalent in the cell	Parental care
No obvious equivalent in the cell	Affection, love, empathy
No obvious equivalent in the cell	Education, Teacher
No obvious equivalent in the cell	School
No obvious equivalent in the cell	Morality, empathy, humanism
Various solutions to criminality	
Inflammation	Establishment of a state of emergency
Fever	Recruitment of law enforcement agents
Excessive inflammation	Abolition of civil rights
Emission of reactive oxygen species (ROS)	Smoking, drugs abuse, gambling, self-destructive behaviour
Peroxidases, catalases, glutathione	Drug detoxification and rehabilitation programmes
Apoptosis	Collective suicide
Compacting aggregates by the aggresome	Active imprisonment of criminals
Compacted amyloids and fibres	Partially neutralized criminals (prisoners)
Exocytosis, secretion of toxic aggregates	Eviction of criminals
Secretion of aggregates to the lysosome	Deportation of criminals to death camps
Molecular chaperones	Law enforcement officers
Binding chaperone (sHsps, Hsp40)	Social worker, community educator, counsellor, priest
Unfolding chaperones (Hsp70, Hsp100)	Psychotherapist, rehabilitation services, probation officers, correctional treatment specialists.
E3-Ubiquitin ligase, lid of the proteasome	Judges
ClpA, ClpX, HslU	Judges
ATP	Salaries of policemen, social workers and judges
Proteasome chamber, ClpP, HslV	Execution chamber
Tricorn proteases, aminopeptidases	Funeral and burial services

step, must deliver these delinquent molecules to the death chamber of the proteasome.[20] This high level of scrutiny is essential to prevent costly judicial errors in the cell.[12] A blind executioner cannot be allowed to ram freely within a dense crowd of law-abiding citizens, waving his axe indiscriminately. Again, the molecular chaperones, in particular Hsp70, as zealous police officers, are the first to point out the damaged delinquent proteins and, in close collaboration with molecular "judges", direct them either to reinsertion programs or to elimination. Poly-Ubiquitin proteins which will be destined to proteasomal degradation. Poly-Ubiquitin proteins would be the social equivalent to the infamous nazi mark of the yellow star. The ATPase proteins guarding the lid of the proteolytic chamber of the proteasome, like the E3-Ubiquitin ligases, which designate the proteins doomed to degradation, can be considered as "molecular judges" (Table 1).

Figure 1 summarizes the various lines of defences that can be used by the eukaryotic cell to prevent the formation of toxic protein conformers and to detoxify those which have already formed. Stress or mutations may cause the partial unfolding of labile native proteins. Unfolding can be prevented by the presence of stabilizing osmolytes, such as glycine betaine, proline and trehalose.[28] Partially unfolded species may then convert into cytotoxic, misfolded species, which may further associate into toxic aggregates. Binding chaperones, such as Hsp27, can prevent aggregation. ATP-hydrolysing chaperones, such as Hsp70 and Hsp100 can actively unfold toxic misfolded and aggregated conformers, converting them into nontoxic native proteins. Alternatively, irreversibly damaged, toxic conformers are degraded by ATPase-gated proteases. Failing that, they can be compacted into the cytoplasm by the aggresome, or secreted outside the cell, to be concentrated into compact, less toxic amyloids. Others are targeted to the lysosome and degraded into amino acids, to be recycled into the energy-consuming resynthesis of functional proteins.

Aging and Conformational Diseases: Failures of Law Enforcement Leading to Lawlessness

In aging mammalian cells, the levels of molecular chaperones and proteases are significantly decreased and, in parallel, irreversibly damaged proteins accumulate.[29,30] In addition to their general cytotoxic effect, irreversibly damaged proteins can inhibit the activity of the remaining minority of active chaperones and proteases. At this stage, old or mutant cells often choose suicide, which may be advantageous to some forms of life, as in the case of cancer cells, or of infected plant tissues that prevent the propagation of the pathogen to the whole plant.

Interestingly, Hsp70 has been shown to protect against cell death by directly interfering with the mitochondrial apoptosis pathway.[31] This is exemplified in the case of the acute respiratory distress syndrome (ARDS), which is an inflammatory response in the lungs culminating in necrosis and fatal apoptosis. Sepsis-induced ARDS in rats correlates with the specific failure of Hsp70 expression in alveolar tissues.[32] In contrast, in a rat model of sepsis-induced ARDS, adenovirus-mediated transient expression of Hsp70 in the lungs effectively prevented apoptosis, lung failure and dramatically improved survival.[33] The overproduction of molecular chaperones observed following treatments with various nonsteroidal anti-inflammatory drugs (NSAIDs), such as sodium salicylate,[13] ibuprofen,[34] and less classical Hsp-inducers such as celastrol,[35] resveratrol (the French paradox),[36] geranylgeranylacetone,[37] may be responsible for the observed reduction of damages related to reactive oxygen species and induced programmed cell death in various damaging contexts, such as ARDS,[38] post-ischemic reperfusions,[39] excessive inflammation, toxic protein aggregation and protein-conformation neurodegenerative diseases and aging.[40]

Human societies seldom choose collective suicide in response to lawlessness, which is a noticeable difference between cellular and social levels of crime responses. Yet, as in the cell, a massive build up of law-enforcement forces, when energetically still possible, may gradually decrease lawlessness and restore cellular functions in societies, as well as in cells otherwise doomed to self destruction. This is observed in the case of apoptotic tissues treated with NSAIDs or low doses of Hsp-inducing poisons.[24,40] Another major difference between criminal molecules in the cell and social criminals is the extreme dependence of human newborns toward their parents and siblings. Unlike proteins, the successful formation of a functional member of the human society demands a massive, long-term investment from the progenitors and the education system. This phenomenon, which has no obvious parallel in the cell, could not withstand selective pressures of evolution, unless the existence of powerful ties between relatives in the human society, namely affective ties, that do not readily disappear once the person has become a functional unit of the society, or a criminal. In contrast to children, nascent proteins exit the ribosome with all the information necessary for their spontaneous folding into functional units of the cell. Unlike parental education, no new information is to be transmitted to a folding protein by the molecular chaperones. Chaperones may only passively prevent proteins from leaving, or actively return astrayed proteins on their innate native folding path.

The resilient affective ties between humans, especially from the same family or clan, have gradually influenced policies of crime management by human societies, especially in recent history, with the appearance of humanism. Whereas ribosomes have apparently no particular feelings towards the proteins they synthesize, siblings of convicted criminals and less-related humanists have become increasingly reluctant to accept irreversible radical solutions to social criminality, such as torture, amputations and executions, and encouraged solutions whereby criminals may be reeducated, pardoned, reinserted as functional members of the society, and in the worst case, incarcerated for a half life period, at most.

Nonetheless, many remarkable parallels, sometimes ethically questionable, but didactically appealing, do exist between the proteins in the cell, and humans in the society (see Table 1). These parallels were initially expressed by the adequate, albeit incomplete, social term "chaperone", suggesting only a rather passive defence mechanism against toxic protein aggregations in the cell. They can inspire scientists and politicians alike, in their commons search for solutions to crime: may it be molecular crime, as in the case of aggregate-induced neurodegeneration and aging, or social crime, as in the case of mobs burning foreign embassies, or terrorists waging war on the democracies of the world.

Acknowledgements

This chapter was inspired, in part, from a review in French by the same authors: Hinault, M-P. and Goloubinoff, P. L'agrégation toxique des protéines: une forme de "délinquance moléculaire" activement combattue dans la cellule par les chaperones moléculaires et les protéases. Annales de cardiologie et d'angéiologie 2006. In Press.

References

1. Anfinsen CB. Principles that govern the folding of protein chains. Science 1973; 181:223-230.
2. Uversky VN. Protein folding revisited. A polypeptide chain at the folding-misfolding-nonfolding cross-roads: Which way to go? Cell Mol Life Sci 2003; 60:1852-1871.
3. Dobson CM. Protein misfolding, evolution and disease. Trends Biochem Sci 1999; 24:329-332.
4. Torok Z, Goloubinoff P, Horvath I et al. Synechocystis HSP17 is an amphitropic protein that stabilizes heat-stressed membranes and binds denatured proteins for subsequent chaperone-mediated refolding. Proc Natl Acad Sci USA 2001; 98:3098-3103.
5. Muchowski PJ, Wacker JL. Modulation of neurodegeneration by molecular chaperones. Nat Rev Neurosci 2005; 6:11-22.
6. Lashuel HA, Hartley D, Petre BM et al. Neurodegenerative disease: Amyloid pores from pathogenic mutations. Nature 2002; 418:291.
7. Dawson TM, Dawson VL. Molecular pathways of neurodegeneration in Parkinson's disease. Science 2003; 302:819-822.
8. Prusiner SB. Scrapie prions. Annu Rev Microbiol 1989; 43:345-374.
9. Ellis RJ. Macromolecular crowding: Obvious but underappreciated. Trends Biochem Sci 2001; 26:597-604.
10. Satyal SH, Schmidt E, Kitagawa K et al. Polyglutamine aggregates alter protein folding homeostasis in Caenorhabditis elegans. Proc Natl Acad Sci USA 2000; 97:5750-5755.
11. Ben-Zvi AP, Goloubinoff P. Proteinaceous infectious behavior in nonpathogenic proteins is controlled by molecular chaperones. J Biol Chem 2002; 277:49422-49427.

12. Tomoyasu T, Mogk A, Langen H et al. Genetic dissection of the roles of chaperones and proteases in protein folding and degradation in the Escherichia coli cytosol. Mol Microbiol 2001; 40:397-413.

13. Westerheide SD, Morimoto RI. Heat shock response modulators as therapeutic tools for diseases of protein conformation. J Biol Chem 2005; 280:33097-33100.

14. Chatellier J, Hill F, Fersht AR. From minichaperone to GroEL 2: Importance of avidity of the multisite ring structure. J Mol Biol 2000; 304:883-896.

15. Mogk A, Tomoyasu T, Goloubinoff P et al. Identification of thermolabile Escherichia coli proteins: Prevention and reversion of aggregation by DnaK and ClpB. Embo J 1999; 18:6934-6949.

16. Goloubinoff P, Christeller JT, Gatenby AA et al. Reconstitution of active dimeric ribulose bisphosphate carboxylase from an unfoleded state depends on two chaperonin proteins and Mg-ATP. Nature 1989; 342:884-889.

17. Goloubinoff P, Mogk A, Zvi AP et al. Sequential mechanism of solubilization and refolding of stable protein aggregates by a bichaperone network. Proc Natl Acad Sci USA 1999; 96:13732-13737.

18. Ben-Zvi AP, Goloubinoff P. Review: Mechanisms of disaggregation and refolding of stable protein aggregates by molecular chaperones. J Struct Biol 2001; 135:84-93.

19. Mayer MP, Bukau B. Hsp70 chaperones: Cellular functions and molecular mechanism. Cell Mol Life Sci 2005; 62:670-684.

20. Wolf DH, Hilt W. The proteasome: A proteolytic nanomachine of cell regulation and waste disposal. Biochim Biophys Acta 2004; 1695:19-31.

21. Dingwall C, Laskey RA. Nucleoplasmin: The archetypal molecular chaperone. Semin Cell Biol 1990; 1:11-17.

22. Ellis RJ, van der Vies SM. Molecular chaperones. Annu Rev Biochem 1991; 60:321-347.

23. Shtilerman M, Lorimer GH, Englander SW. Chaperonin function: Folding by forced unfolding. Science 1999; 284:822-825.

24. Hinault MP, Goloubinoff P. Molecular chaperones and proteases: Cellular fold-controlling factors of proteins in neurodegenerative diseases and aging. J Mol Neurosci 2006, (In Press).

25. Kopito RR. Aggresomes, inclusion bodies and protein aggregation. Trends Cell Biol 2000; 10:524-530.

26. Fevrier B, Vilette D, Archer F et al. Cells release prions in association with exosomes. Proc Natl Acad Sci USA 2004; 101:9683-9688.

27. Turner RS, Suzuki N, Chyung AS et al. Amyloids beta40 and beta42 are generated intracellularly in cultured human neurons and their secretion increases with maturation. J Biol Chem 1996; 271:8966-8970.

28. Diamant S, Eliahu N, Rosenthal D et al. Chemical chaperones regulate molecular chaperones in vitro and in cells under combined salt and heat stresses. J Biol Chem 2001; 276:39586-39591.

29. Heydari AR, You S, Takahashi R et al. Age-related alterations in the activation of heat shock transcription factor 1 in rat hepatocytes. Exp Cell Res 2000; 256:83-93.

30. Soti C, Csermely P. Aging and molecular chaperones. Exp Gerontol 2003; 38:1037-1040.

31. Saleh A, Srinivasula SM, Balkir L et al. Negative regulation of the Apaf-1 apoptosome by Hsp70. Nat Cell Biol 2000; 2:476-483.

32. Weiss YG, Bouwman A, Gehan B et al. Cecal ligation and double puncture impairs heat shock protein 70 (HSP-70) expression in the lungs of rats. Shock 2000; 13:19-23.

33. Weiss YG, Maloyan A, Tazelaar J et al. Adenoviral transfer of HSP-70 into pulmonary epithelium ameliorates experimental acute respiratory distress syndrome. J Clin Invest 2002; 110:801-806.

34. Wang Q, Mosser DD, Bag J. Induction of HSP70 expression and recruitment of HSC70 and HSP70 in the nucleus reduce aggregation of a polyalanine expansion mutant of PABPN1 in HeLa cells. Hum Mol Genet 2005; 14:3673-3684.

35. Westerheide SD, Bosman JD, Mbadugha BN et al. Celastrols as inducers of the heat shock response and cytoprotection. J Biol Chem 2004; 279:56053-56060.

36. Delmas D, Jannin B, Latruffe N. Resveratrol: Preventing properties against vascular alterations and ageing. Mol Nutr Food Res 2005; 49:377-395.

37. Ooie T, Takahashi N, Saikawa T et al. Single oral dose of geranylgeranylacetone induces heat-shock protein 72 and renders protection against ischemia/reperfusion injury in rat heart. Circulation 2001; 104:1837-1843.

38. Haddad JJ. Science review: Redox and oxygen-sensitive transcription factors in the regulation of oxidant-mediated lung injury: Role for nuclear factor-kappaB. Crit Care 2002; 6:481-490.

39. Lopez-Neblina F, Toledo AH, Toledo-Pereyra LH. Molecular biology of apoptosis in ischemia and reperfusion. J Invest Surg 2005; 18:335-350.

40. Hinault MP, Goloubinoff P. L'agrégation toxique des protéines: Une forme de "délinquance moléculaire" activement combattue dans la cellule par les chaperones moléculaires et les protéases. Annales de cardiologie et d'angéiologie 2006, (In Press).

CHAPTER 6

Chaperones as Parts of Cellular Networks

Peter Csermely,* Csaba Söti and Gregory L. Blatch

Abstract

The most important interactions between cellular molecules have a high affinity, are unique and specific, and require a network approach for a detailed description. Molecular chaperones usually have many first and second neighbors in protein-protein interaction networks and they play a prominent role in signaling and transcriptional regulatory networks of the cell. Chaperones may uncouple protein, signaling, membranous, organellar and transcriptional networks during stress, which gives an additional protection for the cell at the network-level. Recent advances uncovered that chaperones act as genetic buffers stabilizing the phenotype of various cells and organisms. This chaperone effect on the emergent properties of cellular networks may be generalized to proteins having a specific, central position and low affinity, weak links in protein networks. Cellular networks are preferentially remodeled in various diseases and aging, which may help us to design novel therapeutic and anti-aging strategies.

Introduction: Cellular Networks and Chaperones

Most of the molecular interactions of our cells have a low affinity, are rather unspecific, and can be described in general terms. An example for this is the self-association of lipids to membranes. However, the most important interactions between cellular molecules have a high affinity, are unique and specific, and require a network approach for a better understanding and prediction of their changes after various environmental changes, like stress.[1-3] One of the good examples for the network description of unique cellular interactions between molecules is the protein-protein interaction network (Fig. 1), where the elements of the network are proteins, and the links between them are permanent or transient bonds.[4-6] At a higher level of complexity we have networks of protein complexes (sometimes built together with lipid membranes), where the individual complexes are modules of the protein-protein network. The cytoskeletal network and the membranous, organellar network are good examples of these, larger networks. In the cytoskeletal network, we have individual cytoskeletal filaments, like actin, tubulin filaments, or their junctions as the elements of the network, and the bonds between them are the links. In the membranous, organellar network various membrane segments (membrane vesicles, domains, rafts, of cellular membranes) and cellular organelles (mitochondria, lysosomes, segments of the endoplasmic reticulum, etc.) are the elements, and they are linked by protein complexes and/or membrane channels. Both the membranes and the organelles contain large protein-protein interaction networks. In signaling networks the elements are proteins or protein complexes and the links are highly specific interactions between them, which undergo a profound change (either activation or inhibition), when a

*Corresponding Author: Peter Csermely—Department of Medical Chemistry, Semmelweis University, Puskin Street 9, H-1088 Budapest, Hungary. Email: csermely@puskin.sote.hu

Molecular Aspects of the Stress Response: Chaperones, Membranes and Networks, edited by Peter Csermely and László Vígh. ©2007 Landes Bioscience and Springer Science+Business Media.

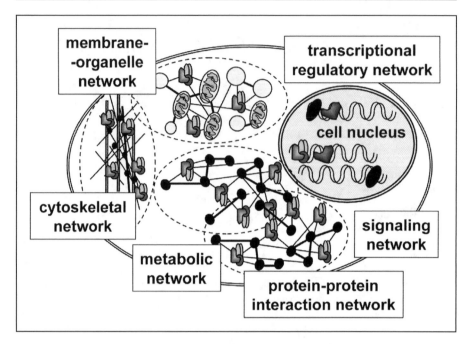

Figure 1. Cellular networks. The figure illustrates the most important networks in our cells. The protein-protein interaction network, the cytoskeletal network and the membranous, organellar networks provide a general scaffold of the cell containing the physical interactions between cellular proteins. Other networks, like the signaling, transcriptional, or metabolic networks are functionally defined. In the signaling network elements of various signaling pathways are linked by the interactions between them. In the transcriptional regulatory network the elements are the transcription factors and the genes, and the connecting links are functional interactions between them. In the metabolic network we have the various metabolites as elements and the enzyme reactions as links. All these networks highly overlap with each other, and some of them contain modules of other networks.

specific signal reaches the cell.[7] In metabolic networks the network elements are metabolites, such as glucose, or adenine, and the links between them are the enzyme reactions, which transform one metabolite from the other.[8] Finally, gene transcription networks have two types of elements, transcriptional factor-complexes and the DNA gene sequences, which they regulate. Here the transcriptional factor-complexes may initiate or block the transcription of the gene's messenger RNA. The links between these elements are the functional (and physical) interactions between the proteins (sometimes RNAs) and various parts of the gene sequences in the cellular DNA.[9]

Cellular networks are often small worlds, where from a given element any other elements of the network can be reached via only a few other elements. Networks of our cells usually have a scale-free degree distribution, which means that these networks have hubs, i.e., elements, which have a large number of neighbors. These networks are rich in motifs, which are regularly appearing combinations of a few adjacent network elements, and contain hierarchical modules, or in other words: are forming hierarchical communities.[1-3] The complex architecture of cellular networks is needed to fulfill four simultaneous tasks (Fig. 2). (1) The first task is the local dissipation of the perturbations/noise coming from outside the cell, and from the stochastic elements of intracellular reactions. (2) The second task is the efficient and reliable global transmission of signals from one distant element of the cell to another. (3) The third task is the

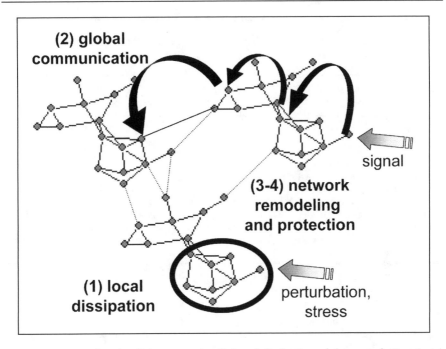

Figure 2. Major tasks of cellular networks. 1) Local dissipation of the perturbations/noise coming from outside the cell, and from the stochastic elements of intracellular reactions. 2) Efficient and reliable global transmission of signals from one element of the cell to another. 3) Discrimination between signals and noise via the continuous remodeling of these networks during the evolutionary learning process of the cell. 4) Protection against the continuous random damage of free radicals and other harmful effects during stress and aging.

discrimination between signals and noise via the continuous remodeling of these networks during the evolutionary learning process of the cell. (4) The fourth task is the protection against the continuous random damage of free radicals and other harmful effects during stress and aging. During the execution of these tasks the assembly of network elements produces a vast number of emergent properties of networks, which can only be understood, if we study the whole network and cannot be predicted knowing the behavior of any of its elements.

The cellular networks use all their features mentioned above to solve their tasks. As a relatively simplified view, hubs help to confine most of the perturbations to a local environment, while the small world character allows the global propagation of signals. Motifs and hierarchical modules help both the discrimination between the two, and provide stability at the network level (which is helped by a number of repair functions at the molecular level).[3] However, this summary of the major features of cellular networks is largely a generalization, and needs to be validated through critical scrutiny of the datasets, sampling procedures and methods of data analysis at each network examined.[10-12]

Molecular chaperones mostly form low affinity, dynamic temporary interactions (weak links) in cellular networks (Table 1).[13,14] Chaperones generally have a large number of partners, thus they behave like hubs of protein-protein interaction or transcriptional regulatory networks.[15,16] Moreover, many chaperone effects (like cell survival, changes in the phenotype diversity, etc.) are typical emergent network properties, which can rarely be understood by studying exclusively the individual chaperone/client interactions. Thus the network approach is a promising tool to explain some key aspects of chaperone function.[17]

Table 1. High and low confidence chaperone neighbors in the yeast protein-protein interaction network

Chaperone Class	High Confidence Partners	Low Confidence Partners
Hsp70	0	28
Hsp90	0	30
Average	1	19

The mean of the partners of the respective chaperone classes (containing the cochaperones as well) was calculated using the annotated yeast protein-protein interaction database of reference 4. High confidence partners are enriched in high affinity chaperone-neighbor interactions. Low confidence partners may contain a considerable amount of artifacts, but may also be enriched in low affinity chaperone-neighbor interactions. The average is the average number of neighbors of all proteins in the database. The differences between the chaperone class values and the average values were not significant due to the high S.D.

Chaperones in Cellular Networks

Chaperones form large complexes and have a large number of cochaperones to regulate their activity, binding properties and function.[18-21] These chaperone complexes regulate local protein networks, such as the mitochondrial protein transport apparatus as well as the assembly and substrate specificity of the major cytoplasmic proteolytic system, the proteasome.[22-24] Chaperones may be important elements to promote the cross-talk between various signaling processes. The Hsp90 chaperone complex promotes the maturation of over a hundred kinase substrates.[14,15] Chaperones have a large number of second neighbors in the yeast protein-protein interaction network (Table 2). The large proportion of hubs in the close vicinity of chaperones gives a central position to these proteins in the protein-protein network, which may help the chaperone-mediated cross talk between signaling pathways.

Chaperones have an important role in membrane stabilization.[25-27] Their membrane association links chaperones to the membrane network of the cell integrating the plasma membrane, the endoplasmic reticulum (ER), the Golgi apparatus, various vesicles, the nuclear membrane and mitochondria together.[28-30]

Table 2. First and second neighbors of molecular chaperones in the yeast protein-protein interaction network

Chaperone Class	First Neighbors	Second Neighbors	% of Hub Neighbors
Hsp70	31	93	41
Hsp90	33	137	49
Average	25	60	28

The mean of the number of first and second neighbors of the respective chaperone classes (containing the cochaperones as well) was calculated using the annotated yeast protein-protein interaction database of reference 4. Hubs are neighbors having more than a 100 interacting proteins. The average is the average number of neighbors of all proteins in the database. The differences between the chaperone class values and the average values were not significant due to the high S.D.

At the end of seventies Porter and coworkers reported various and often quite poorly identifiable, cytoplasmic filamentous structures and called them the microtrabecular network of the cytoplasm.[31,32] Although a rather energetic debate has developed about the validity of the electron microscopic evidence of the microtrabeculae, several independent findings support the existence of the cytoplasmic macromolecular organization.[33-37] This cytoplasmic meshwork not only serves as a scaffold to organize and direct macromolecular traffic, but may also behave as a mesh modulating cytoplasmic streaming assumed to be in the range of 1 to 80 μm/sec.[35] The major cytoplasmic chaperones (TCP1/Hsp60, Hsp70 and Hsp90 and their associated proteins) may well form a part of this cytoplasmic macromolecular network.[38,39]

Molecular chaperones translocate to the cell nucleus, and protect it after stress by a direct protection and repair of damaged proteins, and by changing the intranuclear traffic and nuclear organization.[40,41] Stress-induced nuclear translocation of chaperones may preserve the nuclear remodeling capacity during environmental damage protecting the integrity of DNA. Additionally, chaperones regulate both the activation and the disassembly of numerous transcriptional complexes, thus chaperones emerge as key regulators of the transcriptional network.[42,43]

De-coupling of network elements and modules is a generally used method to stop the propagation of damage.[1-3] When the cell experiences stress, chaperones become increasingly occupied by damaged proteins. This, together with the stress-induced translocation of chaperones to the nucleus mentioned before, might lead to an "automatic" de-coupling of all chaperone-mediated networks including protein-protein, signaling, transcriptional regulatory as well as membranous, organellar networks providing an additional safety measure for the cell.[17]

Chaperone-Mediated Emergent Properties of Cellular Networks

As we have seen before, chaperones are involved in the regulation of signaling, membranous, organellar, cytoskeletal and transcriptional networks. However, relatively little is known on the chaperone-mediated, emergent properties of cellular functions. As mentioned before, these emergent properties are properties of the cell, or the whole organism, which can not be linked to the behavior of any of their particular elements, but emerge as a concerted action of the whole cellular network. One of the best examples of chaperone-mediated emergent network properties was shown by Susannah Rutherford and Susan Lindquist, when they discovered that Hsp90 acts as a buffer to conceal the phenotype of the genetic changes in *Drosophila melanogaster*.[44] Chaperone-induced genetic buffering is released upon stress, which causes the sudden appearance of the phenotype of previously hidden mutations, helps population survival and gives a possible molecular mechanism for fast evolutionary changes. On the other hand, the stress-induced appearance of genetic variation at the level of the phenotype cleanses the genome of the population by allowing the exposure and gradual disappearance of disadvantageous mutations by natural selection. After the initial report of Rutherford and Lindquist in reference 44 on Hsp90, the effect was extended to other chaperones and to *Escherichia coli, Arabidopsis thaliana* and the evolution of resistance in fungi.[45-47] The Hsp90-mediated buffering might have an epigenetic origin due to the Hsp90-induced heritable changes in the chromatin structure.[48]

Chaperones are highly conserved proteins, therefore similar mechanisms might operate in humans.[49] Moreover, the tremendous advance of medicine and the profound changes in human lifestyle in the last two hundred years significantly decreased natural selection, and potentially helped the accumulation of hidden mutations in the human genome. In the first approximation this is not a problem, since we have a large amount of chaperones and other buffering systems to hide these disadvantageous mutations. However, the amount of damaged or newly folded proteins and the available chaperone capacity are two sides of a carefully balanced system in our cells. An excess of chaperone substrates or diminished chaperone content might both induce a "chaperone-overload", i.e., a relative deficit of available active, unloaded chaperones.[50] Chaperone overload becomes especially large in aged subjects, where protein damage is abundant, and both chaperone induction and chaperone function are impaired. A special case of chaperone-overload occurs in folding diseases including various forms of neurodegeneration, such as Alzheimer's disease, Parkinson's disease or Hungtington's disease, where a misfolded and usually not degraded protein sequesters most chaperones.[50-53] In consequence of the chaperone overload, the protein products harboring the "hidden mutations" may be released, and may contribute to the development of civilization diseases, such as cancer, atherosclerosis and diabetes. Since relatively few generations have been passed from the beginning of the medical revolution and lifestyle changes in the 19th century, this effect is most probably negligible today. However, it increases with each generation. Still, we probably have many hundreds of years to think about a possible solution, which gives us time to learn much more and reconcile the serious ethical concerns with a possible solution.[50]

In recent years the scientific community has became increasingly aware of the idea that not only chaperones but a large number of other proteins may also regulate the diversity of the phenotype.[3,14,54-56] Though a relatively small number of regulators were uncovered yet, a common molecular mechanism, such as the involvement in signaling or modifications of histones and DNA structure seems to be an unlikely explanation for all the effects observed. If a general explanation is sought, it is more likely to be related to the network properties of the cell. In this context, chaperones are typical weak linkers, providing low affinity, low probability contacts with other proteins (for a grossly simplified illustration, see Table 1). Weak links are known to help system stability in a large variety of networks from macromolecules to social networks and ecosystems, which may be a general network-level phenomenon explaining many of the genetic buffering effects. Currently we do not know, what position is required, if any, in the network for these 'weak links' besides their low affinity and transient interactions. The central position of chaperones demonstrated in Table 2 may be an additional hallmark of stabilizing weak links.[3,14]

Chaperone Therapies

Cellular networks are remodeled in various diseases and after stress. Proper interventions to push the equilibrium towards the original state may not be limited to single-target drugs, which have a well-designed, high affinity interaction with one of the cellular proteins. In agreement with this general assumption, several examples show that multi-target therapy may be superior to the usual single-target approach. The best known examples of multi-target drugs include Aspirin, Metformin or Gleevec as well as combinatorial therapy and natural remedies, such as herbal teas.[57] Due to the multiple regulatory roles of chaperones, chaperone-modulators provide additional examples for multi-target drugs. Indeed, chaperone substitution (in the form of chemical chaperones), the help of chaperone induction and chaperone inhibition are all promising therapeutic strategies.[58-61]

Conclusion

Chaperones regulate cellular functions at two levels. In several cases they interact with a specific target protein, and become mandatory to its folding as well as for the assistance in the formation of specific protein complexes (and in the prevention of the assembly of others). These specific interactions make chaperones important parts of the core of cellular networks, such as the protein net, the signaling network, the membranous and organellar network as well as the transcriptional network. However, in most cases chaperones have only a low affinity, temporary interactions, i.e., 'weak links' with most of their targets. Changes of these interactions do not affect the general behavior of the whole network, the cell. However, an inhibition of these weak links might lead to a rise in cellular noise, the destabilization and des-integration of the whole network. By this complex version of the 'error-catastrophe', chaperone inhibitors help us to combat cancer. In contrast, chaperone activation may decrease cellular noise, stabilize and integrate cells and thus give a general aid against aging and diseases. Thus, besides slowing the development or reversing protein folding diseases, chaperone-therapies may also generally benefit the aging organism by stabilizing its cells and functions. Properly working chaperones may be key players to help us to reach improved life conditions in an advanced age. The assessment of the multiple roles of chaperones in the context of cellular networks is only just beginning.

Acknowledgements

Work in the authors' laboratory was supported by research grants from the EU (FP6506850, FP6-016003), Hungarian Science Foundation (OTKA-T37357 and OTKA-F47281), Hungarian Ministry of Social Welfare (ETT-32/03), Hungarian National Research Initiative (1A/056/2004 and KKK-0015/3.0) and from the South African National Research Foundation (2067467 and the South African-Hungarian Collaborative Program 2053542). C. Söti is a Bolyai research Scholar of the Hungarian Academy of Sciences.

References

1. Barabasi AL, Oltvai ZN. Network biology: Understanding the cell's functional organization. Nat Rev Genet 2004; 5:101-113.
2. Albert R. Scale-free networks in cell biology. J Cell Sci 2005; 118:4947-4957.
3. Csermely P. Weak links: Stabilizers of complex systems from proteins to social networks. Heidelberg: Springer Verlag, 2006.
4. von Mering C, Krause R, Snel B et al. Comparative assessment of large-scale data sets of protein-protein interactions. Nature 2002; 417:399-403.
5. Rual JF, Venkatesan K, Hao T et al. Towards a proteome-scale map of the human protein-protein interaction network. Nature 2005; 437:1173-1178.
6. Stelzl U, Worm U, Lalowski M et al. A human protein-protein interaction network: A resource for annotating the proteome. Cell 2005; 122:957-968.
7. White MA, Anderson RG. Signaling networks in living cells. Annu Rev Pharmacol Toxicol 2005; 45:587-603.
8. Borodina I, Nielsen J. From genomes to in silico cells via metabolic networks. Curr Opin Biotechnol 2005; 16:350-355.
9. Blais A, Dynlacht BD. Constructing transcriptional regulatory networks. Genes Dev 2005; 19:1499-1511.
10. Arita M. The metabolic world of Escherichia coli is not small. Proc Natl Acad Sci USA 2004; 101:1543-1547.
11. Ma HW, Zeng AP. Reconstruction of metabolic networks from genome data and analysis of their global structure for various organisms. Bioinformatics 2003; 19:220-277.
12. Tanaka R, Yi TM, Doyle J. Some protein interaction data do not exhibit power law statistics. FEBS Lett 2005; 579:5140-5144.
13. Tsigelny IF, Nigam SK. Complex dynamics of chaperone-protein interactions under cellular stress. Cell Biochem Biophys 2004; 40:263-276.
14. Csermely P. Strong links are important - But weak links stabilize them. Trends Biochem Sci 2004; 29:331-334.
15. Zhao R, Davey M, Hsu YC et al. Navigating the chaperone network: An integrative map of physical and genetic interactions mediated by the hsp90 chaperone. Cell 2005; 120:715-727.
16. Nardai G, Vegh E, Prohaszka Z et al. Chaperone-related immune dysfunctions: An emergent property of distorted chaperone-networks. Trends Immunol 2006; 27:74-79
17. Soti C, Pal C, Papp B et al. Chaperones as regulatory elements of cellular networks. Curr Op Cell Biol 2005; 17:210-215.
18. Frydman J. Folding of newly translated proteins in vivo: The role of molecular chaperones. Annu Rev Biochem 2001; 70:603-647.
19. Kleizen B, Braakman I. Protein folding and quality control in the endoplasmic reticulum. Curr Opin Cell Biol 2004; 16:343-349.
20. Young JC, Agashe VR, Siegers K et al. Pathways of chaperone-mediated protein folding in the cytosol. Nat Rev Mol Cell Biol 2004; 5:781-791.
21. Blatch GL, ed. Networking of Chaperones by Co-Chaperones. Georgetown: Landes Bioscience, 2006.
22. Young JC, Hoogenraad NJ, Hartl FU. Molecular chaperones Hsp90 and Hsp70 deliver preproteins to the mitochondrial import receptor Tom70. Cell 2003; 112:41-50.
23. Imai J, Maruya M, Yashiroda H et al. The molecular chaperone Hsp90 plays a role in the assembly and maintenance of the 26S proteasome. EMBO J 2003; 22:3557-3567.

24. Whittier JE, Xiong Y, Rechsteiner MC et al. Hsp90 enhances degradation of oxidized calmodulin by the 20 S proteasome. J Biol Chem 2004; 279:46135-46142.
25. Tsvetkova NM, Horvath I, Torok Z et al. Small heat-shock proteins regulate membrane lipid polymorphism. Proc Natl Acad Sci USA 2002; 99:13504-13509.
26. Torok Z, Horvath I, Goloubinoff P et al. Evidence for a lipochaperonin: Association of active protein-folding GroESL oligomers with lipids can stabilize membranes under heat shock conditions. Proc Natl Acad Sci USA 1997; 94:2192-2197.
27. Torok Z, Goloubinoff P, Horvath I et al. Synechocystis HSP17 is an amphitropic protein that stabilizes heat-stressed membranes and binds denatured proteins for subsequent chaperone-mediated refolding. Proc Natl Acad Sci USA 2001; 98:3098-3103.
28. Filippin L, Magalhaes PJ, Di Benedetto G et al. Stable interactions between mitochondria and endoplasmic reticulum allow rapid accumulation of calcium in a subpopulation of mitochondria. J Biol Chem 2003; 278:39224-39234.
29. Aon MA, Cortassa S, O'Rourke B. Percolation and criticality in a mitochondrial network. Proc Natl Acad Sci USA 2004; 101:4447-4452.
30. Szabadkai G, Simoni AM, Chami M et al. Drp-1-dependent division of the mitochondrial network blocks intraorganellar Ca2+ waves and protects against Ca2+-mediated apoptosis. Mol Cell 2004; 16:59-68.
31. Wolosewick JJ, Porter KR. Microtrabecular lattice of the cytoplasmic ground substance. Artifact or reality. J Cell Biol 1979; 82:114-139.
32. Schliwa M, van Blerkom J, Porter KR. Stabilization of the cytoplasmic ground substance in detergent-opened cells and a structural and biochemical analysis of its composition. Proc Natl Acad Sci USA 1981; 78:4329-4333.
33. Clegg JS. Properties and metabolism of the aqueous cytoplasm and its boundaries. Am J Physiol 1984; 246:R133-R151.
34. Luby-Phelps K, Lanni F, Taylor DL. The submicroscopic properties of cytoplasm as a determinant of cellular function. Annu Rev Biophys Biophys Chem 1988; 17:369-396.
35. Hochachka PW. The metabolic implications of intracellular circulation. Proc Natl Acad Sci USA 1999; 96:12233-12239.
36. Verkman AS. Solute and macromolecule diffusion in cellular aqueous compartments. Trends Biochem Sci 2002; 27:27-33.
37. Spitzer JJ, Poolman B. Electrochemical structure of the crowded cytoplasm. Trends Biochem Sci 2005; 30:536-541.
38. Csermely P. A nonconventional role of molecular chaperones: Involvement in the cytoarchitecture. News Physiol Sci 2001; 16:123-126.
39. Sreedhar AS, Mihaly K, Pato B et al. Hsp90 inhibition accelerates cell lysis. Anti-Hsp90 ribozyme reveals a complex mechanism of Hsp90 inhibitors involving both superoxide- and Hsp90-dependent events. J Biol Chem 2003; 278:35231-35240.
40. Michels AA, Kanon B, Konings AW et al. Hsp70 and Hsp40 chaperone activities in the cytoplasm and the nucleus of mammalian cells. J Biol Chem 1997; 272:33283-33289.
41. Nollen EA, Salomons FA, Brunsting JF et al. Dynamic changes in the localization of thermally unfolded nuclear proteins associated with chaperone-dependent protection. Proc Natl Acad Sci USA 2001; 98:12038-12043.
42. Guo Y, Guettouche T, Fenna M et al. Evidence for a mechanism of repression of heat shock factor 1 transcriptional activity by a multichaperone complex. J Biol Chem 2001; 276:45791-45799.
43. Freeman BC, Yamamoto KR. Disassembly of transcriptional regulatory complexes by molecular chaperones. Science 2002; 296:2232-2235.
44. Rutherford SL, Lindquist S. Hsp90 as a capacitor for morphological evolution. Nature 1998; 396:336-342.
45. Fares MA, Ruiz-González MX, Moya A et al. Endosymbiotic bacteria: GroEL buffers against deleterious mutations. Nature 2002; 417:398.
46. Queitsch C, Sangster TA, Lindquist S. Hsp90 as a capacitor of phenotypic variation. Nature 2002; 417:618-624.
47. Cowen LE, Lindquist S. Hsp90 potentiates the rapid evolution of new traits: Drug resistance in diverse fungi. Science 2005; 309:2185-2189.
48. Sollars V, Lu X, Xiao L et al. Evidence for an epigenetic mechanism by which Hsp90 acts as a capacitor for morphological evolution. Nat Genet 2003; 33:70-74.
49. Whitesell L, Lindquist SL. HSP90 and the chaperoning of cancer. Nat Rev Cancer 2005; 5:761-772.
50. Csermely P. Chaperone-overload as a possible contributor to "civilization diseases": Atherosclerosis, cancer, diabetes. Trends Genet 2001; 17:701-704.

51. Nardai G, Csermely P, Söti C. Chaperone function and chaperone overload in the aged. Exp Gerontol 2002; 37:1255-1260.
52. Söti C, Csermely P. Chaperones and aging: Their role in neurodegeneration and other civilizational diseases. Neurochem International 2002; 41:383-389.
53. Papp E, Száraz P, Korcsmáros T et al. Changes of endoplasmic reticulum chaperone complexes, redox state and impaired protein disulfide reductase activity in misfolding alpha-1-antitrypsin transgenic mice. FASEB J 2006, (in press).
54. Bergman A, Siegal ML. Evolutionary capacitance as a general feature of complex gene networks. Nature 2003; 424:549-552.
55. True HL, Berlin I, Lindquist SL. Epigenetic regulation of translation reveals hidden genetic variation to produce complex traits. Nature 2004; 431:184-187.
56. Sangster TA, Lindquist S, Queitsch C. Under cover: Causes, effects and implications of Hsp90-mediated genetic capacitance. Bioessays 2004; 26:348-362.
57. Csermely P, Agoston V, Pongor S. The efficiency of multi-target drugs: The network approach might help drug design. Trends Pharmacol Sci 2005; 26:178-182.
58. Vigh L, Literati PN, Horvath I et al. Bimoclomol: A nontoxic, hydroxylamine derivative with stress protein-inducing activity and cytoprotective effects. Nat Med 1997; 3:1150-1154.
59. Bernier V, Lagace M, Bichet DG et al. Pharmacological chaperones: Potential treatment for conformational diseases. Trends Endocrinol Metab 2004; 15:222-228.
60. Neckers L, Neckers K. Heat-shock protein 90 inhibitors as novel cancer chemotherapeutics - An update. Expert Opin Emerg Drugs 2005; 10:137-149.
61. Söti C, Nagy E, Giricz Z et al. Heat shock proteins as emerging therapeutic targets. Br J Pharmacol 2005; 146:769-780.

CHAPTER 7

Chaperones as Parts of Organelle Networks

György Szabadkai* and Rosario Rizzuto

Abstract

The efficiency, divergence and specificity of virtually all intracellular metabolic and signalling pathways largely depend on their compartmentalized organization. A corollary of the requirement of compartmentalization is the dynamic structural partition of the intracellular space by endomembrane systems. A branch of these membranes communicate with the extracellular space through the endo- and exocytotic processes. Others, like the mitochondrial and endoplasmic reticulum networks accomplish a further role, being fundamental for the maintenance of cellular energy balance and for determination of cell fate under stress conditions. Recent structural and functional studies revealed that the interaction of these networks and the connectivity state of mitochondria controls metabolic flow, protein transport, intracellular Ca^{2+} signalling, and cell death. Moreover, reflecting the fact that the above processes are accomplished in a microdomain between collaborating organelle membranes, the existence of macromolecular complexes at their contact sites have also been revealed. Being not only assistants of nascent protein folding, chaperones are proposed to participate in assembling and maintaining the function of the above complexes. In this chapter we discuss recently found examples of such an assembly of protein interactions driven by chaperone proteins, and their role in regulating physiological and pathological processes.

Introduction

The functionality of eukaryotic cells lies on their complex and highly dynamic endomembrane system, composed of the endocytotic machinery, the secretory membranes and the mitochondrial network. Membranes at the interface of these systems display dynamic interactions with each other, ensuring protein transport, signal transmission between compartments, and regulation of network shape/connectivity. A long time known function of chaperone proteins is to regulate the assembly of local protein complexes involved in specific organelle function (e.g., mitochondrial or ER protein import). Recently, a novel concept arose, pointing to a more complex, integrating role of these proteins, i.e., to stabilize larger intracellular morphological and functional networks through weak but numerous links.[1-3] One of the most important predictions of such kind of organization is that while under normal conditions chaperoning elements ensure interactions between cellular organelles, during cellular stress their hyperactivation is needed for coping with stress, leading to more intense communication in the regulated networks. Along these lines, exhaustion of this 'network chaperoning' function will lead to decomposition of normal cellular function, thus it will lie at the base of a wide range of pathophysiological conditions. Indeed, chaperones in the endoplasmic reticulum (ER) have now a well established role in sensing and signalling stress conditions and executing defence

*Corresponding Author: György Szabadkai—Department of Experimental and Diagnostic Medicine, University of Ferrara, Via Borsari 46, Ferrara, 44100, Italy. Email: szg@unife.it

Molecular Aspects of the Stress Response: Chaperones, Membranes and Networks, edited by Peter Csermely and László Vígh. ©2007 Landes Bioscience and Springer Science+Business Media.

responses such as the unfolded protein response (UPR).[4-7] A similar, but less characterized stress response has been also observed in the mitochondrial network, following insults leading impairment of protein folding in the organelle. Moreover, following ER stress, mitochondria seems to be directly activated by ER-derived signals and they are the target of numerous pro- and anti- cell death mechanisms.

To discuss the potential role of chaperones in the interaction between organellar networks we will use the above described paradigm as an example and we will refer to eventual similarities to mechanisms described in other interacting organelle networks. First, we will briefly describe the regulation of dynamics how ER-mitochondria interaction evolved, presumably by the help of numerous chaperoning activities; then, through the example of recent data on ER-mitochondrial Ca^{2+} metabolic and cell death signalling, we will discuss the eventual specific functions which chaperons might fulfil in microdomains between these organelles; finally, we will describe the perspectives related to chaperone assisted organellar interactions under pathophysiological conditions.

Biogenesis of the ER and Mitochondrial Networks: A Role for Chaperones in Interorganellar Communication?

The endoplasmic reticulum (ER), adjoining to the nuclear envelope, forms a continuous network in virtually all cell types that is the major site of lipid biosynthesis, protein folding and the entry point into the secretory pathway. Thus the ER can be considered the common ancestor of all membranes downstream in this pathway, including the Golgi, secretory vesicles, the lysosomes, and the plasma membrane.[8-11] The biogenesis of the ER is a result of coordinate activation of protein and lipid biosynthesis, as deduced from experiments showing the interdependency of the unfolded protein response (UPR) and inositol response pathways.[8] These works also provided the clue that chaperoning inside the ER lumen plays an important role in organelle biogenesis and network formation. Indeed, during UPR, accumulation of unfolded proteins (e.g., by a disequilibrium between the amount of proteins processed in the secretory pathway and the actual requirement of protein synthesis at a given developmental stage of the cell) leads to activation of a transcriptome including ER chaperones (to ensure proper folding in the ER lumen), and concerted activation of genes involved in phospholipid synthesis can also be observed during this process.[12,13] Similar mechanisms operate during 'ER stress' conditions, which now seems to be involved in pathogenesis of several diseases (see below).[4-7]

In contrast to the ER, the mitochondrial network* is apparently shaped in a more variable and complex way, depending on cell type and metabolic state. These features of mitochondrial morphology seems to reflect a complex evolutionary path evolving from their α-proteobacterial ancestor, since, to the best of our knowledge, it had a quite standard ovoid shape.[14-16] The mitochondrial proteome consists of about 500-1500 proteins of which only a minor fraction (8 proteins in yeast and 13 in humans) is encoded by mitochondrial DNA. Thus, during mitochondrial biogenesis, virtually all mitochondrial proteins are synthesized as precursor polypeptides in the cytosol and are imported post-translationally into the organelle.[17] Defective protein folding and processing in the organelle, e.g., due to the impairment of the assembly of multi-subunit mitochondrial complexes or downregulation of genes encoding mitochondrial chaperones or proteases, lead to a mitochondria-specific (ER independent) 'stress' response, comprising transcriptional upregulation of matrix chaperones such as Hsp60, mtHsp70 (also known as Grp-75/mortalin).[18-20]

---

* Whether mitochondria form a 'real' network, is still a matter of debate. e.g., from a structural point of view, there has never been documentation of a network of mitochondria in the neurons or glia of brain tissue. Although, a mitochondrial network is not disputed in lower organisms, such as yeast or in most cultured cell lines. On the other hand, from a functional point of view we can define mitochondria as a communication "network" wherein the mitochondria of a cell respond in a concerted fashion to a stimulus, then "network" is probably the right term to use for mitochondria of all eukaryotic cells.

In this manner, the ER and mitochondrial network appear to function in an analogous way in utilizing chaperones to maintain correct organelle function. Moreover, the biogenesis of both organelles relies on protein import from the cytosolic compartment, marking out a cross-road for newly synthesized proteins in the cytosol.[21] The proper protein sorting into the organelles is driven by a complex protein interaction network.[22] The general model for protein recognition and transport at the mitochondrial outer membrane and the ER are fascinatingly similar and relies on a series of chaperones interacting with the ER and mitochondrial membranes: first, either cytosolic chaperones of the Hsp70 family, or occupancy of the C-terminal region of nascent chains in the ribosomal exit site prevents premature stable folding of preproteins. Membrane-bound receptors (such as Tom70 and Tom20 of the mitochondrial outer membrane translocase complex [TOM], and the Signal recognition particle [Srp] of the ER membrane) bind preproteins and transfer them into the translocation channel (Tom40 of the OMM and the Sec61 complex of the ER). Mitochondria have to operate with further complexes of IMM to ensure translocation and insertion of matrix and IMM proteins (Tim22 and Tim23, respectively).[17,23] Finally, protein motors and chaperones in the matrix and the ER lumen (mtHsp70/Grp-75 and BiP/Grp-78 in the ER and mitochondria, respectively) complete the transport task. This protein sorting task of the cytosolic chaperone network requires several other elements, such as Hsp90 and its co-chaperone interactors.[24]

A recent meticulous proteomic analysis of the different mitochondrial compartments shed light on the existence of a conserved subset of matrix preproteins, which reside also on the OMM surface.[25] Among them, mitochondrial matrix chaperones (Grp-75, Hsp60) were also found, indicating that they may localize along the whole protein translocation pathway, and that they could play a role in interacting with the cytosolic or ER protein sorting machinery, thus also ER membrane proteins. A corollary of the latter assumption is that the ER and mitochondria are located close enough to mediate these interactions. Indeed, recently significant progress has been achieved in exploring the morphological characteristics of their interaction. This topic is extensively covered and illustrated in recent reviews (see refs. 26-28), here we just mention that a close, synaptic-like apposition of the organelles was clearly demonstrated in a broad variety of cell types using light and electron microscopy studies (Fig. 1).[29-32]

It is now generally accepted that even if the ER forms a structurally cohesive network, it contains functionally and structurally separate sub-domains, specialized to a defined function (for a recent see review ref. 9). Indeed, the ER-mitochondria interface seems to be one these specialized domains, containing smooth ER membrane, surrounded by attached ribosomes.[33] This specific ER compartment involved in interactions with mitochondria can be isolated as a light mitochondrial fraction by density gradient centrifugation and is called as mitochondria-associated membranes (MAM).[34] Originally it was recognized as a fraction enriched of phospholipid synthetic and transfer activities, but now it has been shown to have a role in other signaling pathways, like Ca^{2+} and apoptotic signaling following ER stress.[35,36]

A further implication of chaperones at the ER-mitochondria interface came from the description of an alternative protein sorting pathway, independent of the one described above. In this path, Ydj1, the yeast homologue of HDJ2 (a DJ-domain containing co-chaperone, known to function as regulator of ATPase activity of hsp70 family proteins), assists the Hsc70 (Ssa in yeast)-mediated post-translational targeting of some proteins to both mitochondria and the endoplasmic reticulum by transiently associating with ER and mitochondria membranes through its covalently attached lipid moiety.[37] Similarly, other reports, by analysing alternative mitochondrial and ER targeting sequences (e.g., tail anchors), showed that the two organelles may share particular proteins, including also integral membrane proteins.[38-40]

In summary, the biogenesis of the ER and mitochondria requires a continuous flow of their protein components in a tight microdomain between the organelles, through alternative routes which are delineated by a cluster of chaperones. Along these lines, a further avenue to explore is how these processes might provide a dynamic framework to maintain organelle interactions. In the next section, we will discuss this prospect, through the example of ER-mitochondrial Ca^{2+} signalling machinery.

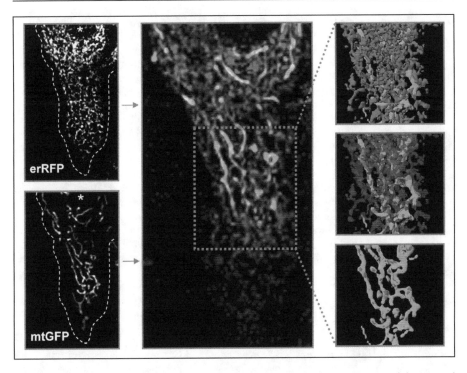

Figure 1. Visualization of the three dimensional structure and colocalization of the ER and mitochondrial network. HeLa cells were transfected with ER targeted mRFP1 and mitochondrially targeted GFP (as described in refs. 76, 101), and digital microscopy, 3D deconvolution and image reconstruction was applied.[76] The left panel shows the original images of one focal plane, overlayed on the middle panel (ER—red; mitochondria—green). The right panels show 3D surface rendered images using the same colour code. The top right panel shows the interconnected network, rendered transparent on the right middle panel. The bottom right panel shows the mitochondrial network with the surface colocalizing with the ER (shown in red). A color version of this figure is available online at www.Eurekah.com.

ER-Mitochondrial Ca^{2+} Transfer: A Major Example of Organelle Interactions

A significant part of our current view on ER-mitochondrial interactions arose from the investigation of Ca^{2+} signal transmission between these organelles; an affirmation which might reflect its importance in cellular signalling but also the availability of technical approaches to study Ca^{2+} signals. Introduction of protein based Ca^{2+} sensitive probes like aequorins and GFP/YFP fusion proteins with genetically engineered Ca^{2+} binding sites, targeted to different subcellular locations facilitated profoundly the development of this field.[41,42]

The outer mitochondrial membrane (OMM) is permeable to ions of a limited size, owing to the abundance of the large conductance channel VDAC (for the role of the channel in metabolic channelling, see refs. 43, 44). Indeed, recent data shows that the availability and selective placement of VDAC channels at ER/mitochondria contact sites facilitates mitochondrial Ca^{2+} accumulation, in keeping with the idea that the latter process requires the fast and efficient transfer of Ca^{2+} microdomains from the mouth of the Ca^{2+} channels located in neighbouring ER to the transporters of the inner mitochondrial membrane (IMM).[45,46] The IMM is an ion impermeable membrane, forming foldings into the internal space, known as *cristae*.[32] The activity of respiratory chain complexes allows the translocation of H$^+$ into the intermembrane space (IMS), with consequent generation of an electrochemical gradient ($\Delta\mu_H$), composed of a chemical (ΔpH) and electrical ($\Delta\Psi_m$) component, the latter being about -180 mV, negative inside, thus providing a huge driving force for Ca^{2+} entry into the organelle. As to the uptake route, its molecular identity is still unknown. It is suggested to be a gated channel rather than a carrier, showing a Ca^{2+} activated second order kinetics, with a separate Ca^{2+} activation and a Ca^{2+} transport site.[47,48] Based on these properties, the rate of mitochondrial Ca^{2+} uptake through the channel was shown to be significant only above an 'activatory' threshold (200- 300 nM), also known as 'set point'. A known inhibitor of Ca^{2+} uptake into mitochondria is ruthenium red (RuR), of which the binding site however, was shown to be on the OMM located VDAC channel.[46,49] This latter finding indicates that, similarly to the protein translocation mechanisms, Ca^{2+} transport through the two mitochondrial membranes is mediated by interacting protein complexes of both the OMM and IMM, presumably at the contact sites between the two membranes.[50] Interestingly, at these contact sites a non specific ion- and metabolite-channel activity of major pathophysiological interest has been characterized in the past years, commonly referred to the permeability transition pore (PTP).[51] This high-conductance channel was shown to be a multiprotein complex (the putatively essential components being VDAC, the adenine-nucleotide translocase [ANT] of the IMM and cyclophilin D from the matrix side), and is activated by $[Ca^{2+}]$ increase in the mitochondrial matrix and oxidation of critical cysteins of its protein components.[52-54] Since this channel was shown to display several conductance state, and the Ca^{2+} uniporter mechanism is sensitive to RuR acting on VDAC, one can imagine that this or a similar protein complex mediates Ca^{2+} uptake under physiological conditions. Importantly, reflecting the relatively low affinity of Ca^{2+} activation of the uniporter, efficient mitochondrial Ca^{2+} uptake in intact cells was shown to be dependent on the close apposition of mitochondria to Ca^{2+} release sites of the ER or the sarcoplasmic reticulum.[55,56] Here, at mouth of Ca^{2+} release channels (inositol 1,4,5 trisphosphate receptor, IP_3R or the ryanodine receptor, RyR, respectively), during Ca^{2+} release, a high $[Ca^{2+}]$ microdomain is formed, which in turn is instantly transferred to the mitochondrial matrix.

Taken together, the ER-mitochondria Ca^{2+} transmission machinery represents a transport system for which necessary is (i) the interaction of two organelles, (ii) and the interaction of channelling proteins of three adjacent membranes. Augmenting the complexity of the system, these interactions do not seem to be static, considering the continuous dynamic structural remodelling of both the ER and mitochondrial networks in living cells.[9,57] Moreover, ER-mitochondrial interactions and mitochondrial mobility were shown to be regulated by Ca^{2+} itself.[58]

Chaperone Control of ER-Mitochondrial Interaction along the Ca^{2+} Signal Transmission Pathway

The morphological and functional description of the above detailed complex machineries logically led to the question: what are the molecular components that bring together this pathway? To answer this challenge, first, biochemical studies, exploring protein interactions with the already known components provided some clues. Surprisingly, a two-hybrid study, using VDAC as a bait (see ref. 59), has identified mtHsp70/Grp75, the above mentioned mitochondrial matrix chaperone as a VDAC interactor, which not only attached to VDAC, located in the OMM, but was able also to change its conductance properties, proving the physiological significance of the interaction. This study called to mind that several lines of recent evidence showed the presence of different classes of chaperone proteins along the ER-mitochondrial Ca^{2+} transfer route. Indeed, the IP_3R and RyR Ca^{2+} release channels, as well as VDAC were shown to be important nodes of protein interaction networks (described in detail in the reviews refs. 43, 60-63), among which several exhibit chaperone activity.[61] The broad interaction modules of the IP_3R and VDAC, and the particular role of chaperones in that network are

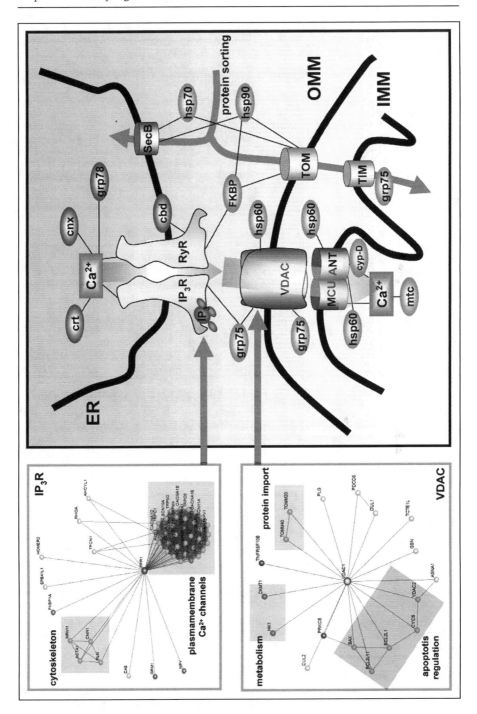

Figure 2. Please see figure legend on next page.

Figure 2, viewed on previous page. Components of the Ca^{2+} signal transmission axis between the ER and mitochondria, and its interaction points with chaperone networks. The left panels show the cellular interaction network of the IP_3R (upper) and VDAC proteins (lower) as browsed from the HiMAP human protein-protein interaction map (www.himap.org), based on two-hybrid datasets, literature-confirmed interactions (Human Protein Reference Database, www.hprd.org) and prediction algorithms.[102] Gene names are shown as defined in the OMIM database (www.ncbi.nlm.nih.gov). Highlighted genes represent subsets of proteins with similar function or subcellular localization. The right panel shows literature based interactions of chaperone proteins (showed in ovals, blue: ER origin; red: cytosolic origin; green: mitochondrial origin) with the Ca^{2+} signal transmission machinery from the endoplasmic reticulum (ER) to the mitochondria (OMM: outer mitochondrial membrane; IMM: inner mitochondrial membrane). Abbreviations and citations proving interaction not mentioned in the text: cbd: calbindin; cnx: calnexin;[103] crt: calreticulin;[104] cyp-D: cyclophilin-D;[105] FKBP: FK506 binding protein;[69] mtc: mitocalcin.[106] Note the multiple localization of Hsp60 and Grp75.[88,107] A color version of this figure is available online at www.Eurekah.com.

shown on (Fig. 2) and are described in more detail in the figure legend. Seeing the vast presence of chaperones in this system, again an apparent question arises: what can be their function in the Ca^{2+} signal transmission machinery, apparently unrelated to protein folding and transfer?

The first clue emerges from an analogous system at the plasmamembrane-ER interface, described in hippocampal neurons. In their postsynaptic density the Homer and Shank proteins play a scaffolding role to maintain the interaction between the glutamatergic receptors of the postsynaptic membrane and the Ca^{2+} signalling components of the ER: the IP_3R and ER Ca^{2+} ATPases (SERCAs).[64-66] Strikingly, overexpression of Homer1b and Shank1b induces relocation to plasmamembrane not only of the above Ca^{2+} transport proteins but also of calreticulin and calbindin, Ca^{2+} binding chaperones of the ER lumen and membrane, respectively.[65] Similarly, we can propose that the chaperones found at the ER-mitochondrial interface may structurally bring together the building blocks of the Ca^{2+} transfer system. A noticeable difference between the two systems would be the ATP dependent dynamic binding/scaffolding of the Hsp70/90 family chaperones, which might represent a target for unlocking the interaction under conditions when the ATP supply is not sufficient, i.e., under stress conditions (see also below).

A second plausible role for chaperone association to ER and mitochondrial Ca^{2+} channels is the modulation of the ion channel function itself. Indeed, both the peptydil-prolyl isomerase/ immunophilin FK506 binding protein (FKBP12) binding to the IP_3Rs and RyRs, and mtHsp70/ Grp75 binding to VDAC was shown to modify their conductance.[59] While the effect on the RyRs of FKBP12 was shown not to be related to its enzyme activity (ref. 67), it appears to stabilize the receptor homo- and heterodimers through a conformational effect.[68] Direct, durable binding of FKBP12 to the IP_3R was failed to be shown (ref. 69), still it is able to inhibit Ca^{2+} release through both Ca^{2+} channels.[70] Another member of the FKBP family of proteins, FKBP52 binds to Hsp90 (see protein sorting to mitochondria above), and through binding to the dynein motor on the microtubule network it directs translocation of glucocorticoid receptors and p53 to the nucleus.[71,72] The finding that mtHsp70/Grp75 also interacts with the dynein light chain component TCTEL and p53, and that VDAC docks both former proteins (refs. 59, 73), raise the possibility that also the presence of FKBP (family member 12) close to the IP_3R and RyR, plays a similar role, i.e., it recruits regulatory factors on the surface of mitochondria and the ER. In this case, mtHsp70/Grp75 can be a regulatory protein by its plain interaction, since its ATPase domain was shown to be dispensable for modifying VDAC conductance.[59]

Our recent findings along the ER-mitochondria Ca^{2+} signal transmission axis provided further confirmation of chaperone mediated coupling between these two organelles. We have found that the IP_3R and VDAC-1 reside in a common macromolecular complex, localized in the above described mitochondria-associated subfraction of the ER (MAM). Moreover, in this complex the N-terminal ligand binding domain of the IP_3R was found to directly enhance mitochondrial Ca^{2+} accumulation. Since the downregulation of the OMM localized mtHsp70/Grp75 abolished this effect, we concluded that a chaperone-mediated direct coupling between the organelles is mediated by these three proteins (Szabadkai, Bianchi et al, manuscript submitted).

To summarize the above described scenario, we can embed these findings in the concept that chaperones, through their interaction with the elements of a signal transduction pathway between the ER and mitochondria, build up and maintain the integrity of the interacting surface the two organelles. Pending on results and new experimental approaches that will clarify the nature and strength of interaction of chaperones with the Ca^{2+} signalling protein components, they might also underlie the emerging concept of stabilization of networks by weak, transient links.[1,2]

Here we considered Ca^{2+} merely as an example of a substance transferred from the ER to mitochondria, but we have also to mention that Ca^{2+}, as a messenger released from the ER, was shown to determine functional properties and integrity of the mitochondrial network. Elevated $[Ca^{2+}]$ in the cytosol and the ER-mitochondrial interface may induce two different responses from the mitochondrial remodelling apparatus. The Ca^{2+} induced translocation of the mitochondrial fission factor Drp-1 induces mitochondrial fragmentation, i.e., disintegration of the continuous mitochondrial network.[74,75] The functional consequence of the breakdown of the network is well illustrated by the block of intramitochondrial Ca^{2+} waves, which in the normal mitochondrial network was shown to be initiated close to the ER Ca^{2+} release sites and to be responsible for the complete upload of the integral mitochondrial network.[76,77]

Thus, in summary, the chaperone mediated interaction between the ER and mitochondria appears as a novel way to integrate organellar networks and by regulating signalling between the organelles, might be responsible for controlling organellar and cellular function. Apart from massive biochemical work to identify interacting partners of the chaperons located at the organellar interface, detailed mathematical modelling of the network properties will be necessary to show the exact role of these proteins.

Perspectives: The Role of Chaperone Mediated ER-Mitochondria Coupling in Cell Death

Mitochondrial Ca^{2+} overload, originating from Ca^{2+} release from the ER, has been shown to play a crucial role in apoptosis induction caused by certain proapoptotic stimuli, such as C_2 ceramide.[78-80] C_2 ceramide was shown to directly induce Ca^{2+} release from the ER Ca^{2+} store (ref. 79) and also to sensitize mitochondria to Ca^{2+} impulses from $InsP_3$ or ryanodine receptor ($InsP_3R$, RyR) mediated Ca^{2+} release, leading to OMM permeabilization (OMP), mitochondrial permeability transition (MPT) and depolarization.[78] This is followed by the release of proapoptotic factors (e.g., cytochrome *c*) activating the effector caspases and finally triggering apoptotic cell death.[81] Importantly, mitochondrial depolarization in these cases is propagated as a wave throughout the cells, pointing to a fundamental role of mitochondrial network integrity in apoptotic signaling.[82] The overall picture emerging from a series of recent studies is that an important mechanism underlying the sensitivity of cells to apoptosis is the up- and downregulation of steady state $[Ca^{2+}]_{er}$ by pro- and antiapoptotic factors, respectively. The resulting increase or decrease of Ca^{2+} release from the ER and Ca^{2+} uptake into mitochondria in turn is pivotal in triggering apoptotic signals through the mitochondrial pathway.[83] Indeed, Bcl-2 was shown to reduce steady state $[Ca^{2+}]_{er}$, leading to a marked inhibition in apoptotic death and preservation of the mitochondrial structure.[79] Similarly, following overexpression of calreticulin (an abundant luminal ER Ca^{2+} buffer) which does not raise $[Ca^{2+}]_{er}$, but does increase the releasable Ca^{2+} pool, cell survival is drastically reduced upon C_2-ceramide treatment.[79] Furthermore, cell lines derived from calreticulin knock-out mice are more resistant to apoptosis, indicating that the crucial requirement is the amount of released Ca^{2+} and not $[Ca^{2+}]_{er}$.[84] Scorrano and coworkers provided further evidence in favor of the hypothesis that Ca^{2+} movement from the ER to mitochondria is a key process in the activation of apoptosis by a number of stimuli.[85] They showed that mouse embryonic fibroblasts deficient of the two proapoptotic proteins Bax and Bak double knock out (DKO) cells, are markedly resistant to a variety of apoptotic stimuli and have a much reduced Ca^{2+} concentration in the ER. If the ER Ca^{2+} levels are restored by recombinantly overexpressing the sarco-endoplasmic reticulum Ca^{2+} ATPase (SERCA), not only mitochondrial Ca^{2+} uptake in response to stimulation is reestablished, but the cells regain sensitivity to apoptotic stimuli such as arachidonic acid, C_2-ceramide and oxidative stress. An important positive feed-back loop was recently added to this picture, by showing that cytochrome c, released from the mitochondria, binds to the IP$_3$R and further stimulates Ca^{2+} release from the ER even at low IP$_3$ concentrations, leading to PTP opening, ulterior breakdown of mitochondrial network, and metabolic failure.[86]

Several heat shock proteins have been demonstrated to directly interact with various components of the cell death machinery, both upstream and downstream of the above described mitochondrial events. Indeed, as extensively (reviewed in refs. 87-89), chaperones (the cytosolic Hsp40, Hsp70 and Hsp90) promote cell survival by preventing the Bax/Bad and tBid induced OMP, or by promoting the antiapoptotic action of Bcl-2 (FKPB38, ref. 90). In a broader context it means that they help to maintain the rheostat between the action of pro- and antiapoptotic Bcl-2 family members.[91] As for the pathways downstream of mitochondria, Hsp70 and Hsp27 were shown to bind to proapoptotic factors (AIF and cytochrome c, respectively) released from the intermembrane space, thus inhibiting the activation of the apoptosome.[92,93]

Apart from identifying the role of the individual chaperone proteins in the complex apoptotic pathways, a more general question arises: how the above described role of chaperones in integrating organellar networks could be applied to the cell death process? The answer appears to be rather complex. Above we concluded that the links provided by different chaperones are necessary for efficient signalling between the ER and mitochondria. Indeed, downregulation of any of these proteins (Grp78, Grp75, Hsp60) leads to increased cell death.[88,89,94] A reasonable explanation for this effect is the uncoupling of organelle interaction, disrupting normal organelle biogenesis, e.g., due to reduced protein import and metabolic disturbances, potential mechanisms, still awaiting experimental demonstration. On the other hand, several line of evidence show that massive signaling from the ER to mitochondria is necessary also to induce the mitochondrial pathway of cell death. This include the above described Ca^{2+} signaling pathway (refs. 83, 95), but unrelated pathways were also demonstrated, such as Bax translocation following cleavage of the ER membrane protein Bap31 (ref. 75), or ROS or RNS signaling during ER stress.[96] In these cases, overexpression of chaperones are mostly protective. For instance, Grp75-transfected cells show resistance to cell death induction by different proapoptotic agents (refs. 97, 98), leading to extended life span and in certain cases cell transformation and tumorigenesis.[97] Similar protective mechanisms were described for other chaperones (Grp78, Hsp60), participating in organelle communication.[87,89]

Thus, to rationalize the impact of chaperone linkage between the ER and mitochondria on cell death we can propose two scenarios. First, chaperones may not only maintain the communication between these organelles, but they can have a buffering role, i.e., they can also protect from excessive death signal transduction. This hypothesis is validated by the finding that Grp75, apart from maintaining contact between the IP$_3$R and VDAC (Szabadkai, Bianchi et al., manuscript in preparation) reduces also the cation selectivity (is that also of Ca^{2+}?) of VDAC, thus protecting from Ca^{2+} overload and mitochondrial permeability transition.[59] An increased colocalization between the ER and mitochondria and Ca^{2+} induced cell death was shown by applying recombinant linkers between the ER and OMM, or by inducing ER stress (Csordas and Hajnoczky; Chami and Paterlini-Brechot, manuscripts in preparation). Moreover, to substantiate the role of chaperones in this ER-mitochondrial signalling process, Grp78, which normally reside in the ER lumen, during ER stress was shown to translocate even to the mitochondria.[99] The exact role of Grp78 in mitochondria was not described in this work, but our finding that Grp78 interacts with VDAC (in a yeast two-hybrid screen) indicates, that it might also control VDAC conductance.

A second scenario may envisage that chaperones, by recognizing multiple targets in the diverse death signalling pathways, discriminate between 'normal' and 'dangerous' targets, and thus they can counter the vicious effect of death inducing signals. Hsp27, a small cytoplasmic chaperone, binds to cytochrome *c* released from the mitochondria, preventing cytochrome *c* induced apoptosome activation (ref. 93). In the context of ER-mitochondria communication, it might also inhibit the positive feedback loop induced by cytochrome *c* on the ER surface by interacting with the IP_3R (ref. 86, see also above). Similarly, as recently proposed by Lemasters and He, chaperones (e.g., Hsp25 of the mitochondrial matrix, ref. 100) may recognize the unfolded state of the protein components of permeability transition pore, and thus might be able protect mitochondria by blocking conductance through these misfolded protein clusters.[52]

Conclusions

In summary, the authors of this chapter, by performing an uncountable number of experiments exploring the role of Grp75 on mitochondrial Ca^{2+} signaling and cell death, recognized that chaperones, by interacting the actual bait (VDAC) in the actual two-hybrid screen, may not only represent an artefact of the experimental approach, but can also have a significant contribution to the mechanisms determining cell signalling and cell fate under stress conditions. During this route, Katiuscia Bianchi and Diego De Stefani gave us an immense contribution, and we hope that conserving our interactions, we will describe the above outlined picture in more details, leading to innovations to alleviate stress induced impairment of organelle, cell and organism function.

Acknowledgements

This work was supported by grants from Telethon-Italy, the Italian Association for Cancer Research (AIRC), the Italian University Ministry (PRIN, FIRB and local research grants), the Emilia-Romagna PRRIITT program and the Italian Space Agency (ASI) to R. Rizzuto Part of the work by G. Szabadkai was supported by a Marie-Curie individual fellowship.

References

1. Csermely P. Strong links are important, but weak links stabilize them. Trends Biochem Sci 2004; 29:331-334.
2. Soti C, Pal C, Papp B et al. Molecular chaperones as regulatory elements of cellular networks. Curr Opin Cell Biol 2005; 17:210-215.
3. Barabasi AL, Oltvai ZN. Network biology: Understanding the cell's functional organization. Nat Rev Genet 2004; 5:101-113.
4. Szczesna-Skorupa E, Chen CD, Liu H et al. Gene expression changes associated with the endoplasmic reticulum stress response induced by microsomal cytochrome p450 overproduction. J Biol Chem 2004; 279:13953-13961.
5. Xu C, Bailly-Maitre B, Reed JC. Endoplasmic reticulum stress: Cell life and death decisions. J Clin Invest 2005; 115:2656-2664.
6. Rao RV, Bredesen DE. Misfolded proteins, endoplasmic reticulum stress and neurodegeneration. Curr Opin Cell Biol 2004; 16:653-662.
7. Paschen W, Mengesdorf T. Endoplasmic reticulum stress response and neurodegeneration. Cell Calcium 2005; 38:409-415.
8. Federovitch CM, Ron D, Hampton RY. The dynamic ER: Experimental approaches and current questions. Curr Opin Cell Biol 2005; 17:409-414.
9. Levine T, Rabouille C. Endoplasmic reticulum: One continuous network compartmentalized by extrinsic cues. Curr Opin Cell Biol 2005; 17:362-368.
10. Baumann O, Walz B. Endoplasmic reticulum of animal cells and its organization into structural and functional domains. Int Rev Cytol 2001; 205:149-214.
11. Voeltz GK, Rolls MM, Rapoport TA. Structural organization of the endoplasmic reticulum. EMBO Rep 2002; 3:944-950.
12. Cox JS, Chapman RE, Walter P. The unfolded protein response coordinates the production of endoplasmic reticulum protein and endoplasmic reticulum membrane. Mol Biol Cell 1997; 8:1805-1814.
13. Powell KS, Latterich M. The making and breaking of the endoplasmic reticulum. Traffic 2000; 1:689-694.
14. Bereiter-Hahn J. Behavior of mitochondria in the living cell. Int Rev Cytol 1990; 122:1-63.

15. Yaffe MP. The machinery of mitochondrial inheritance and behavior. Science 1999; 283:1493-1497.
16. Gray MW, Burger G, Lang BF. Mitochondrial evolution. Science 1999; 283:1476-1481.
17. Truscott KN, Brandner K, Pfanner N. Mechanisms of protein import into mitochondria. Curr Biol 2003; 13:R326-R337.
18. Yoneda T, Benedetti C, Urano F et al. Compartment-specific perturbation of protein handling activates genes encoding mitochondrial chaperones. J Cell Sci 2004; 117:4055-4066.
19. Zhao Q, Wang J, Levichkin IV et al. A mitochondrial specific stress response in mammalian cells. EMBO J 2002; 21:4411-4419.
20. Martinus RD, Garth GP, Webster TL et al. Selective induction of mitochondrial chaperones in response to loss of the mitochondrial genome. Eur J Biochem 1996; 240:98-103.
21. Deshaies RJ, Koch BD, Werner-Washburne M et al. A subfamily of stress proteins facilitates translocation of secretory and mitochondrial precursor polypeptides. Nature 1988; 332:800-805.
22. Wickner W, Schekman R. Protein translocation across biological membranes. Science 2005; 310:1452-1456.
23. Neupert W, Brunner M. The protein import motor of mitochondria. Nat Rev Mol Cell Biol 2002; 3:555-565.
24. Young JC, Agashe VR, Siegers K et al. Pathways of chaperone-mediated protein folding in the cytosol. Nat Rev Mol Cell Biol 2004; 5:781-791.
25. Zahedi RP, Sickmann A, Boehm AM et al. Proteomic analysis of the yeast mitochondrial outer membrane reveals accumulation of a subclass of preproteins. Mol Biol Cell 2006; E05-E08.
26. Frey TG, Mannella CA. The internal structure of mitochondria. Trends Biochem Sci 2000; 25:319-324.
27. Mannella CA, Pfeiffer DR, Bradshaw PC et al. Topology of the mitochondrial inner membrane: Dynamics and bioenergetic implications. IUBMB Life 2001; 52:93-100.
28. Hajnoczky G, Csordas G, Yi M. Old players in a new role: Mitochondria-associated membranes, VDAC, and ryanodine receptors as contributors to calcium signal propagation from endoplasmic reticulum to the mitochondria. Cell Calcium 2002; 32:363-377.
29. Rizzuto R, Pinton P, Carrington W et al. Close contacts with the endoplasmic reticulum as determinants of mitochondrial Ca2+ responses. Science 1998; 280:1763-1766.
30. Mannella CA, Buttle K, Rath BK et al. Electron microscopic tomography of rat-liver mitochondria and their interaction with the endoplasmic reticulum. Biofactors 1998; 8:225-228.
31. Marsh BJ, Mastronarde DN, Buttle KF et al. Organellar relationships in the Golgi region of the pancreatic beta cell line, HIT-T15, visualized by high resolution electron tomography. Proc Natl Acad Sci USA 2001; 98:2399-2406.
32. Frey TG, Renken CW, Perkins GA. Insight into mitochondrial structure and function from electron tomography. Biochim Biophys Acta 2002; 1555:196-203.
33. Wang HJ, Guay G, Pogan L et al. Calcium regulates the association between mitochondria and a smooth subdomain of the endoplasmic reticulum. J Cell Biol 2000; 150:1489-1498.
34. Vance JE. Phospholipid synthesis in a membrane fraction associated with mitochondria. J Biol Chem 1990; 265:7248-7256.
35. Simmen T, Aslan JE, Blagoveshchenskaya AD et al. PACS-2 controls endoplasmic reticulum-mitochondria communication and Bid-mediated apoptosis. EMBO J 2005.
36. Filippin L, Magalhaes PJ, Di Benedetto G et al. Stable interactions between mitochondria and endoplasmic reticulum allow rapid accumulation of calcium in a subpopulation of mitochondria. J Biol Chem 2003; 278:39224-39234.
37. Becker J, Walter W, Yan W et al. Functional interaction of cytosolic hsp70 and a DnaJ-related protein, Ydj1p, in protein translocation in vivo. Mol Cell Biol 1996; 16:4378-4386.
38. Colombo S, Longhi R, Alcaro S et al. N-myristoylation determines dual targeting of mammalian NADH-cytochrome b5 reductase to ER and mitochondrial outer membranes by a mechanism of kinetic partitioning. J Cell Biol 2005; 168:735-745.
39. Miyazaki E, Kida Y, Mihara K et al. Switching the sorting mode of membrane proteins from cotranslational endoplasmic reticulum targeting to posttranslational mitochondrial import. Mol Biol Cell 2005; 16:1788-1799.
40. van Herpen RE, Oude Ophuis RJ, Wijers M et al. Divergent mitochondrial and endoplasmic reticulum association of DMPK splice isoforms depends on unique sequence arrangements in tail anchors. Mol Cell Biol 2005; 25:1402-1414.
41. Chiesa A, Rapizzi E, Tosello V et al. Recombinant aequorin and green fluorescent protein as valuable tools in the study of cell signalling. Biochem J 2001; 355:1-12.
42. Bianchi K, Rimessi A, Prandini A et al. Calcium and mitochondria: Mechanisms and functions of a troubled relationship. Biochim Biophys Acta 2004; 1742:119-131.

43. Colombini M. VDAC: The channel at the interface between mitochondria and the cytosol. Mol Cell Biochem 2004; 256-257:107-115.
44. Lemasters JJ, Holmuhamedov E. Voltage-dependent anion channel (VDAC) as mitochondrial governator-Thinking outside the box. Biochim Biophys Acta 2006; 1762:181-190.
45. Rapizzi E, Pinton P, Szabadkai G et al. Recombinant expression of the voltage-dependent anion channel enhances the transfer of Ca2+ microdomains to mitochondria. J Cell Biol 2002; 159:613-624.
46. Gincel D, Zaid H, Shoshan-Barmatz V. Calcium binding and translocation by the voltage-dependent anion channel: A possible regulatory mechanism in mitochondrial function. Biochem J 2001; 358:147-155.
47. Bernardi P. Mitochondrial transport of cations: Channels, exchangers, and permeability transition. Physiol Rev 1999; 79:1127-1155.
48. Kirichok Y, Krapivinsky G, Clapham DE. The mitochondrial calcium uniporter is a highly selective ion channel. Nature 2004; 427:360-364.
49. Israelson A, Arzoine L, Abu-hamad S et al. A photoactivable probe for calcium binding proteins. Chem Biol 2005; 12:1169-1178.
50. Crompton M, Barksby E, Johnson N et al. Mitochondrial intermembrane junctional complexes and their involvement in cell death. Biochimie 2002; 84:143-152.
51. Bernardi P, Colonna R, Costantini P et al. The mitochondrial permeability transition. Biofactors 1998; 8:273-281.
52. He L, Lemasters JJ. Regulated and unregulated mitochondrial permeability transition pores: A new paradigm of pore structure and function? FEBS Lett 2002; 512:1-7.
53. Halestrap AP, McStay GP, Clarke SJ. The permeability transition pore complex: Another view. Biochimie 2002; 84:153-166.
54. Crompton M. Mitochondrial intermembrane junctional complexes and their role in cell death. J Physiol 2000; 529(Pt 1):11-21.
55. Rizzuto R, Brini M, Murgia M et al. Microdomains with high Ca2+ close to IP3-sensitive channels that are sensed by neighboring mitochondria. Science 1993; 262:744-747.
56. Csordas G, Thomas AP, Hajnoczky G. Calcium signal transmission between ryanodine receptors and mitochondria in cardiac muscle. Trends Cardiovasc Med 2001; 11:269-275.
57. Yaffe MP. Dynamic mitochondria. Nat Cell Biol 1999; 1:E149-E150.
58. Yi M, Weaver D, Hajnoczky G. Control of mitochondrial motility and distribution by the calcium signal: A homeostatic circuit. J Cell Biol 2004; 167:661-672.
59. Schwarzer C, Barnikol-Watanabe S, Thinnes FP et al. Voltage-dependent anion-selective channel (VDAC) interacts with the dynein light chain Tctex1 and the heat-shock protein PBP74. Int J Biochem Cell Biol 2002; 34:1059-1070.
60. Bosanac I, Michikawa T, Mikoshiba K. Structural insights into the regulatory mechanism of IP3 receptor. 2004; 1742:89-102.
61. Patterson RL, Boehning D, Snyder SH. Inositol 1,4,5-trisphosphate receptors as signal integrators. Annu Rev Biochem 2004; 73:437-465.
62. Rostovtseva TK, Tan W, Colombini M. On the role of VDAC in apoptosis: Fact and fiction. J Bioenerg Biomembr 2005; 37:129-142.
63. Vyssokikh M, Brdiczka D. VDAC and peripheral channelling complexes in health and disease. Mol Cell Biochem 2004; 256-257:117-126.
64. Brakeman PR, Lanahan AA, O'Brien R et al. Homer: A protein that selectively binds metabotropic glutamate receptors. Nature 1997; 386:284-288.
65. Sala C, Roussignol G, Meldolesi J et al. Key role of the postsynaptic density scaffold proteins shank and homer in the functional architecture of Ca2+ homeostasis at dendritic spines in hippocampal neurons. J Neurosci 2005; 25:4587-4592.
66. Xiao B, Tu JC, Worley PF. Homer: A link between neural activity and glutamate receptor function. Curr Opin Neurobiol 2000; 10:370-374.
67. Timerman AP, Onoue H, Xin HB et al. Selective binding of FKBP12.6 by the cardiac ryanodine receptor. J Biol Chem 1996; 271:20385-20391.
68. Lehnart SE, Huang F, Marx SO et al. Immunophilins and coupled gating of ryanodine receptors. Curr Top Med Chem 2003; 3:1383-1391.
69. Bultynck G, de Smet P, Rossi D et al. Characterization and mapping of the 12 kDa FK506-binding protein (FKBP12)-binding site on different isoforms of the ryanodine receptor and of the inositol 1,4,5-trisphosphate receptor. Biochem J 2001; 354:413-422.
70. MacMillan D, Currie S, Bradley KN et al. In smooth muscle, FK506-binding protein modulates IP3 receptor-evoked Ca2+ release by mTOR and calcineurin. J Cell Sci 2005; 118:5443-5451.

71. Galigniana MD, Radanyi C, Renoir JM et al. Evidence that the peptidylprolyl isomerase domain of the hsp90-binding immunophilin FKBP52 is involved in both dynein interaction and glucocorticoid receptor movement to the nucleus. Journal of Biological Chemistry 2001; 276:14884-14889.
72. Galigniana MD, Harrell JM, O'Hagen HM et al. Hsp90-binding immunophilins link p53 to dynein during p53 transport to the nucleus. Journal of Biological Chemistry 2004; 279:22483-22489.
73. Wadhwa R, Yaguchi T, Hasan MK et al. Hsp70 family member, mot-2/mthsp70/GRP75, binds to the cytoplasmic sequestration domain of the p53 protein. Exp Cell Res 2002; 274:246-253.
74. Bleazard W, McCaffery JM, King EJ et al. The dynamin-related GTPase Dnm1 regulates mitochondrial fission in yeast. Nat Cell Biol 1999; 1:298-304.
75. Breckenridge DG, Stojanovic M, Marcellus RC et al. Caspase cleavage product of BAP31 induces mitochondrial fission through endoplasmic reticulum calcium signals, enhancing cytochrome c release to the cytosol. J Cell Biol 2003; 160:1115-1127.
76. Szabadkai G, Simoni AM, Chami M et al. Drp-1 dependent division of the mitochondrial network blocks intraorganellar Ca2+ waves and protects against Ca2+ mediated apoptosis. Mol Cell 2004; 16:59-68.
77. Gerencser AA, Adam-Vizi V. Mitochondrial Ca2+ dynamics reveals limited intramitochondrial Ca2+ diffusion. Biophys J 2005; 88:698-714.
78. Szalai G, Krishnamurthy R, Hajnoczky G. Apoptosis driven by IP(3)-linked mitochondrial calcium signals. EMBO J 1999; 18:6349-6361.
79. Pinton P, Ferrari D, Rapizzi E et al. The Ca2+ concentration of the endoplasmic reticulum is a key determinant of ceramide-induced apoptosis: Significance for the molecular mechanism of Bcl-2 action. EMBO J 2001; 20:2690-2701.
80. Walter L, Hajnoczky G. Mitochondria and endoplasmic reticulum: The lethal interorganelle cross-talk. J Bioenerg Biomembr 2005; 37:191-206.
81. Orrenius S, Zhivotovsky B, Nicotera P. Regulation of cell death: The calcium-apoptosis link. Nat Rev Mol Cell Biol 2003; 4:552-565.
82. Pacher P, Hajnoczky G. Propagation of the apoptotic signal by mitochondrial waves. EMBO J 2001; 20:4107-4121.
83. Szabadkai G, Rizzuto R. Participation of endoplasmic reticulum and mitochondrial calcium handling in apoptosis: More than just neighborhood? FEBS Lett 2004; 567:111-115.
84. Nakamura K, Bossy-Wetzel E, Burns K et al. Changes in endoplasmic reticulum luminal environment affect cell sensitivity to apoptosis. J Cell Biol 2000; 150:731-740.
85. Scorrano L, Oakes SA, Opferman JT et al. BAX and BAK regulation of endoplasmic reticulum Ca2+: A control point for apoptosis. Science 2003; 300:135-139.
86. Boehning D, Patterson RL, Sedaghat L et al. Cytochrome c binds to inositol (1,4,5) trisphosphate receptors, amplifying calcium-dependent apoptosis. Nat Cell Biol 2003; 5:1051-1061.
87. Beere HM. Death versus survival: Functional interaction between the apoptotic and stress-inducible heat shock protein pathways. J Clin Invest 2005; 115:2633-2639.
88. Gupta S, Knowlton AA. HSP60, Bax, apoptosis and the heart. J Cell Mol Med 2005; 9:51-58.
89. Sreedhar AS, Csermely P. Heat shock proteins in the regulation of apoptosis: New strategies in tumor therapy: A comprehensive review. Pharmacol Ther 2004; 101:227-257.
90. Shirane M, Nakayama KI. Inherent calcineurin inhibitor FKBP38 targets Bcl-2 to mitochondria and inhibits apoptosis. Nat Cell Biol 2003; 5:28-37.
91. Korsmeyer SJ, Shutter JR, Veis DJ et al. Bcl-2/Bax: A rheostat that regulates an anti-oxidant pathway and cell death. Semin Cancer Biol 1993; 4:327-332.
92. Ravagnan L, Gurbuxani S, Susin SA et al. Heat-shock protein 70 antagonizes apoptosis-inducing factor. Nat Cell Biol 2001; 3:839-843.
93. Bruey JM, Ducasse C, Bonniaud P et al. Hsp27 negatively regulates cell death by interacting with cytochrome c. Nat Cell Biol 2000; 2:645-652.
94. Wadhwa R, Takano S, Taira K et al. Reduction in mortalin level by its antisense expression causes senescence-like growth arrest in human immortalized cells. J Gene Med 2004; 6:439-444.
95. Szabadkai G, Simoni AM, Rizzuto R. Mitochondrial Ca2+ uptake requires sustained Ca2+ release from the endoplasmic reticulum. J Biol Chem 2003; 278:15153-15161.
96. Xu W, Liu L, Charles IG et al. Nitric oxide induces coupling of mitochondrial signalling with the endoplasmic reticulum stress response. Nat Cell Biol 2004; 6:1129-1134.
97. Kaul SC, Yaguchi T, Taira K et al. Overexpressed mortalin (mot-2)/mthsp70/GRP75 and hTERT cooperate to extend the in vitro lifespan of human fibroblasts. Exp Cell Res 2003; 286:96-101.
98. Liu Y, Liu L, Song XD et al. Effect of GRP75/mthsp70/PBP74/mortalin overexpression on intracellular ATP level, mitochondrial membrane potential and ROS accumulation following glucose deprivation in PC12 cells. Mol Cell Biochem 2005; 268:45-51.

99. Sun FC, Wei S, Li CW et al. Localization of GRP78 to mitochondria in response to unfolded protein response. Biochem J 2006.
100. He L, Lemasters JJ. Heat shock suppresses the permeability transition in rat liver mitochondria. J Biol Chem 2003; 278:16755-16760.
101. Varnai P, Balla A, Hunyady L et al. Targeted expression of the inositol 1,4,5-triphosphate receptor (IP3R) ligand-binding domain releases Ca2+ via endogenous IP3R channels. Proc Natl Acad Sci USA 2005; 102:7859-7864.
102. Rhodes DR, Tomlins SA, Varambally S et al. Probabilistic model of the human protein-protein interaction network. Nat Biotechnol 2005; 23:951-959.
103. Zuppini A, Groenendyk J, Cormack LA et al. Calnexin deficiency and endoplasmic reticulum stress-induced apoptosis. Biochemistry 2002; 41:2850-2858.
104. Enyedi P, Szabadkai G, Krause KH et al. Inositol 1,4,5-trisphosphate binding sites copurify with the putative Ca-storage protein calreticulin in rat liver. Cell Calcium 1993; 14:485-492.
105. Crompton M, Virji S, Ward JM. Cyclophilin-D binds strongly to complexes of the voltage-dependent anion channel and the adenine nucleotide translocase to form the permeability transition pore. Eur J Biochem 1998; 258:729-735.
106. Tominaga M, Kurihara H, Honda S et al. Molecular characterization of mitocalcin, a novel mitochondrial Ca2+-binding protein with EF-hand and coiled-coil domains. J Neurochem 2006; 96:292-304.
107. Ran Q, Wadhwa R, Kawai R et al. Extramitochondrial localization of mortalin/mthsp70/PBP74/GRP75. Biochem Biophys Res Commun 2000; 275:174-179.

CHAPTER 8

Heat Shock Factor 1 as a Coordinator of Stress and Developmental Pathways

Julius Anckar and Lea Sistonen*

Abstract

The transition from normal growth conditions to stressful conditions is accompanied by a robust upregulation of heat shock proteins, which dampen the cytotoxicity caused by misfolded and denatured proteins. The most prominent part of this transition occurs on the transcriptional level. In mammals, protein-damaging stress leads to the activation of heat shock factor 1 (HSF1), which binds to upstream regulatory sequences in the promoters of heat shock genes. The activation of HSF1 proceeds through a multi-step pathway, involving a monomer-to-trimer transition, nuclear accumulation and extensive posttranslational modifications. In addition to its established role as the main regulator of heat shock genes, new data link HSF1 to developmental pathways. In this chapter, we examine the established stress-related functions and prospect the intriguing role of HSF1 as a developmental coordinator.

Introduction

One year after the identification of mRNA in 1961, Ferruccio Ritossa reported a different puffing pattern in the polytene chromosomes of *Drosophila* after an accidental upshift in temperature of an incubator containing *Drosophila* larvae.[1,2] Subsequent analysis of the heat-induced puffs verified that the loci were transcriptionally active regions, in which mRNAs coding for heat shock proteins (Hsps) were synthesized.[2] The transcriptional response to hyperthermia and other protein-damaging stresses has later been shown to be a highly regulated process in which the transcription of most nonstress genes is repressed, but that of heat shock genes rapidly upregulated.[3] Hence, the cell possesses a machinery for the privileged transcription and translation of stress-related genes and mRNA. In invertebrates such as yeast, nematode and fruit fly, the transcriptional activation of the heat shock response is regulated by a single heat shock factor (HSF). Vertebrates, on the other hand, have evolved a family of HSF members, HSF1-4, of which HSF3 is specific for avian species and HSF4 for mammals.[4] HSF1 is the mammalian counterpart of the single invertebrate HSF and cannot be replaced by other HSF members, since the heat shock response is obliterated in mice and fibroblasts lacking *hsf1*.[5,6] In contrast, HSF2 and HSF4 are known to play important roles in differentiation and development. For example, *hsf2* knockout mice have an intact heat shock response, but disrupted spermatogenesis and neuronal development.[7,8] HSF4, on the other hand, is expressed in only a few tissues and

*Corresponding Author: Lea Sistonen—Department of Biology, Åbo Akademi University and Turku Centre for Biotechnology, University of Turku and Åbo Akademi University, P.O. Box 123 FI-20521 Turku, Finland. Email: lea.sistonen@btk.fi

Molecular Aspects of the Stress Response: Chaperones, Membranes and Networks, edited by Peter Csermely and László Vígh. ©2007 Landes Bioscience and Springer Science+Business Media.

required for proper eye development.[9,10] These differences notwithstanding, the different HSFs may have overlapping roles in target gene regulation, as discussed later in this chapter.

Functional Domains of HSF1

Although different HSFs play distinct biological roles, they share certain conserved functional domains, of which the N-terminal helix-loop-helix DNA-binding domain is best preserved (see Fig. 1). HSFs bind to DNA as trimers, in which an individual DNA-binding domain recognizes the pentameric sequence nGAAn in the major groove of the DNA double helix.[11] The concurrent binding of all DNA-binding domains of the trimer to three adjacent nGAAn repeats is required for stable binding.[11] Thus, the minimal functional upstream regulatory elements recognized by HSF1 contain at least three nGAAn repeats, and are called heat shock elements (HSEs). The promoters of most *hsp* genes contain more than one HSE, allowing for multiple HSF1 trimers to bind simultaneously. It has been convincingly shown that HSF1 binds to HSEs in a cooperative manner, in which one DNA-bound trimer facilitates the binding of the next trimer.[12,13] Such cooperativity may be involved in regulating the magnitude of HSF1-mediated transcription, although some HSEs are not required for downstream gene expression.[11] Furthermore, the precise architecture of the HSE is an important determinant in gene-specific expression. For example, yeast HSF adopts different conformations when bound to typical and atypical HSEs and displays distinct requirements for temperature and posttranslational modifications depending on HSE architecture.[14,15]

The regions involved in regulating HSF1 trimerization have been thoroughly investigated in mutagenesis studies. The oligomerization domain is found adjacent to the DNA-binding domain and is characterized by hydrophobic amino acids at every seventh residue.[16] In the trimeric HSF1, these heptad repeats (HR-A/B; Fig. 1) are proposed to form an unusual triple-stranded coiled-coil configuration through the interactions of the hydrophobic residues.[17] In addition, most members of the HSF family contain an additional C-terminal heptad repeat (HR-C). Whereas a deletion of HR-A/B abolishes the HSF1 trimer formation, the deletion of HR-C generates a constitutively trimeric HSF1, suggesting that HR-C negatively regulates HSF1 trimerization.[18] In accordance, HSF of *Saccharomyces cerevisiae* and *Kluyveromyces lactis* as well as mammalian HSF4, which lack a conserved HR-C, are constitutively trimeric.[19,20] These data have impelled a model in which the HR-C represses HSF1 trimerization by folding back and directly interacting with the HR-A/B. The hydrophobic contacts between HR-A/B and HR-C are thought to restrain the HR-A/B, forming an intramolecular coiled-coil and thereby maintaining the monomeric HSF1 in a closed inactive state in the absence of stress.[11]

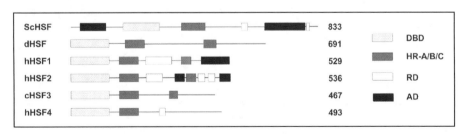

Figure 1. Functional regions of HSF family members. The helix-loop-helix DNA-binding domain (DBD) is the most conserved region within the HSF family. DNA-binding activity is achieved through homo- or heterotrimerization, and regulated by three leucin-zipper-like heptad-repeat domains (HR-A/B/C). The centrally located regulatory domain (RD) of HSF1 is extensively modified posttranslationally and regulates HSF1 transactivation capacity, conveyed by the C-terminal activation domain (AD). HSF of *S. cerevisiae* has both an N-terminal and a C-terminal activation domain. Note that the regulatory and activation domains of dHSF1, cHSF3 and hHSF4 are yet poorly characterized. The numbers indicate the last amino acid of the protein. Abbreviations: h: human; c: chicken; Sc: *S. cerevisiae*; d: *Drosophila*. Modified from reference 83.

The last 150 C-terminal residues of mammalian HSF1 encompass the transactivation domain, which is rich in hydrophobic and acidic amino acids (Fig. 1).[21,22] As discussed below, the transactivation domain interacts with several regulatory proteins as well as with components of the preinitiation complex. Under nonstressful conditions, the transactivation domain is kept inactive by a regulatory domain, which is found between the HR-A/B and HR-C domains (Fig. 1).[23] Whereas the isolated HSF1 transactivation domain possesses constitutive transcriptional activity, the fusion of the regulatory domain to the transactivation domain leads to stress-inducible HSF1 activation.[22,23] Thus, the regulatory domain can work as a gatekeeper, on which stress stimuli act to unleash the function of the transactivation domain.

Activation Mechanisms of HSF1

Under normal growth conditions, HSF1 is found mainly as transcriptionally inactive monomers in the nucleoplasm and, to a minor extent, in the cytosol.[24,25] The inactive state is thought to be maintained through both intermolecular and intramolecular interactions as well as by posttranslational modifications (Fig. 2).[26] Although the signals regulating HSF1 can be of diverse origin, including pathophysiological conditions, developmental cues and environmental stresses, they are similar in that they generate an increase in nonnative proteins.[26] This has led to a widely accepted model in which the elevated level of misfolded protein intermediates liberates Hsps from HSF1, thereby allowing HSF1 to be converted into an active state (reviewed in Chapter 8). Indeed, injection of denatured, but not native proteins into *Xenopus* oocytes is enough to trigger HSF1 activation.[27] It is also known that different stressors can have additive or synergistic effects on *hsp* gene transcription, which could reflect a common signaling pathway from protein damage to HSF1 activation.[26]

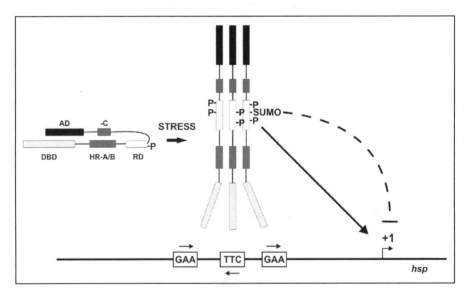

Figure 2. Multistep activation mechanism of HSF1. When exposed to stress, HSF1 acquires DNA-binding activity through a monomer-to-trimer transition and nuclear accumulation. In the nucleus, HSF1 binds to HSEs, consisting of repetitive nGAAn sequences, in the promoters of target genes such as *hsps*. During activation, HSF1 is hyperphosphorylated (P) at several serine residues, mostly in the RD. For simplicity, only a few phosphorylation sites are indicated. Phosphorylation of serine 303 is required for the stress-inducible SUMO modification of lysine 298, which represses the transcriptional activity of HSF1. +1 denotes the transcription start site. Abbreviations for the functional domains of HSF1 are as indicated in Figure 1.

The activation mechanism seems to be more multifaceted than depicted above, since HSF1 also possesses an intrinsic ability to sense stresses such as elevated temperatures and changed redox states. Recombinant HSF1 from different species can be activated by various stressors, such as heat, H_2O_2 and low pH.[28-31] Ahn and Thiele noted that two conserved cysteine residues in the DNA-binding domain of HSF1 were crucial for HSF1 trimer formation in vitro, and suggested that disulfide bonds between these cysteines would function as a stress sensor for HSF1 trimerization.[31] This built-in sensor capacity of HSF1 might also explain the extreme rapidity of HSF1 activation in response to severe stress. In fact, binding of HSF to the *Drosophila hsp70* promoter is detected already within seconds after the onset of heat shock, and the number of HSF molecules on the promoter reaches saturated levels in about one minute.[32] Although little is known about how HSF1 is converted from a latent monomer to a transcriptionally active trimer, the rapidity of the process strongly implies the involvement of posttranslational modifications in the transcriptional onset of HSF1 target genes. Concurrently with the cloning of HSF from yeast, HSF was found to be extensively phosphorylated, and a recent mass spectrometric analysis revealed that human HSF1 is phosphorylated on at least 12 sites, most of which are located in the regulatory domain.[33,34] Therefore, stress-induced phosphorylation of the regulatory domain likely acts as a switch that relieves the inhibitory effect of the regulatory domain on HSF1 transactivation capacity.

It is interesting to note that although HSF1 is rapidly hyperphosphorylated upon various stresses and in a manner that correlates with its transactivation capacity, only a few studies have been able to directly link phosphorylation to HSF1 transcriptional activity.[34-38] On the contrary, most phosphorylation events have been shown to repress the transcriptional activity of HSF1. A particularly interesting example is the phosphorylation of serines 303 and 307, located in the regulatory domain. The phosphorylation of these residues maintains HSF1 in a transcriptionally inactive state under nonstress conditions, but this inhibitory effect can be overridden by heat shock.[35,39] A putative mechanism for this regulatory role was presented by Wang and coworkers, who reported that phosphorylation of S303 and S307 mediated an interaction between HSF1 and the scaffolding protein 14-3-3ε followed by a sequestration of HSF1 in the cytoplasm.[40,41] The phosphorylation of the serines might be a hierarchical event, in which ERK-primed phosphorylation of S307 would be required for GSK3β-mediated phosphorylation of S303.[42,43] Nevertheless, mutation of S303 alone leads to a dramatic derepression of HSF1 activity, demonstrating a functional independence between the two phosphorylation events.[35,44] The mechanism behind the repressive function of S303 phosphorylation was revealed when it was shown to be a prerequisite for yet another posttranslational modification, the covalent attachment of a protein called SUMO (small ubiquitin-like modifier) to the adjacent lysine 298.[45] SUMO is a member of the ubiquitin-like protein superfamily of modifiers and an established regulator of transcription.[46] For most transcriptional regulators, such as Sp3, Elk-1 and HSF2, sumoylation negatively regulates the transactivation capacity of the substrate.[47,48] Accordingly, the phosphorylation-dependent sumoylation inhibits the transcriptional activity of HSF1 in vivo (Fig. 2).[44] SUMO modification usually requires the consensus tetrapeptide (I/V/L)-K-X-E, in which X denotes any amino acid and K the target lysine to which SUMO becomes attached.[46] Interestingly, the identification of phosphorylation-dependent sumoylation of HSF1 was the preamble to the recent discovery that several transcriptional regulators contain a corresponding phosphorylation-dependent sumoylation motif (IVL)-K-X-E-X-X-S-P, called PDSM, and that this regulatory mechanism is a conserved feature for a number of transcriptional players.[44]

Regulation of *hsp* Gene Transcription by HSF1

The *hsp70* promoter is an established model system for inductive transcriptional responses and by far the best studied among the HSF/HSF1-responsive promoters. The details of how HSF1 triggers *hsp70* transcription are, however, not fully understood. The uninduced *hsp70* promoter is primed for transcription in that it is occupied by a transcriptionally engaged, but stalled RNA polymerase II (RNAP II).[49,50] Upon heat shock, RNAP II is released from its dormant state and the transcript is rapidly elongated. In vitro, the stalling of RNAP II is greatly enhanced by nucleosome formation, suggesting that chromatin remodeling is pertinent for the release of the paused RNAP II.[51] Subsequently, certain hydrophobic residues in the HSF1 activation domain were shown to stimulate RNAP II release, and directly interact with BRG1, the ATPase subunit of the chromatin-remodeller SWI/SNF.[52-54] The maturation of RNAP II into an active elongating complex is accompanied by site-specific hyperphosphorylation of the C-terminal domain of RNAP II. In *Drosophila*, heat shock induces the recruitment of P-TEFb kinase to heat shock loci. The P-TEFb kinase is composed of a Cdk9 subunit and a cyclin T subunit, and known to hyperphosphorylate the C-terminal domain of RNAP II.[55] When P-TEFb is artificially recruited to the *hsp70* promoter, *hsp70* is transcribed also under nonstressful conditions, suggesting that heat-induced recruitment of P-TEFb to heat shock loci could act as a switch in stress-induced transcription of *hsp* genes.[56] Accordingly, the translocation of P-TEFb to heat shock loci is abrogated in the absence of *Drosophila* HSF, but the recruitment does not seem to be due to a direct interaction between these two proteins.[56] Interestingly, inhibition of P-TEFb kinase activity does not affect the maturation of the stalled RNAP II complex, but instead leads to defect processing of the 3' end of *hsp70* and *hsp26* transcripts.[57] In addition, mammalian HSF1 interacts with symplekin, a scaffold for polyadenylation factors, and this interaction is required for efficient polyadenylation of *hsp70* transcripts.[58] Together, these intriguing findings that HSF1 not only activates *hsp* gene transcription but also participates in mRNA processing broaden the repertoire of mechanisms by which HSF1 regulates the expression of heat shock genes.

Upon heat shock, HSF recruits the Mediator coactivator to heat shock loci.[59] Mediator complexes regulate a wide range of genes through relaying activating signals from transcription factors to the basal transcription machinery. In contrast to P-TEFb, the recruitment of Mediator occurs through a direct interaction with HSF, since the activation domain of *Drosophila* HSF binds to the TRAP80 subunit of the Mediator complex.[59] HSF1 activity may further be conveyed to the preinitiation complex more directly, which is supported by findings that the activation domain of HSF1 binds to the TATA-binding protein TBP as well as to other components of the preinitiation complex, including TFIIB in vitro.[60,61] These results may indicate that a spatial constraint regulates the interaction between HSF1 and the general transcription machinery and possibly explain why most HSEs lie in close proximity to the transcription start site.[11]

Stress-Specific Activation of HSF1

Activation of HSF1 by stresses that are greatly different in nature raises the important question whether the HSF1-activating stressors simply signal through a common pathway or whether distinct stimuli differentially modulate HSF1-dependent transcriptional processes. Several reports have provided evidence for the latter, showing that the mechanism of HSF1 activation and HSF1-dependent transcription can proceed in a stimulus-specific manner. For example, studies in *S. cerevisiae* have demonstrated that both glucose starvation and heat shock lead to HSF-mediated transcription of the *CUP1* gene, but through distinct mechanisms.[62] Specifically, during glucose starvation HSF interacts with and is phosphorylated by the Snf1 kinase, which is required for HSF activation. However, HSF activation by heat shock is not Snf1 dependent, demonstrating the involvement of stress-specific signal transduction pathways in HSF activation.[63] Similarly, HSF activated by oxidative stress shows a phosphorylation pattern distinct from HSF activated by heat.[64] Although phosphorylation is involved in a stress-specific activation mechanism of HSF, it may also specify the subset of HSF target genes that are to be transcribed in response to a particular stress. A mutation that inhibits HSF phosphorylation impairs transcription of HSF targets which have short or mildly dispersed nGAAn elements in their promoter regions, but does not significantly affect transcription of genes with multiple copies of nGAAn pentamers.[15]

In mammals, little is known how HSF1 activation can be adjusted according to different signals. By introduction of dominant-negative forms of the SWI/SNF ATPases BRG1 and BRM, de la Serna and coworkers noticed a differential SWI/SNF dependency of HSF1-mediated transcription. Whereas the transcriptional activation of *hsp70* by arsenite or cadmium was severely impaired by the disruption of SWI/SNF activity, *hsp70* transcription by heat shock was unaffected.[65] Further studies are necessary, since chromatin immunoprecipitation and restriction enzyme accessibility of the *hsp70* promoter demonstrated that the direct recruitment of SWI/SNF by HSF1 is essential for chromatin accessibility and efficient *hsp70* transcription during heat shock.[54] The differential activation of *hsp70* by heat and arsenite was further elucidated when Thomson and coworkers noticed that *hsp70* induction by arsenite, but not by heat shock, was dependent on the p38 MAP kinase pathway, perhaps due to specific histone phosphorylation mediated by p38 signaling.[66] Moreover, during both stresses HSF1 was found responsible for targeting histone H4 acetylation to *hsp70* chromatin, highlighting the importance of epigenetic modifications in HSF1-mediated stimulus-specific transcription.[66]

HSF1 as a Developmental Regulator

Albeit HSF1 plays its most prominent role in protecting cells and organisms from proteotoxic stresses, several lines of genetic evidence support a versatile role for HSF1 in regulation of development. In yeast, HSF is essential for viability also under normal growth conditions, and more recently, a genome-wide analysis of *S. cerevisiae* HSF targets showed that HSF participates in diverse biological pathways, such as energy metabolism, synthesis of cytoskeletal components and vesicular transport.[33,67,68] In higher invertebrates and vertebrates, HSF/HSF1 is not required for survival, but its absence causes severe reproductive disturbances, a reduction in body size and deficient embryonic development. For example, the deletion of the *hsf* gene leads to defect oogenesis and larvae development in *Drosophila*.[69] *hsf1* knockout mice undergo normal oogenesis but have malformed chorioallantoic placenta, resulting in high prenatal lethality. Moreover, *hsf1*-/- females are infertile, which is due to HSF1 being a maternal factor.[70] In the testis, HSF1 plays a dual role in the protection against harmful signals. Whereas HSF1 acts as a survival factor of more immature germ cells in testes exposed to hyperthermia, it is also required for apoptosis of pachytene spermatocytes in response to thermal insults.[71] Accordingly, the expression of a constitutively active form of HSF1 inhibits the progression of spermatogenesis and leads to apoptosis of spermatocytes at the pachytene stage.[72] It is interesting to note that the threshold for HSF1 activation is reduced in pachytene spermatocytes relative to somatic cells, implying that HSF1 could play a role in monitoring the quality control of male germ cells.[73,74]

HSF1-Mediated Expression of Cytokines

The requirement for HSF1 in developmental pathways does not seem to be explicitly due to altered Hsp levels, emphasizing the importance of other HSF1 targets. An interesting exception is seen in the mouse heart, where the disruption of *hsf1* causes decreased constitutive expression of several Hsps, which is intimately associated with impaired redox homeostasis and mitochondrial damage.[75] Transcriptional profiling of *hsf1*-deficient fibroblasts has revealed that HSF1 regulates several genes also under nonstressful conditions. Surprisingly, the transcription of many of these genes is not enhanced by heat shock, suggesting that the stress-dependent activation of HSF1 may favor the expression of only a subset of its target genes.[76,77] Among the nonstress HSF1 targets, locally acting signal molecules, including cytokines and chemokines, make up a significant part.[76] One of these targets, *IL-6* was shown to be under direct control of HSF1 in spleen cells. IL-6 is a multi-functional cytokine that participates in immune responses and inflammation and is important for B cell differentiation and maturation into plasma cells.[78] Accordingly, in response to immunization, the *hsf1* knockout mice show an impaired production of immunoglobulins, especially IgG2a.[76]

In the olfactory epithelium of pubescent mice, HSF1 acquires DNA-binding activity during post-natal week 4. One of the targets of this activation is another member of the IL-6 family, leukemia inhibitory factor (LIF), a pleiotropic cytokine implicated in numerous developmental pathways.[79,80] Interestingly, HSF1 seems to repress the expression of LIF, thereby allowing the differentiation of the olfactory epithelium.[80] Negative regulation by HSF1 is also involved in modulating the expression of the proinflammatory tumor necrosis factor alpha (TNFα). When *hsf1-/-* mice were injected with an LD$_{50}$ dose of bacterial lipopolysaccharide LPS, they displayed an increased mortality and abnormally high expression of TNFα.[6] The fascinating concept of HSF1 functioning both as a transcriptional activator and as a transcriptional repressor in a context-dependent manner markedly broadens the biological role of HSF1. The repressive function of HSF1 was originally introduced by Cahill and coworkers, who reported that the LPS-induced *interleukin 1β* gene expression could be inhibited by HSF1 activation.[81] This phenomenon was subsequently shown to be mediated through the direct interaction between HSF1 and C/EBPβ, a regulator of *IL-1β* transcription.[82] In most cases, however, the mechanism of HSF1-mediated repression is poorly understood. Moreover, possible indirect effects of HSF1 activity may be emphasized through the complexity of cytokine signaling networks, accentuating the need for further studies to elucidate the intriguing function of HSF1 in cytokine expression.

Heat Shock Factors Working Together

Since the DNA-binding and oligomerization domains are particularily well conserved within the HSF family, it is not surprising that different HSFs display similar fundamental properties. This is especially evident in vitro, where all HSFs are able to form trimers and bind to the same HSEs, strongly suggesting that different members of the HSF family could share a subset of target genes.[83] Nevertheless, as HSF1 cannot be substituted by other HSF members, and the *hsf1*, *hsf2* and *hsf4* knockout mice have clearly distinct phenotypes, the HSFs might be restricted to different fields of action in vivo.[6-8,10,84] Recently, however, HSF1 was found to converge with HSF4 in the regulation of eye development. In certain Chinese and Danish families, mutations in the HSF4 DNA-binding domain were found to be the cause of autosomal dominant lamellar and Marnier cataract, and shortly thereafter, *hsf4* null mice were shown to develop cataract during the early postnatal period.[9,10,84] The *hsf4-/-* lens contains inclusion-like structures of protein aggregates, perhaps due to an increased expression of fibroblast growth factors (FGFs) and a decrease in protective γ-crystallins and Hsps, particularly Hsp27, in lens cells.[10,84] Surprisingly, HSF1 and HSF4 have opposing effects on FGF expression in the lens. Both HSF1 and HSF4 bind to the *FGF7* promoter, but whereas the expression of FGF7 is increased in *hsf4-/-*, it is decreased in *hsf1-/-* and returned to normal levels in the *hsf1/hsf4* double knockout mice.[10] Similarly, the increased LIF expression in the absence of *hsf1* is partially relieved by the simultaneous disruption of *hsf4*, demonstrating an intimate relationship between these two factors in the maintenance of sensory systems.[80]

The generation of double knockout mice has provided valuable information for a functional relationship also between HSF1 and HSF2. While spermatogenesis is normal in *hsf1-/-* mice and only modestly impaired in *hsf2-/-* mice, the simultaneous disruption of both *hsf1* and *hsf2* leads to male infertility.[7,8,85] A gene expression analysis of the *hsf1/hsf2* double knockout testis revealed that many genes involved in spermatogenesis were downregulated, suggesting that HSF1 and HSF2 could have complementary or interdependent functions in the regulation of male fertility.[85] The notion that HSF1 and HSF2 could share a subset of target genes was initially proposed when HSF2 was found to bind to the *hsp70* promoter upon hemin-induced erythroid differentiation of K562 erythroleukemia cells.[86] More recently, an extensive promoter analysis in K562 cells by Trinklein and coworkers demonstrated that both HSF1 and HSF2 are recruited to several *hsp* promoters in response to hemin treatment, implying that HSF2 could be involved in regulating the expression of a broad repertoire of *hsp* genes.[87] An elegant mechanism by which HSF2 could influence the transcription of heat shock genes was recently presented by Xing and coworkers, who reported that HSF2 maintains chromatin accessibility of the inducible *hsp70* gene during the M phase in mitosis, a phenomenon called bookmarking, and that this function is required for HSF1-mediated transcription of *hsp70* in the G1 phase.[88] The putative modulatory role of HSF2 in HSF1-mediated transcription could also result from heterocomplex formation, since the direct interaction between HSF1 and HSF2 as well as the colocalization of HSF1 and HSF2 in nuclear stress bodies is dependent on intact oligomerization domains.[89,90] This hypothesis is further supported by the recent observation that upon proteasome inhibition, both HSF1 and HSF2 bind to the *clusterin* promoter in the context of chromatin and to oligonucleotides containing binding sites for only one HSF trimer.[91] Although the fundamental question of how HSF2 modulates HSF1-mediated transcription remains to be answered, the prospect of cooperative heterogenous HSF complexes certainly adds a novel flavor to the transcriptional regulation of heat shock genes as well as other targets.

Future Perspectives

Although the heat shock response was initially described already four decades ago, the molecular details of this strictly regulated process are only beginning to emerge. For example, the essential questions how cells are able to sense stress signals and how these signals are subsequently conveyed to HSF1 for transcription of *hsp* genes need to be explored more thoroughly. While genome-wide analyses may provide further understanding of target gene selectivity in response to specific stress stimuli, it is becoming evident that distinct stressors employ different signaling pathways for activation of the same HSF1 target genes. The elucidation of how stress-specific networks are organized poses an intriguing challenge for future studies. Moreover, the involvement of HSF1 in several developmental pathways brings unforeseen complexity to the mechanisms of HSF1 action. The recent discoveries of HSF1 functioning in tight cooperation with other HSFs during development propose that such cross talk might occur also upon exposure to stress stimuli. To this end, analyses of the posttranscriptional regulation, to which HSF2 and HSF4 are subjected under stressful conditions, likely provide novel insights into how these factors contribute to the heat shock response.

Acknowledgements

We apologize to our many colleagues whose work could only be cited indirectly due to space limitations. We are grateful to Aura Kaunisto, Annika Meinander, Pia Roos-Mattjus and Anton Sandqvist for their valuable comments on the manuscript. The work done in our laboratory is financially supported by the Academy of Finland, the Finnish Cancer Organizations, the Sigrid Jusélius Foundation, the Finnish Life and Pension Insurance Companies, and Åbo Akademi University. J. Anckar is supported by the Turku Graduate School for Biomedical Sciences (TuBS).

References

1. Brenner S, Jacobson F, Meselson M. An unstable intermediate carrying information from genes to ribosomes for protein synthesis. Nature 1961; 190:576-577.
2. Ritossa F. A new puffing pattern induced by temperature shock and DNP in Drosophila. Experimentia 1962; 18:571-573.
3. Lindquist S. The heat-shock response. Annu Rev Biochem 1986; 55:1151-1191.
4. Nakai A. New aspects in the vertebrate heat shock factor system: Hsf3 and Hsf4. Cell Stress Chap 1999; 2:86-93.
5. McMillan DR, Xiao X, Shao L et al. Targeted disruption of heat shock transcription factor 1 abolishes thermotolerance and protection against heat-inducible apoptosis. J Biol Chem 1998; 273:7523-7528.
6. Xiao X, Zuo X, Davis AA et al. HSF1 is required for extra-embryonic development, postnatal growth and protection during inflammatory responses in mice. EMBO J 1999; 18:5943-5952.
7. Kallio M, Chang Y, Manuel M et al. Brain abnormalities, defective meiotic chromosome synapsis and female subfertility in HSF2 null mice. EMBO J 2002; 21:2591-2601.

8. Wang G, Zhang J, Moskophidis D et al. Targeted disruption of the heat shock transcription factor (hsf)-2 gene results in increased embryonic lethality, neuronal defects, and reduced spermatogenesis. Genesis 2003; 36:48-61.

9. Bu L, Jin Y, Shi Y et al. Mutant DNA-binding domain of HSF4 is associated with autosomal dominant lamellar and Marner cataract. Nat Genet 2002; 31:276-278.

10. Fujimoto M, Izu H, Seki K et al. HSF4 is required for normal cell growth and differentiation during mouse lens development. EMBO J 2004; 23:4297-4306.

11. Wu C. Heat shock transcription factors: Structure and regulation. Annu Rev Cell Dev Biol 1995; 11:441-469.

12. Xiao H, Perisic O, Lis JT. Cooperative binding of Drosophila heat shock factor to arrays of a conserved 5 bp unit. Cell 1991; 64:585-593.

13. Kroeger PE, Morimoto RI. Selection of new HSF1 and HSF2 DNA-binding sites reveals difference in trimer cooperativity. Mol Cell Biol 1994; 14:7592-7603.

14. Santoro N, Johansson N, Thiele DJ. Heat shock element architecture is an important determinant in the temperature and transactivation domain requirements for heat shock transcription factor. Mol Cell Biol 1998; 18:6340-6352.

15. Hashikawa N, Sakurai H. Phosphorylation of the yeast heat shock transcription factor is implicated in gene-specific activation dependent on the architecture of the heat shock element. Mol Cell Biol 2004; 24:3648-3659.

16. Sorger PK, Nelson HCM. Trimerization of a yeast transcriptional activator via a coiled-coil motif. Cell 1989; 59:807-813.

17. Peteranderl R, Nelson HCM. Trimerization of the heat shock transcription factor by a triple-stranded alpha-helical coiled-coil. Biochemistry 1992; 31:12272-12276.

18. Rabindran SK, Haroun RI, Clos J et al. Regulation of heat shock factor trimer formation: Role of a conserved leucine zipper. Science 1993; 259:230-234.

19. Chen Y, Barlev NA, Westergaard O et al. Identification of the C-terminal activator domain in yeast heat shock factor: Independent control of transient and sustained transcriptional activity. EMBO J 1993; 12:5007-5018.

20. Nakai A, Tanabe M, Kawazoe Y et al. HSF4, a new member of the human heat shock factor family which lacks properties of a transcriptional activator. Mol Cell Biol 1997; 17:469-481.

21. Shi Y, Kroeger PE, Morimoto RI. The carboxyl-terminal transactivation domain of heat shock factor 1 is negatively regulated and stress responsive. Mol Cell Biol 1995; 15:4309-4318.

22. Newton EM, Knauf U, Green M et al. The regulatory domain of human heat shock factor 1 is sufficient to sense heat stress. Mol Cell Biol 1996; 16:839-846.

23. Green M, Schuetz TJ, Sullivan EK et al. A heat shock-responsive domain of human HSF1 that regulates transcription activation domain function. Mol Cell Biol 1995; 15:3354-3362.

24. Mercier PA, Winegarden NA, Westwood JT. Human heat shock factor 1 is predominantly a nuclear protein before and after heat stress. J Cell Sci 1999; 112:2765-2774.

25. Vujanac M, Fenaroli A, Zimarino V. Constitutive nuclear import and stress-regulated nucleocytoplasmic shuttling of mammalian heat-shock factor 1. Traffic 2005; 6:214-229.

26. Morimoto RI. Regulation of the heat shock transcriptional response: Cross talk between a family of heat shock factors, molecular chaperones, and negative regulators. Genes Dev 1998; 12:3788-3796.

27. Ananthan J, Goldberg AL, Voellmy R. Abnormal proteins serve as eukaryotic stress signals and trigger the activation of heat shock genes. Science 1986; 232:522-524.

28. Goodson ML, Sarge KD. Heat-inducible DNA binding of purified heat shock transcription factor 1. J Biol Chem 1995; 270:2447-2450.

29. Farkas T, Kutskova YA, Zimarino V. Intramolecular repression of mouse heat shock factor 1. Mol Cell Biol 1998; 18:906-918.

30. Zhong M, Orosz A, Wu C. Direct sensing of heat and oxidation by Drosophila heat shock transcription factor. Mol Cell 1998; 2:101-108.

31. Ahn SG, Thiele DJ. Redox regulation of mammalian heat shock factor 1 is essential for Hsp gene activation and protection from stress. Genes Dev 2003; 17:516-528.

32. Boehm AK, Saunders A, Werner J et al. Transcription factor and polymerase recruitment, modification, and movement on dhsp70 in vivo in the minutes following heat shock. Mol Cell Biol 2003; 23:7628-7637.

33. Sorger PK, Pelham HRB. Yeast heat shock factor is an essential DNA-binding protein that exhibits temperature-dependent phosphorylation. Cell 1988; 54:855-864.

34. Guettouche T, Boellmann F, Lane WS et al. Analysis of phosphorylation of human heat shock factor 1 in cells experiencing a stress. BMC Biochem 2005; 6:4.

35. Kline MP, Morimoto RI. Repression of the heat shock factor 1 transcriptional activation domain is modulated by constitutive phosphorylation. Mol Cell Biol 1997; 17:2107-2115.

36. Holmberg CI, Hietakangas V, Mikhailov A et al. Phosphorylation of serine 230 promotes inducible transcriptional activity of heat shock factor 1. EMBO J 2001; 20:3800-3810.
37. Holmberg CI, Tran SEF, Eriksson JE et al. Multisite phosphorylation provides sophisticated regulation of transcription factors. Trends Biochem Sci 2002; 27:619-627.
38. Boellmann F, Guettouche T, Guo Y et al. DAXX interacts with heat shock factor 1 during stress activation and enhances its transcriptional activity. Proc Natl Acad Sci USA 2004; 101:4100-4105.
39. Knauf U, Newton EM, Kyriakis J et al. Repression of human heat shock factor 1 activity at control temperature by phosphorylation. Genes Dev 1996; 10:2782-2793.
40. Wang X, Grammatikakis N, Siganou A et al. Regulation of molecular chaperone gene transcription involves the serine phosphorylation, 14-3-3 epsilon binding, and cytoplasmic sequestration of heat shock factor 1. Mol Cell Biol 2003; 23:6013-6026.
41. Wang X, Grammatikakis N, Siganou A et al. Interactions between extracellular signal-regulated protein kinase 1, 14-3-3epsilon, and heat shock factor 1 during stress. J Biol Chem 2004; 279:49460-49469.
42. Chu B, Soncin F, Price BD et al. Sequential phosphorylation by mitogen-activated protein kinase and glycogen synthase kinase 3 represses transcriptional activation by heat shock factor-1. J Biol Chem 1996; 271:30847-20857.
43. Chu B, Zhong R, Soncin F et al. Transcriptional activity of heat shock factor 1 at 37 degrees C is repressed through phosphorylation on two distinct serine residues by glycogen synthase kinase 3 and protein kinases Calpha and Czeta. J Biol Chem 1998; 273:18640-18646.
44. Hietakangas V, Anckar J, Blomster HA et al. PDSM, a motif for phosphorylation-dependent SUMO modification. Proc Natl Acad Sci USA 2006; 103:45-50.
45. Hietakangas V, Ahlskog JK, Jakobsson AM et al. Phosphorylation of serine 303 is a prerequisite for the stress-inducible SUMO modification of heat shock factor 1. Mol Cell Biol 2003; 23:2953-68.
46. Johnson ES. Protein modification by SUMO. Annu Rev Biochem 2004; 73:355-382.
47. Gill G. Something about SUMO inhibits transcription. Curr Opin Genet Dev 2005; 15:536-541.
48. Anckar J, Hietakangas V, Denessiouk K et al. Inhibition of DNA binding by differential sumoylation of heat shock factors. Mol Cell Biol 2006; 26:955-964.
49. Rougvie AE, Lis JT. The RNA polymerase II molecule at the 5' end of the uninduced hsp70 gene of D. melanogaster is transcriptionally engaged. Cell 1988; 54:795-804.
50. Lis JT. Promoter-associated pausing in promoter architecture and postinitiation transcriptional regulation. Cold Spring Harb Symp Quant Biol 1998; 63:347-56.
51. Brown SA, Imbalzano AN, Kingston RE. Activator-dependent regulation of transcriptional pausing on nucleosomal templates. Genes Dev 1996; 10:1479-1490.
52. Brown SA, Weirich CS, Newton EM et al. Transcriptional activation domains stimulate initiation and elongation at different times and via different residues. EMBO J 1998; 17:3146-3154.
53. Sullivan EK, Weirich CS, Guyon JR et al. Transcriptional activation domains of human heat shock factor 1 recruit human SWI/SNF. Mol Cell Biol 2001; 21:5826-5837.
54. Corey LL, Weirich CS, Benjamin IJ et al. Localized recruitment of a chromatin-remodeling activity by an activator in vivo drives transcriptional elongation. Genes Dev 2003; 17:1392-1401.
55. Marshall, NF, Peng J, Xie Z et al. Control of RNA polymerase II elongation potential by a novel carboxyl-terminal domain kinase. J Biol Chem 1996; 271:27176-27183.
56. Lis JT, Mason P, Peng J et al. P-TEFb kinase recruitment and function at heat shock loci. Genes Dev 2000; 14:792-803.
57. Ni Z, Schwartz BE, Werner J et al. Coordination of transcription, RNA processing, and surveillance by P-TEFb kinase on heat shock genes. Mol Cell 2004; 13:55-65.
58. Xing H, Mayhew CN, Cullen KE et al. HSF1 modulation of Hsp70 mRNA polyadenylation via interaction with symplekin. J Biol Chem 2004; 279:10551-10555.
59. Park JM, Werner J, Kim JM et al. Mediator, not holoenzyme, is directly recruited to the heat shock promoter by HSF upon heat shock. Mol Cell 2001; 8:9-19.
60. Mason Jr PB, Lis JT. Cooperative and competitive protein interactions at the hsp70 promoter. J Biol Chem 1997; 272:33227-33233.
61. Yuan CX, Gurley WB. Potential targets for HSF1 within the preinitiation complex. Cell Stress Chap 2000; 5:229-242.
62. Tamai KT, Liu X, Silar P et al. Heat shock transcription factor activates yeast metallothionein gene expression in response to heat and glucose starvation via distinct signalling pathways. Mol Cell Biol 1994; 14:8155-8165.
63. Hahn JS, Thiele DJ. Activation of the Saccharomyces cerevisiae heat shock transcription factor under glucose starvation conditions by Snf1 protein kinase. J Biol Chem 2004; 279:5169-5176.
64. Liu XD, Thiele DJ. Oxidative stress induced heat shock factor phosphorylation and HSF-dependent activation of yeast metallothionein gene transcription. Genes Dev 1996; 10:592-603.

65. de la Serna IL, Carlson KA, Hill DA et al. Mammalian SWI-SNF complexes contribute to activation of the hsp70 gene. Mol Cell Biol 2000; 20:2839-2851.
66. Thomson S, Hollis A, Hazzalin CA et al. Distinct stimulus-specific histone modifications at hsp70 chromatin targeted by the transcription factor heat shock factor-1. Mol Cell 2004; 15:585-594.
67. Gallo GJ, Prentice H, Kingston RE. Heat shock factor is required for growth at normal temperatures in the fission yeast Schizosaccharomyces pombe. Mol Cell Biol 1993; 13:749-761.
68. Hahn JS, Hu Z, Thiele DJ et al. Genome-wide analysis of the biology of stress responses through heat shock transcription factor. Mol Cell Biol 2004; 24:5249-5256.
69. Jedlicka P, Mortin MA, Wu C. Multiple functions of Drosophila heat shock transcription factor in vivo. EMBO J 1997; 16:2452-2462.
70. Christians E, Davis AA, Thomas SD et al. Maternal effect of Hsf1 on reproductive success. Nature 2000; 407:693-694.
71. Izu H, Inouye S, Fujimoto M et al. Heat shock transcription factor 1 is involved in quality-control mechanisms in male germ cells. Biol Reprod 2004; 70:18-24.
72. Nakai A, Suzuki M, Tanabe M. Arrest of spermatogenesis in mice expressing an active heat shock transcription factor 1. EMBO J 2000; 19:1545-1554.
73. Sarge KD. Male germ cell-specific alteration in temperature set point of the cellular stress response. J Biol Chem 1995; 270:18745-18748.
74. Sarge KD, Bray AE, Goodson ML. Altered stress response in testis. Nature 1995; 374:126.
75. Yan LJ, Christians ES, Liu L et al. Mouse heat shock transcription factor 1 deficiency alters cardiac redox homeostasis and increases mitochondrial oxidative damage. EMBO J 2002; 21:5164-5172.
76. Inouye S, Izu H, Takaki E et al. Impaired IgG production in mice deficient for heat shock transcription factor 1. J Biol Chem 2004; 279:38701-38709.
77. Trinklein ND, Murray JI, Hartman SJ et al. The role of heat shock transcription factor 1 in the genome-wide regulation of the mammalian heat shock response. Mol Biol Cell 2004; 15:1254-1261.
78. Heinrich PC, Behrmann I, Haan S et al. Principles of interleukin (IL)-6-type cytokine signalling and its regulation. Biochem J 2003; 374:1-20.
79. Metcalf D. The unsolved enigmas of leukemia inhibitory factor. Stem Cells 2003; 21:5-14.
80. Takaki E, Fujimoto M, Sugahara K et al. Maintenance of olfactory neurogenesis requires HSF1, a major heat shock transcription factor in mice. J Biol Chem 2006; 281:4931-4937.
81. Cahill CM, Waterman WR, Xie Y et al. Transcriptional repression of the prointerleukin 1beta gene by heat shock factor 1. J Biol Chem 1996; 271:24874-24879.
82. Xie Y, Chen C, Stevenson MA et al. Heat shock factor 1 represses transcription of the IL-1beta gene through physical interaction with the nuclear factor of interleukin 6. J Biol Chem 2002; 277:11802-11810.
83. Pirkkala L, Nykänen P, Sistonen L. Roles of the heat shock transcription factors in regulation of the heat shock response and beyond. FASEB J 2001; 15:1118-1131.
84. Min JN, Zhang Y, Moskophidis D et al. Unique contribution of heat shock transcription factor 4 in ocular lens development and fiber cell differentiation. Genesis 2004; 40:205-217.
85. Wang G, Ying Z, Jin X et al. Essential requirement for both hsf1 and hsf2 transcriptional activity in spermatogenesis and male fertility. Genesis 2004; 38:66-80.
86. Sistonen L, Sarge KD, Phillips B et al. Activation of heat shock factor 2 during hemin-induced differentiation of human erythroleukemia cells. Mol Cell Biol 1992; 12:4104-4111.
87. Trinklein ND, Chen WC, Kingston RE et al. Transcriptional regulation and binding of heat shock factor 1 and heat shock factor 2 to 32 human heat shock genes during thermal stress and differentiation. Cell Stress Chap 2004; 9:21-28.
88. Xing H, Wilkerson DC, Mayhew CN et al. Mechanism of hsp70i gene bookmarking. Science 2005; 307:421-423.
89. Alastalo TP, Hellesuo M, Sandqvist A et al. Formation of nuclear stress granules involves HSF2 and coincides with the nucleolar localization of Hsp70. J Cell Sci 2003; 116:3557-3570.
90. He H, Soncin F, Grammatikakis N et al. Elevated expression of heat shock factor (HSF) 2A stimulates HSF1-induced transcription during stress. J Biol Chem 2003; 278:35465-35475.
91. Loison F, Debure L, Nizard P et al. Up-regulation of the clusterin gene after proteotoxic stress: implication of HSF1-HSF2 heterocomplexes. Biochem J 2006; 395:223-231.

Chaperone Regulation of the Heat Shock Protein Response

Richard Voellmy* and Frank Boellmann

Abstract

The heat shock protein response appears to be triggered primarily by nonnative proteins accumulating in a stressed cell and results in increased expression of heat shock proteins (HSPs). Many heat shock proteins prevent protein aggregation and participate in re-folding or elimination of misfolded proteins in their capacity as chaperones. Even though several mechanisms exist to regulate the abundance of cytosolic and nuclear chaperones, activation of heat shock transcription factor 1 (HSF1) is an essential aspect of the heat shock protein response. HSPs and co-chaperones that are assembled into multichaperone complexes regulate HSF1 activity at different levels. HSP90-containing multichaperone complexes appear to be the most relevant repressors of HSF1 activity. Because HSP90-containing multichaperone complexes interact not only specifically with client proteins including HSF1 but also generically with nonnative proteins, the concentration of nonnative proteins influences assembly on HSF1 of HSP90-containing complexes that repress activation, and may play a role in inactivation, of the transcription factor. Proteins that are unable to achieve stable tertiary structures and remain chaperone substrates are targeted for proteasomal degradation through polyubiquitination by co-chaperone CHIP. CHIP can activate HSF1 to regulate the protein quality control system that balances protection and degradation of chaperone substrates.

Introduction

The term "heat shock protein response" relates to the induction of heat shock or stress protein (HSP) synthesis that occurs in eukaryotic cells subsequent to heat treatment or exposure to other proteotoxic stress. The transcriptional response to heat and other proteotoxic stress is mediated by so-called heat shock transcription factors (HSF).[1,2] Vertebrate animals and plants express several different but related HSFs. Among the different vertebrate HSFs, the factor termed HSF1 is essentially required for stress regulation of HSP expression.[3-6] However, it is noted that other HSF family members also may contribute and, in some cases, play critical roles in regulating the response.[7-9] Disruption of HSF in invertebrates abolishes the heat shock response in a similar way as disruption of HSF1 in vertebrates.[10-15]

In this chapter we are primarily concerned with the regulation of vertebrate, in particular human, HSF1 and the single HSF species of invertebrate organisms such as *Saccharomyces cerevisiae* and *Drosophila melanogaster* that perform the function of vertebrate HSF1 in these organisms (also referred to herein as HSF1 for simplicity). Sequence and functional features of

*Corresponding Author: Richard Voellmy—HSF Pharmaceuticals SA, Avenue des Cerisiers 39B, 1009 Pully, Switzerland. Email: rvoellmy@hsfpharma.com

Molecular Aspects of the Stress Response: Chaperones, Membranes and Networks, edited by Peter Csermely and László Vígh. ©2007 Landes Bioscience and Springer Science+Business Media.

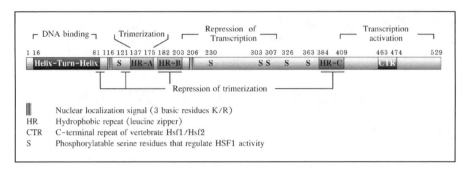

Figure 1. Linear map of human HSF1. Numbers refer to positions in the sequence of the HSF1 protein. Functions of individual domains/regions are indicated above and below the map.

a typical HSF1 are outlined in Figure 1, using human HSF1 as the example. A DNA-binding domain comprising a winged, helix-turn-helix motif is located near the amino terminus.[16-19] The domain is capable of interacting specifically with so-called heat shock element (HSE) sequences present in the promoters of heat shock protein (*hsp*) genes.[20,21] A long, interrupted, hydrophobic repeat sequence (HR-A/B) is situated adjacent to the DNA-binding domain. The presence of at least a portion of this sequence is required for oligomerization of the factor.[16,22,23] HSF1 includes an additional hydrophobic repeat sequence (HR-C) in the carboxy-terminal third of its sequence. Studies on mammalian and *Drosophila* HSF1 identified transactivation domains near the carboxy ends of these factors.[24-27] Regulation of HSF1 activity is largely controlled posttranslationally and not at the level of synthesis/degradation of the transcription factor.[26,28-31]

Feedback Regulation of the Heat Shock Protein Response by Stress-Inducible Chaperones

In their early studies of cultured *Drosophila* cells, Lindquist and colleagues observed that, under conditions of moderate heat stress, the heat shock protein response was self-regulated, meaning that HSP expression increased rapidly after initiation of a heat treatment, continued for some time and then decreased to a low rate approximately corresponding to the prestress rate (a phenomenon later called attenuation).[32,33] Reduction of rates of heat-induced transcription or translation resulted in an extension of the period of elevated *hsp* gene transcription during stress recovery, suggesting that one or more newly synthesized HSPs feedback-regulated the heat shock protein response. Since activation of *hsp* gene transcription did not require new protein synthesis, it was also recognized that the heat shock protein response is regulated both transcriptionally and post-transcriptionally. Attenuation of the stress protein response was subsequently also observed and studied in mammalian cells.[34] The signal that triggers the heat shock protein response appears to be accumulation of nonnative proteins resulting from chemical or physical denaturation.[35-41] Many HSPs can bind such nonnative proteins, assist in their refolding or target them for degradation, preventing cytotoxic protein aggregation. Proteins that display these capabilities are known as molecular chaperones.

Hsps and Co-Chaperones Repress Activation of HSF1

Homotrimerization of HSF1 was found to be essential for DNA binding and, therefore, absolutely required for its function as a transcriptional activator.[3,4,23,42-44] Heat or hydrogen peroxide exposure can induce this trimerization in vitro, arguing for an inherent ability of HSF1 to sense these stresses.[45-47] Several observations, however, suggested the existence of a cellular mechanism(s) that keeps HSF1 in a nonhomooligomeric state in the absence of a stress. When human HSF1 was introduced into *Drosophila* or *Xenopus laevis* cells, the factor was induced to trimerize at the low heat shock temperatures characteristic of the latter cells, demonstrating that it is not an absolute temperature that results in activation of the transcription factor.[26,48] Another observation in favor of a cellular mechanism negatively regulating the transcription factor was that overexpression of HSF1 from transduced genes resulted in accumulation of trimeric HSF1 even in the absence of a stress. This observation suggested the existence of a cellular repressor of trimerization that was titrated in these experiments.[26] When it was found that the transcriptional activity of homotrimeric HSF1 derived from overexpression was negligible compared to that of stress-activated factor, it became obvious that additional layers of regulation exist. It is noted that regulation at multiple levels had been postulated earlier based on observations that compounds such as salicylates, menadione and hydrogen peroxide induced HSE DNA-binding activity but not expression of inducible *hsp* genes.[49,50] Even prior to the latter observations, heat exposure of an erythroleukemia cell line was reported to result in HSE DNA-binding activity but not in increased HSP expression.[51]

The search for factors that repress HSF1 homooligomerization initially focused on HSP70 after heterocomplexes with monomeric HSF1 could be shown to exist in unstressed cells.[52,53] However, several eukaryotic species contain very low levels of endogenous HSP70 and are perfectly capable of suppressing HSF1 activity in the absence of a stress.[54-56] To develop a general model of HSF1 regulation, the possible involvement of other chaperones was examined as well. Several observations that will be reviewed in more detail in the next section pointed to HSP90 and HSP90-containing multichaperone complexes as the key regulators of HSF1 activity.

HSP90-Containing Multichaperone Complexes Regulate HSF1 Oligomeric Status and Transcriptional Competence

Repression of Oligomerization of HSF1 in the Absence of a Stress

Exposure of cells to benzoquinone ansamycins herbimycin A and geldanamycin was found to induce homotrimerization of HSF1 and result in an increased rate of expression of HSPs.[41,57-60] Geldanamycin is known to bind in the ATP-binding pocket of HSP90 with apparent specificity and to alter HSP90 function.[61-63] Together, these findings suggested that HSP90, alone or in concert with other factors, repressed HSF1 oligomerization in the absence of a stress. The observations of Nadeau et al and Nair et al of in vitro interactions between HSF1 and HSP90 were consistent with this hypothesis.[64,65] To obtain more definitive information about the involvement of HSP90 in HSF1 regulation, an in vitro system derived from HeLa cells was developed by Zou et al that was capable of reproducing important aspects of the in vivo regulation of HSF1 oligomeric status: HSF1 oligomerization and HSE DNA-binding activity were induced upon exposure to elevated temperature, or addition of geldanamycin or chemically denatured proteins.[60] Immunodepletion of HSP90 induced specific HSE DNA-binding activity in this system. Back addition of purified HSP90 immediately following HSP90 immunodepletion prevented an increase in HSE DNA-binding activity. Ali et al carried out corresponding experiments in vivo, using *Xenopus* oocytes that were microinjected with antibodies and/or HSP90.[66] Results obtained essentially mirrored those of the Zou et al study. Genetic evidence for a functional role of HSP90 in HSF1 regulation in yeast was reported by Duina et al.[67] In a complementary type of study, Zhao et al demonstrated that overexpression of HSP90 desensitized heat induction of HSE DNA-binding activity in mammalian cells.[68] Other experiments examined the abundance of HSF1-HSP90 complexes in HeLa cells prior to and during a heat stress using in situ cross-linking techniques.[69] Results revealed that the level of HSF1-HSP90 decreased dramatically in the course of a 15-min exposure of the cells to moderately severe heat. Another study demonstrated that administration to cardiac cells of geldanamycin also reduced the concentration of HSF1-HSP90 complexes.[70] Taken together, these results provide strong evidence that HSF1 oligomerization is inhibited by HSP90 that binds to HSF1 polypeptide in the absence of a stress.

Typically, Hsp90 does not bind its client proteins alone but as part of an HSP90-containing multichaperone complex. In the case of steroid receptors, where this has been studied extensively, "mature" multichaperone complexes comprise HSP90, P23 and an immunophilin.[71] Assembly of this type of complexes appears to be the end result of several intermediate assembly steps, involving initial binding by HSP70 and HSP40, introduction of HSP90 through adapter protein HOP and replacement of the resulting HSP90-containing multichaperone complex by above-mentioned mature HSP90-containing complex. Results obtained by Bharadwaj et al and Duina et al suggested that a similar mature HSP90-containing multichaperone complex assembled on HSF1 polypeptide.[67,72] The study by Baradwaj et al showed that immunodepletion not only of HSP90 but also of P23 enhanced HSE DNA-binding activity in *Xenopus* oocytes.[72] Duina et al provided evidence for an involvement of an immunophilin, a CYP40-like cyclophilin, in repression of HSF1 activity in yeast.[67] Finally, Marchler and Wu obtained data supporting the notion that HSF1 polypeptide undergoes a similar series of chaperone complex assembly reactions as steroid receptors.[73] The latter authors reported that siRNA depletion of HSP70, DROJ1 (HSP40) or HSP90 increased the HSE DNA-binding activity of *Drosophila* HSF1. Codepletion of HSP70 and HSP40 or HSP90 and HSP40 produced synergistic effects. In conclusion, the information available suggests that a mature HSP90-containing multichaperone complex similar to that associating with steroid receptors assembles on HSF1 polypeptide and prevents oligomerization of the HSF1 in the absence of a stress.

Repression of Transcriptional Competence of HSF1 in the Absence of a Stress

Transcriptional competence of HSF1, i.e., the ability of the factor to transactivate *hsp* genes, is regulated independent from oligomeric status. A so-called regulatory domain was defined in the HSF1 polypeptide that is required for repression of transcriptional competence in the absence of a stress.[24,26] Deletion of this sequence or parts thereof results in a factor that effectively transactivates *hsp* genes when overexpressed in the absence of a stress. As discussed before, overexpression of wildtype HSF1 titrates the mechanism that represses oligomerization but not that controlling transcriptional competence. Nair et al and Guo et al found that Hsp90-containing multichaperone complexes (HSP90-P23-FKBP52 and others) assembled on trimeric HSF1 in vitro and that components of these complexes also associated with trimeric HSF1 in vivo.[65,69] Interestingly, FKBP52 appeared to only interact with trimeric but not (or differently) with nontrimeric HSF1 in human HeLa cells. Therefore, FKBP52 could be used as a marker for HSP90-containing complexes associating with trimeric HSF1. Using this tool and working with HeLa cells transfected to overexpress an HSF1 form (a LexA-HSF1 chimera), Guo et al were able to examine in parallel HSF1 transcriptional competence and relative concentration of HSP90-containing complexes associated with trimeric HSF1.[69] The authors reported that, in different experimental situations that caused concentration of chaperone substrates to rise, concentration of HSP90-containing complexes found in association with trimeric HSF1 declined and transcriptional competence of HSF1 increased. Furthermore, they observed that mutations in the HSF1 regulatory domain that increased transcriptional competence impaired the assembly of HSP90-containing complexes. These findings strongly suggest the existence of a mechanism that can prevent inadvertently formed HSF1 trimers from becoming active transcription factors in the absence of a stress and involves assembly of HSP90-containing multichaperone complexes such as HSP90-P23-FKBP52 on such HSF1 trimers. The mechanism may also serve to ensure that HSF1 activity increases proportionally with the level of stress a cell is exposed to. This function of the mechanism becomes plausible if one considers that mammalian HSF1 trimers are relatively stable to dissociation, at least in vitro. Consequently, homooligomerization of HSF1 cannot be expected to maintain proportionality with the stress level, except at relatively low stress intensities. The notion that assembly of HSP90-containing multichaperone complexes on HSF1 trimers serves to downmodulate HSF1 activity under all but the most extreme stress conditions is also supported by the observation that, upon exposure of human cells to a moderately severe stress, the HSF1-FKBP52 interaction increased during the exposure and remained prominent even after essentially all HSF1 molecules had been incorporated in homotrimers.[69] It is conceivable that the mechanism also plays a role in inactivation and, possibly, even dissociation of HSF1 trimers that occurs subsequent to a stress.

Another mechanism for repression of HSF1 transcriptional competence was described by Shi et al.[74] These authors observed that HSP70 and HSP40 were capable of binding to the transcriptional activation domain region of HSF1. Overexpression of HSP70 or HSP40 repressed transactivation by a GAL4-HSF1 transcriptional activation domain chimera. These results support the notion that transcriptional competence of HSF1 may be suppressed during recovery from a stress by binding of HSP70 or HSP40, and, possibly, subsequent assembly of a multichaperone complex in the HSF1 transcriptional activation domain region.

Regulation of HSF1 by CHIP as Part of the Protein Quality Control System

HSPs and chaperones in general are directly responsible for the establishment, protection and restoration of protein functionality. However, when a cell is unable to salvage proteins that are irreparably damaged, trapped in a nonnative conformation or unable to participate in the formation of their native complexes, it needs to eliminate these proteins to prevent unregulated protein aggregation. Chaperones are an integral part of this process. Eukaryotic cells target intracellular proteins for degradation by the ^{26}S proteasome complex through covalent attachment of polyubiquitin side chains.[75,76] The interplay and partitioning of proteins between protection/refolding and proteasomal degradation is known as the protein quality control system.[77] At this date, only one protein of higher eukaryotes is known that is able to target chaperone substrates (i.e., nonnative or misfolded proteins) for proteasomal degradation by facilitating their polyubiquitination. This co-chaperone has been named CHIP (C-terminus of HSP70-interacting protein).[78] CHIP is able to target nonnative proteins to the proteasome because it contains not only a tetratricopeptide repeat (TPR) domain but also a U-Box domain that has E3 ubiquitin ligase activity.[79,80] The TPR domain is responsible for binding the conserved C-terminus of several chaperones (HSC70, HSP70 and HSP90), and the E3 ubiquitin ligase activity of CHIP is able to attach polyubiquitin side chains to a chaperone-bound nonnative protein. The N-terminal TPR domain is separated from the C-terminal U-Box domain by a charged domain that is required for homodimerization of the protein.[81] Inactivation of any of the TPR, charged or U-Box domains abolishes the ability of CHIP to ubiquitinate chaperone-bound proteins.

Under normal growth conditions, the cell apparently achieves a dynamic equilibrium between chaperone-dependent protein protection and chaperone-directed proteasomal protein degradation. This equilibrium manifests itself in the stringency of the protein quality control, or in how many of the naturally occurring chaperone substrates (e.g., spontaneously unfolded proteins) are targeted for degradation. In the case of the cystic fibrosis transmembrane receptor, it has been clearly demonstrated that raising the concentration of CHIP increases the stringency of the protein quality control system.[82,83] Another model substrate used to illustrate the ability of CHIP to influence protein quality control was p53.[84] Not only could endogenous levels of p53 be reduced by CHIP overexpression, but p53 protein levels increased after depletion of CHIP by specific siRNA transfection. The protein level of CHIP therefore seems to determine whether a chaperone substrate either achieves its final conformation and seizes to be a chaperone substrate or whether it will be polyubiquitinated and subsequently degraded. How an appropriate CHIP level and, hence, level of protein quality control is established and maintained is not well understood at this time. One aspect of the mechanism that contributes to a balanced protein quality control is the activation of HSF1 and, consequently, the heat shock protein response by increased levels of CHIP.[85] Activation of HSF1 and the ensuing increase in chaperone concentration occurring at elevated levels of CHIP should counteract an undue increase in stringency of protein quality control. Experimental results suggest direct and/or indirect mechanisms for this functional interaction between HSF1 and CHIP. In order to activate HSF1, CHIP needs to be able to bind chaperones via its TPR domain.[85] Hence, CHIP appears to interfere with the assembly of Hsp90-containing chaperone complexes on HSF1 or may force dissassembly of these chaperone complexes, resulting in activation of the transcription factor. Other results favor a direct involvement of CHIP in HSF1 activation, suggesting a direct physical interaction between HSF1 and CHIP in vivo and in vitro.[85,86] Genomic inactivation of endogenous CHIP in mouse cells led to an impairment of the heat shock protein response.[85] This finding was interpreted as evidence for a direct connection between HSF1 and CHIP. However, cells lacking CHIP can be expected to have a chronically elevated level of nonnative proteins that may attenuate the stress-induced heat shock protein response. Direct and indirect explanations for the activation of HSF1 by CHIP are not mutually exclusive; both may contribute to balancing the stringency of the protein quality control.

The only known transcriptional mechanism that can influence the protein level of CHIP is tissue-specific induction of its mRNA expression. Interestingly, elevated levels of CHIP transcription occur in tissues that are known for their elevated oxygen consumption, such as heart, pancreas and skeletal muscle, and are perhaps more prone to oxidative damage.[85,87] In such tissues, in which both CHIP and heat shock protein levels may be elevated when compared to other tissues, protein protection and degradation should remain balanced, but the cells should have acquired a higher capacity for partitioning damaged proteins between chaperones and the proteasome. Recent evidence demonstrates that CHIP in fact plays an important role in maximal cardioprotection after myocardial infarction.[87]

Synopsis

A graphic representation of interactions between HSF1, chaperone complexes (prominently including HSP90-containing complexes), nonnative proteins and CHIP is provided in Figure 2.

HSP90-containing multichaperone complexes (and, possibly, HSP90 alone) appear to repress oligomerization of nonhomotrimeric and transcriptional competence of homotrimeric HSF1. These complexes exert their repressing effects by dynamically interacting with HSF1. When a cell is exposed to a proteotoxic stress, cellular proteins are denatured at an elevated rate and begin to accumulate in the cell. Because such nonnative proteins are substrates of chaperones, total concentration of client proteins of chaperones increases, and the rate of assembly of HSP90-containing chaperone complexes on HSF1 and, hence, the concentration of chaperone-bound HSF1 decrease. Unbound HSF1 homotrimerizes, acquires transcriptional competence and then transactivates *hsp* genes. HSP concentration in the cell begins to rise. During and subsequent to the proteotoxic stress, HSPs and co-chaperones engage in the disposal of stress-unfolded proteins, either by chaperoning their refolding or their proteolytic degradation. The newly synthesized HSPs accelerate these processes (observed as feedback regulation). The proper balance between the cell's capacities for refolding and degrading nonnative proteins is maintained by CHIP that modulates HSF1 activity prior to (background HSF1 activity) and during the stress. Presumably, when the concentration of nonnative proteins in the cell has returned to a level that is close to the prestress level, HSF1 activity is suppressed, trimeric factor is induced to dissociate, and nonhomooligomeric HSF1 reassociates with HSP90-containing chaperone complex that inhibits its reoligomerization. Although little is known about the type and number of reactions involved in returning active HSF1 to its inactive form, we believe that most of them will be found to be chaperone-mediated. We know this to be true for at the least the final reaction that consists of reassembly of HSP90-containing chaperone complex on HSF1 polypeptide.

Although the interactions with chaperones and chaperone complexes discussed herein are important aspects of the control of HSF1 activity, the transcription factor is exposed to several other regulatory influences. It is sumoylated, phosphorylated, and associates with regulatory

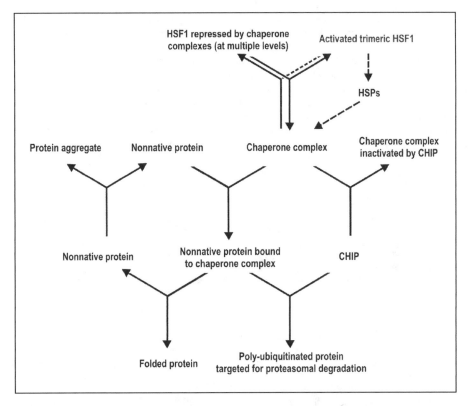

Figure 2. Network of functional interactions relevant to regulation of the heat shock protein response and protein quality control. The illustration shows how chaperone complexes (and/or chaperones) are shared between nonnative proteins and client protein HSF1, and how CHIP affects HSF1 activity through inactivation of chaperone complexes and protein quality through ubiquitination of chaperone complex-bound, nonnative proteins.

factors such as DAXX and 14-3-3epsilon, other transcription factors such as STAT-1 and NF-IL6, chromatin, splicing factors, chromatin-remodeling factors and components of the transcriptional machinery.[88] DAXX is a nuclear protein that in a stressed cell appears to interact with trimeric HSF1 and enhance activation of its transcriptional competence.[89] As discussed before, HSF1 trimers resulting from overexpression from transduced *hsf1* genes possess only marginal transcriptional competence in an unstressed cell. Coexpression of DAXX dramatically enhances the activity of these HSF1 trimers. Considering that transcriptional competence of the latter HSF1 trimers is repressed by dynamically bound HSP90-containing chaperone complexes, the finding suggests the possibility that DAXX may function to assist trimeric HSF1 in shedding its repressive HSP90 complex(es) in a stressed cell.

Functional activation of HSF1 after stress exposure correlates with elevated levels of serine and threonine phosphorylation of the factor.[88] Only a small number of the 12 phosphorylation sites of human HSF1 identified to date (Ser[121], Ser[230], Ser[292], Ser[303], Ser[307], Ser[314], Ser[319], Ser[326], Ser[344], Ser[363], Ser[419] and Ser[444]) appears to play a role in activation of the factor during a stress (Ser[230], Ser[326]) or inactivation subsequent to a stress (Ser[303], Ser[307], Ser[363]).[88,90,91] A generalized connection between chaperone binding or release from binding and phosphorylation of HSF1 was not established. However, Wang et al recently reported results suggesting that phosphorylation of Ser[121] of human HSF1 suppresses activation by stabilizing the interaction between nonhomotrimeric factor and HSP90 complex.[92] Phosphorylation of Ser[121] increases under proinflammatory conditions but not during proteotoxic stress (i.e., heat). It remains to be examined whether stress-induced phosphorylation of HSF1 destabilizes (e.g., in the case of Ser[230] and/or Ser[326]) or stabilizes (e.g., in the case of Ser[303], Ser[307], and/or Ser[363]) interactions between HSF1 and HSP90-containing chaperone complexes. It is noted that such auxiliary roles for HSF1 phosphorylation would be consistent with the finding that phosphorylation of none of the known sites is absolutely required for activation of HSF1 in response to a stress and subsequent inactivation in the wake of the stress.

Much needs to be learned about the interplay between chaperone complexes and other regulatory factors. The small amount of information available today and presented above hopefully will trigger further investigation.

References

1. Wu C. Heat shock transcription factors: Structure and regulation. Annu Rev Cell Dev Biol 1995; 11:441-469.
2. Pirkkala L, Nykanen P, Sistonen L. Roles of the heat shock transcription factors in regulation of the heat shock response and beyond. FASEB J 2001; 15(7):1118-1131.
3. Sarge KD, Murphy SP, Morimoto RI. Activation of heat shock gene transcription by heat shock factor 1 involves oligomerization, acquisition of DNA-binding activity, and nuclear localization and can occur in the absence of stress. Mol Cell Biol 1993; 13(3):1392-1407.
4. Baler R, Dahl G, Voellmy R. Activation of human heat shock genes is accompanied by oligomerization, modification, and rapid translocation of heat shock transcription factor HSF1. Mol Cell Biol 1993; 13(4):2486-2496.
5. McMillan DR, Xiao X, Shao L et al. Targeted disruption of heat shock transcription factor 1 abolishes thermotolerance and protection against heat-inducible apoptosis. J Biol Chem 1998; 273(13):7523-7528.
6. Zhang Y, Huang L, Zhang J et al. Targeted disruption of hsf1 leads to lack of thermotolerance and defines tissue-specific regulation for stress-inducible Hsp molecular chaperones. J Cell Biochem 2002; 86(2):376-393.
7. He H, Soncin F, Grammatikakis N et al. Elevated expression of heat shock factor (HSF) 2A stimulates HSF1-induced transcription during stress. J Biol Chem 2003; 278(37):35465-35475.
8. Tanabe M, Kawazoe Y, Takeda S et al. Disruption of the HSF3 gene results in the severe reduction of heat shock gene expression and loss of thermotolerance. EMBO J 1998; 17(6):1750-1758.
9. Inouye S, Katsuki K, Izu H et al. Activation of heat shock genes is not necessary for protection by heat shock transcription factor 1 against cell death due to a single exposure to high temperatures. Mol Cell Biol 2003; 23(16):5882-5895.
10. Lohmann C, Eggers-Schumacher G, Wunderlich M et al. Two different heat shock transcription factors regulate immediate early expression of stress genes in Arabidopsis. Mol Genet Genomics 2004; 271(1):11-21.
11. Wiederrecht G, Seto D, Parker CS. Isolation of the gene encoding the S. cerevisiae heat shock transcription factor. Cell 1988; 54(6):841-853.
12. Sorger PK, Pelham HR. Yeast heat shock factor is an essential DNA-binding protein that exhibits temperature-dependent phosphorylation. Cell 1988; 54(6):855-864.
13. Walker GA, Lithgow GJ. Lifespan extension in C. elegans by a molecular chaperone dependent upon insulin-like signals. Aging Cell 2003; 2(2):131-139.
14. Gallo GJ, Prentice H, Kingston RE. Heat shock factor is required for growth at normal temperatures in the fission yeast Schizosaccharomyces pombe. Mol Cell Biol 1993; 13(2):749-761.
15. Jedlicka P, Mortin MA, Wu C. Multiple functions of Drosophila heat shock transcription factor in vivo. EMBO J 1997; 16(9):2452-2462.
16. Clos J, Westwood JT, Becker PB et al. Molecular cloning and expression of a hexameric Drosophila heat shock factor subject to negative regulation. Cell 1990; 63(5):1085-1097.
17. Vuister GW, Kim SJ, Orosz A et al. Solution structure of the DNA-binding domain of Drosophila heat shock transcription factor. Nat Struct Biol 1994; 1(9):605-614.
18. Schultheiss J, Kunert O, Gase U et al. Solution structure of the DNA-binding domain of the tomato heat-stress transcription factor HSF24. Eur J Biochem 1996; 236(3):911-921.
19. Damberger FF, Pelton JG, Harrison CJ et al. Solution structure of the DNA-binding domain of the heat shock transcription factor determined by multidimensional heteronuclear magnetic resonance spectroscopy. Protein Sci 1994; 3(10):1806-1821.
20. Amin J, Ananthan J, Voellmy R. Key features of heat shock regulatory elements. Mol Cell Biol 1988; 8(9):3761-3769.

21. Xiao H, Lis JT. Germline transformation used to define key features of heat-shock response elements. Science 1988; 239(4844):1139-1142.
22. Peteranderl R, Rabenstein M, Shin YK et al. Biochemical and biophysical characterization of the trimerization domain from the heat shock transcription factor. Biochemistry 1999; 38(12):3559-3569.
23. Sorger PK, Nelson HC. Trimerization of a yeast transcriptional activator via a coiled-coil motif. Cell 1989; 59(5):807-813.
24. Green M, Schuetz TJ, Sullivan EK et al. A heat shock-responsive domain of human HSF1 that regulates transcription activation domain function. Mol Cell Biol 1995; 15(6):3354-3362.
25. Shi Y, Kroeger PE, Morimoto RI. The carboxyl-terminal transactivation domain of heat shock factor 1 is negatively regulated and stress responsive. Mol Cell Biol 1995; 15(8):4309-4318.
26. Zuo J, Rungger D, Voellmy R. Multiple layers of regulation of human heat shock transcription factor 1. Mol Cell Biol 1995; 15(8):4319-4330.
27. Wisniewski J, Orosz A, Allada R et al. The C-terminal region of Drosophila heat shock factor (HSF) contains a constitutively functional transactivation domain. *Nucleic Acids Res* 1996; 24(2):367-374.
28. Rabindran SK, Giorgi G, Clos J et al. Molecular cloning and expression of a human heat shock factor, HSF1. Proc Natl Acad Sci USA 1991; 88(16):6906-6910.
29. Zimarino V, Wu C. Induction of sequence-specific binding of Drosophila heat shock activator protein without protein synthesis. Nature 1987; 327(6124):727-730.
30. Wu C, Wilson S, Walker B et al. Purification and properties of Drosophila heat shock activator protein. Science 1987; 238(4831):1247-1253.
31. Kline MP, Morimoto RI. Repression of the heat shock factor 1 transcriptional activation domain is modulated by constitutive phosphorylation. Mol Cell Biol 1997; 17(4):2107-2115.
32. Lindquist S. Varying patterns of protein synthesis in Drosophila during heat shock: Implications for regulation. Dev Biol 1980; 77(2):463-479.
33. DiDomenico BJ, Bugaisky GE, Lindquist S. The heat shock response is self-regulated at both the transcriptional and posttranscriptional levels. Cell 1982; 31(3 Pt 2):593-603.
34. Abravaya K, Phillips B, Morimoto RI. Attenuation of the heat shock response in HeLa cells is mediated by the release of bound heat shock transcription factor and is modulated by changes in growth and in heat shock temperatures. Genes Dev 1991; 5(11):2117-2127.
35. Kelley PM, Schlesinger MJ. The effect of amino acid analogues and heat shock on gene expression in chicken embryo fibroblasts. Cell 1978; 15(4):1277-1286.
36. Hightower LE. Cultured animal cells exposed to amino acid analogues or puromycin rapidly synthesize several polypeptides. J Cell Physiol 1980; 102(3):407-427.
37. Ananthan J, Goldberg AL, Voellmy R. Abnormal proteins serve as eukaryotic stress signals and trigger the activation of heat shock genes. Science 1986; 232(4749):522-524.
38. Freeman ML, Borrelli MJ, Syed K et al. Characterization of a signal generated by oxidation of protein thiols that activates the heat shock transcription factor. J Cell Physiol 1995; 164(2):356-366.
39. Liu H, Lightfoot R, Stevens JL. Activation of heat shock factor by alkylating agents is triggered by glutathione depletion and oxidation of protein thiols. J Biol Chem 1996; 271(9):4805-4812.
40. McDuffee AT, Senisterra G, Huntley S et al. Proteins containing nonnative disulfide bonds generated by oxidative stress can act as signals for the induction of the heat shock response. J Cell Physiol 1997; 171(2):143-151.
41. Zou J, Salminen WF, Roberts SM et al. Correlation between glutathione oxidation and trimerization of heat shock factor 1, an early step in stress induction of the Hsp response. Cell Stress Chaperones 1998; 3(2):130-141.
42. Westwood JT, Clos J, Wu C. Stress-induced oligomerization and chromosomal relocalization of heat-shock factor. Nature 1991; 353(6347):822-827.
43. Zuo J, Baler R, Dahl G et al. Activation of the DNA-binding ability of human heat shock transcription factor 1 may involve the transition from an intramolecular to an intermolecular triple-stranded coiled-coil structure. Mol Cell Biol 1994; 14(11):7557-7568.
44. Westwood JT, Wu C. Activation of Drosophila heat shock factor: Conformational change associated with a monomer-to-trimer transition. Mol Cell Biol 1993; 13(6):3481-3486.
45. Goodson ML, Sarge KD. Heat-inducible DNA binding of purified heat shock transcription factor 1. J Biol Chem 1995; 270(6):2447-2450.
46. Larson JS, Schuetz TJ, Kingston RE. In vitro activation of purified human heat shock factor by heat. Biochemistry 1995; 34(6):1902-1911.
47. Zhong M, Orosz A, Wu C. Direct sensing of heat and oxidation by Drosophila heat shock transcription factor. Mol Cell 1998; 2(1):101-108.
48. Clos J, Rabindran S, Wisniewski J et al. Induction temperature of human heat shock factor is reprogrammed in a Drosophila cell environment. Nature 1993; 364(6434):252-255.

49. Jurivich DA, Sistonen L, Kroes RA et al. Effect of sodium salicylate on the human heat shock response. Science 1992; 255(5049):1243-1245.
50. Bruce JL, Price BD, Coleman CN et al. Oxidative injury rapidly activates the heat shock transcription factor but fails to increase levels of heat shock proteins. Cancer Res 1993; 53(1):12-15.
51. Hensold JO, Hunt CR, Calderwood SK et al. DNA binding of heat shock factor to the heat shock element is insufficient for transcriptional activation in murine erythroleukemia cells. Mol Cell Biol 1990; 10(4):1600-1608.
52. Baler R, Zou J, Voellmy R. Evidence for a role of Hsp70 in the regulation of the heat shock response in mammalian cells. Cell Stress Chaperones 1996; 1(1):33-39.
53. Rabindran SK, Wisniewski J, Li L et al. Interaction between heat shock factor and hsp70 is insufficient to suppress induction of DNA-binding activity in vivo. Mol Cell Biol 1994; 14(10):6552-6560.
54. Velazquez JM, Sonoda S, Bugaisky G et al. Is the major Drosophila heat shock protein present in cells that have not been heat shocked? J Cell Biol 1983; 96(1):286-290.
55. Milner CM, Campbell RD. Structure and expression of the three MHC-linked HSP70 genes. Immunogenetics 1990; 32(4):242-251.
56. Jaattela M, Wissing D, Kokholm K et al. Hsp70 exerts its anti-apoptotic function downstream of caspase-3-like proteases. EMBO J 1998; 17(21):6124-6134.
57. Murakami Y, Uehara Y, Yamamoto C et al. Induction of hsp 72/73 by herbimycin A, an inhibitor of transformation by tyrosine kinase oncogenes. Exp Cell Res 1991; 195(2):338-344.
58. Hegde RS, Zuo J, Voellmy R et al. Short circuiting stress protein expression via a tyrosine kinase inhibitor, herbimycin A. J Cell Physiol 1995; 165(1):186-200.
59. Conde AG, Lau SS, Dillmann WH et al. Induction of heat shock proteins by tyrosine kinase inhibitors in rat cardiomyocytes and myogenic cells confers protection against simulated ischemia. J Mol Cell Cardiol 1997; 29(7):1927-1938.
60. Zou J, Guo Y, Guettouche T et al. Repression of heat shock transcription factor HSF1 activation by HSP90 (HSP90 complex) that forms a stress-sensitive complex with HSF1. Cell 1998; 94(4):471-480.
61. Whitesell L, Mimnaugh EG, De Costa B et al. Inhibition of heat shock protein HSP90-pp60v-src heteroprotein complex formation by benzoquinone ansamycins: Essential role for stress proteins in oncogenic transformation. Proc Natl Acad Sci USA 1994; 91(18):8324-8328.
62. Grenert JP, Sullivan WP, Fadden P et al. The amino-terminal domain of heat shock protein 90 (hsp90) that binds geldanamycin is an ATP/ADP switch domain that regulates hsp90 conformation. J Biol Chem 1997; 272(38):23843-23850.
63. Prodromou C, Roe SM, O'Brien R et al. Identification and structural characterization of the ATP/ADP-binding site in the Hsp90 molecular chaperone. Cell 1997; 90(1):65-75.
64. Nadeau K, Das A, Walsh CT. Hsp90 chaperonins possess ATPase activity and bind heat shock transcription factors and peptidyl prolyl isomerases. J Biol Chem 1993; 268(2):1479-1487.
65. Nair SC, Toran EJ, Rimerman RA et al. A pathway of multi-chaperone interactions common to diverse regulatory proteins: Estrogen receptor, Fes tyrosine kinase, heat shock transcription factor Hsf1, and the aryl hydrocarbon receptor. Cell Stress Chaperones 1996; 1(4):237-250.
66. Ali A, Bharadwaj S, O'Carroll R et al. HSP90 interacts with and regulates the activity of heat shock factor 1 in Xenopus oocytes. Mol Cell Biol 1998; 18(9):4949-4960.
67. Duina AA, Kalton HM, Gaber RF. Requirement for Hsp90 and a CyP-40-type cyclophilin in negative regulation of the heat shock response. J Biol Chem 1998; 273(30):18974-18978.
68. Zhao C, Hashiguchi A, Kondoh K et al. Exogenous expression of heat shock protein 90kDa retards the cell cycle and impairs the heat shock response. Exp Cell Res 2002; 275(2):200-214.
69. Guo Y, Guettouche T, Fenna M et al. Evidence for a mechanism of repression of heat shock factor 1 transcriptional activity by a multichaperone complex. J Biol Chem 2001; 276(49):45791-45799.
70. Knowlton AA, Sun L. Heat-shock factor-1, steroid hormones, and regulation of heat-shock protein expression in the heart. Am J Physiol Heart Circ Physiol 2001; 280(1):H455-H464.
71. Pratt WB, Toft DO. Steroid receptor interactions with heat shock protein and immunophilin chaperones. Endocr Rev 1997; 18(3):306-360.
72. Bharadwaj S, Ali A, Ovsenek N. Multiple components of the HSP90 chaperone complex function in regulation of heat shock factor 1 In vivo. Mol Cell Biol 1999; 19(12):8033-8041.
73. Marchler G, Wu C. Modulation of Drosophila heat shock transcription factor activity by the molecular chaperone DROJ1. EMBO J 2001; 20(3):499-509.
74. Shi Y, Mosser DD, Morimoto RI. Molecular chaperones as HSF1-specific transcriptional repressors. Genes Dev 1998; 12(5):654-666.
75. Goldberg AL. Protein degradation and protection against misfolded or damaged proteins. Nature 2003; 426(6968):895-899.

76. Ciechanover A, Iwai K. The ubiquitin system: From basic mechanisms to the patient bed. IUBMB Life 2004; 56(4):193-201.
77. Wickner S, Maurizi MR, Gottesman S. Posttranslational quality control: Folding, refolding, and degrading proteins. Science 1999; 286(5446):1888-1893.
78. Ballinger CA, Connell P, Wu Y et al. Identification of CHIP, a novel tetratricopeptide repeat-containing protein that interacts with heat shock proteins and negatively regulates chaperone functions. Mol Cell Biol 1999; 19(6):4535-4545.
79. Jiang J, Ballinger CA, Wu Y et al. CHIP is a U-box-dependent E3 ubiquitin ligase: Identification of Hsc70 as a Target for Ubiquitylation. J Biol Chem 2001; 276(46):42938-42944.
80. Murata S, Minami Y, Minami M et al. CHIP is a chaperone-dependent E3 ligase that ubiquitylates unfolded protein. EMBO Rep 2001; 2(12):1133-1138.
81. Nikolay R, Wiederkehr T, Rist W et al. Dimerization of the human E3 ligase CHIP via a coiled-coil domain is essential for its activity. J Biol Chem 2004; 279(4):2673-2678.
82. Meacham GC, Patterson C, Zhang W et al. The Hsc70 co-chaperone CHIP targets immature CFTR for proteasomal degradation. Nat Cell Biol 2001; 3(1):100-105.
83. Cyr DM, Hohfeld J, Patterson C. Protein quality control: U-box-containing E3 ubiquitin ligases join the fold. Trends Biochem Sci 2002; 27(7):368-375.
84. Esser C, Scheffner M, Hohfeld J. The chaperone-associated ubiquitin ligase CHIP is able to target p53 for proteasomal degradation. J Biol Chem 2005; 280(29):27443-27448.
85. Dai Q, Zhang C, Wu Y et al. CHIP activates HSF1 and confers protection against apoptosis and cellular stress. EMBO J 2003; 22(20):5446-5458.
86. Kim SA, Yoon JH, Kim DK et al. CHIP interacts with heat shock factor 1 during heat stress. FEBS Lett 2005; 579(29):6559-6563.
87. Zhang C, Xu Z, He XR et al. CHIP, a co-chaperone/ubiquitin ligase that regulates protein quality control, is required for maximal cardioprotection after myocardial infarction in mice. Am J Physiol Heart Circ Physiol 2005; 288(6):H2836-H2842.
88. Voellmy R. Transcriptional regulation of the metazoan stress protein response. Prog Nucleic Acid Res Mol Biol 2004; 78:143-185.
89. Boellmann F, Guettouche T, Guo Y et al. DAXX interacts with heat shock factor 1 during stress activation and enhances its transcriptional activity. Proc Natl Acad Sci USA 2004; 101(12):4100-4105.
90. Holmberg CI, Hietakangas V, Mikhailov A et al. Phosphorylation of serine 230 promotes inducible transcriptional activity of heat shock factor 1. EMBO J 2001; 20(14):3800-3810.
91. Guettouche T, Boellmann F, Lane WS et al. Analysis of phosphorylation of human heat shock factor 1 in cells experiencing a stress. BMC Biochem 2005; 6(1):4.
92. Wang X, Khaleque MA, Zhao MJ et al. Phosphorylation of HSF1 by MAPK-activated protein kinase 2 on Serine 121, inhibits transcriptional activity and promotes HSP90 binding. J Biol Chem 2006; 281(2):782-791.

Mechanisms of Activation and Regulation of the Heat Shock-Sensitive Signaling Pathways

Sébastien Ian Nadeau and Jacques Landry*

Abstract

Heat shock (HS), like many other stresses, induces specific and highly regulated signaling cascades that promote cellular homeostasis. The three major mitogen-activated protein kinases (MAPK) and protein kinase B (PKB/Akt) are the most notable of these HS-stimulated pathways. Their activation occurs rapidly and sooner than the transcriptional upregulation of heat shock proteins (Hsp), which generate a transient state of extreme resistance against subsequent thermal stress. The direct connection of these signaling pathways to cellular death or survival mechanisms suggests that they contribute importantly to the HS response. Some of them may counteract early noxious effects of heat, while others may bolster key apoptosis events. The triggering events responsible for activating these pathways are unclear. Protein denaturation, specific and nonspecific receptor activation, membrane alteration and chromatin structure perturbation are potential initiating factors.

Introduction

The cellular ability to resist and rapidly adapt to hazardous environmental conditions was one of the main prerequisites for the development of living organisms during evolution. Sudden changes in temperature are among the earliest and most ubiquitous insults that cells had to cope with to preserve their structural and enzymatic integrity. Accordingly, the HS response is one of the most ancient and conserved cellular stress responses. This response is characterized by the transcriptional activation and accumulation of a set of proteins known as heat shock proteins (Hsp), which provide a state of extreme thermal resistance to the cells.[1-5] The common view on the protective function of Hsp relies on their chaperone activity, i.e., their capacity to bind unfolded or damaged proteins and prevent their aggregation as well as enhancing refolding or degradation.[6,7]

The accumulation of heat-denatured proteins appears to be the main trigger of the Hsp transcriptional response.[8] At normal temperatures, most Hsp are expressed at a low basal level, keeping the HS transcriptional factor (HSF) in a repressed conformation. Relief of repression occurs via the titration of the Hsp by the stress-induced denaturation of native proteins. This leads to the activation of HSF by homotrimerization and to the accumulation of the Hsp

*Corresponding Author: Jacques Landry—Centre de recherche en cancérologie de l'Université Laval, L'Hôtel-Dieu de Québec, 9, rue McMahon, Québec, Canada G1R 2J6.
Email: jacques.landry@med.ulaval.ca

Molecular Aspects of the Stress Response: Chaperones, Membranes and Networks, edited by Peter Csermely and László Vígh. ©2007 Landes Bioscience and Springer Science+Business Media.

which, in mammals, occurs over a few hours and lasts for 2-3 days. The cellular increase in Hsp enhances cell resistance against subsequent protein-denaturing stresses and turns off HSF in a negative feedback loop.[9-13]

The attention directed towards the transcriptional activation and function of Hsp has overshadowed an aspect of the HS response that might be fundamental for cell adaptation. Prior to the accumulation of Hsp, HS rapidly activates many distinct signaling pathways. The exact mechanisms of activation of these pathways and their cellular functions during thermal stress are still ill-defined. However, studies on other cellular stress responses have indicated that these transducing pathways regulate many aspects of survival or death, and therefore are primordial determinants of cell fates.

Major Signaling Pathways Activate Heat Shock

HS activates in minutes an array of signal-transducing kinases, which comprise the three mitogen-activated protein kinase (MAPK) pathways, namely the extracellular signal-regulated kinase (ERK), c-jun N-terminal kinase (JNK) and p38.[14-18] MAPK cascades are organized hierarchically into three-layered phosphorylation modules, where each MAPK is the specific target of a restricted number of MAPK kinases (MAP2K), which in turn are phosphorylated by stimulus-specific MAP2K kinases (MAP3K).[19-21] Protein kinase B (PKB/Akt) is another important kinase rapidly stimulated by heat stress.[22] The early induction of the pathways leading to MAPK and PKB induction can either serve to trigger adaptive responses or be used to signal cell death. As discussed below, many studies have shown that most of these signaling cascades can have both pro-survival and pro-death functions. This plasticity in mediating diverses responses depends on many factors, such as the nature and intensity of the stimulus applied and the cellular context (i.e., cell-cycle, cell type and extracellular conditions). Such adverse consequences highlight the need for a tight control over the activation of these pathways in order to elicit adequate responses. Current experimental data effectively indicate that HS-induced pathways are highly regulated and rely upon very specific and distinct mechanisms of induction.

Activation of the ERK Pathway (Fig. 1)

The ERK kinases ERK1 and ERK2 (also known as p44/p42) are preferentially activated by growth-promoting stimuli such as growth factors and oncogenes.[23] However, many stressful agents including HS also activate them.[16,24-26] It is still not totally clear how HS induces ERK. It was initially found that ERK activation by HS requires the phosphorylation and activation of the epidermal growth factor receptor (EGFR), probably in conjunction with another tyrosine kinase such as c-Src, which is also induced by HS.[27] Activation of EGFR by HS likely induces the downstream canonical ERK pathways involving the MAP3K Raf-1 and the MAP2K MEK1/2, since the expression of a dominant negative mutant of Raf-1 inhibited HS-induced ERK activity.[27] Interestingly, the activation of EGFR by HS is ligand-independent. This mode of induction resembles the activation of the ERK pathway by ultraviolet (UV) light, hyperosmolarity and hydrogen peroxide (H_2O_2), which all hijack growth factor receptors from their normal physiological function to initiate stress signals.[28-31] In contrast, other reports claimed that ERK activation by HS is independent of both EGFR and Raf-1, but still depends on the action of the MAP2K MEK1/2.[32,33] However, the use of suramin, an extracellular antagonist of several membrane receptors, reduced significantly this activation, therefore confirming that some cell surface receptors are indeed involved in the activation of ERK by HS.[27,33] A possible explanation for these different results was provided from another study that showed that the mechanisms of ERK activation are different depending on the strength of the HS.[34] ERK activation appears to result not only from the stimulation of the MAP2K MEK1/2 and upstream pathways, but also from the inhibition of ERK dephosphorylation. Under mild HS conditions, ERK is strongly activated by MEK1/2. As the heat treatment becomes more severe, the activation of MEK1/2 diminishes gradually, whereas the ERK dual specificity phosphatases

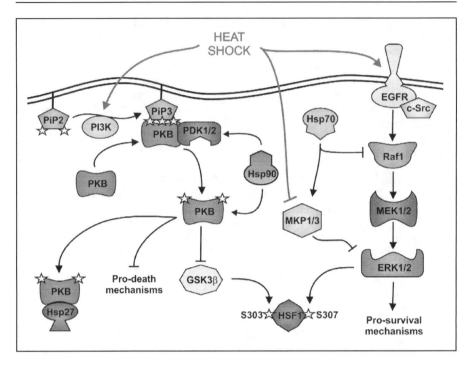

Figure 1. Major players in the HS-activation of the PKB and ERK pathways. See text for details.

MKP3 and MKP1 are rapidly insolubilized and inactivated by heat-aggregation. Thus, during a prolonged or acute HS, the inhibition of phosphatases and the consequent lack of dephosphorylation may be more important than the action of an upstream activator to explain the sustained induction of ERK. Interestingly, Hsp70 overexpression prevents activation of ERK by two distinct mechanisms. It can stabilize the ERK phosphatases through a chaperone dependent-mechanism and block MEK1/2 activation in a chaperone-independent mechanism.[34] The inhibitory effect of Hsp70 on MEK1/2 may be through the regulation of Raf-1 activity by the Hsp70 cofactor Bag-1.[35]

Although a sustained activation of ERK in a variety of stress conditions can lead to cell death, ERK is generally considered as a pro-survival kinase.[36-39] Accordingly, preventing the activation of ERK during HS by the MEK inhibitor PD98059 or overexpression of a dominant-negative construct is sufficient to greatly reduce cell survival.[33,40] The exact mechanisms by which ERK promotes long term cell survival following hyperthermia are not understood. One consequence of ERK activation under both growth and HS conditions is the direct phosphorylation of HSF1 on serine 307.[41] Phosphorylation on some specific residues modulates the transcriptional activity of HSF1.[42] Intriguingly, the HSF1 phosphorylation by ERK rather promotes its repression.[41,43] The phosphorylation of HSF1 on serine 307 is a necessary step for its binding to the scaffold protein 14-3-3ε.[44,45] During serum stimulation, this association negatively regulates HSF1 activity by favoring its cytoplasmic sequestration, thereby insuring a low basal activity under nonstressful conditions.[44] This physical interaction also takes place during HS and might be implicated in the progressive shut-down of HSF1 during the HS recovery period.[45] However, it is not clear how HSF1 phosphorylation/repression can contribute to the protective activity of ERK during HS.

Activation of the JNK Pathway (Fig. 2)

The HS activation of the JNK pathway shows major similarities with the activation of ERK. Although its induction does not initially proceed from a receptor-born signal, it also arises from the heat-inactivation of a JNK phosphatase.[46,47] Indeed, JNK activation is mainly caused by the heat-induced insolubilization and inactivation of the dual specificity JNK phosphatase M3/6 in a manner akin to MKP1 and MKP3 in the ERK pathway.[47,48] The basal activities of the MAP2K MKK4 is nonetheless essential to initiate this pathway since the absence of an activable form of this kinase abrogates JNK activation.[46,49-52] Hence, without any proper phosphatase activity to counteract the accumulation of active JNK, only a slight increase in the activities of upstream signaling elements is sufficient to augment drastically the amount of hyperphosphorylated JNK. This model agrees with the observation that MKK4 is indeed minimally activated by HS, as compared to other stresses such as UV light, but still causes a strong induction of JNK.[46,53] However, little is known about the mechanisms of activation of MKK4 during HS. The small GTPase Rac1 might play a role since the presence of a dominant negative form of this protein reduces partly the heat activation of JNK.[54]

In contrast to the ERK pathway, JNK is generally considered as an apoptosis-promoting pathway. JNK can mediate cell death signals initiated by a variety of stress factors including UV light, TNFα and HS.[55-57] Under these stresses, JNK activation can cause the efflux of cytochrome c from mitochondria, either via the phosphorylation/inactivation of the anti-apoptotic proteins Bcl-2 and Bcl-x or the cleavage/activation of the pro-apoptotic protein Bid.[55,56,58,59] The release of cytochrome c from mitochondria leads to the execution of the apoptotic cascade triggered by activated caspases. Stabilization of the tumor suppressor p53 is another mechanism by which JNK can promote apoptosis. In unstressed cells, JNK can bind p53 and targets it to ubiquitylation and proteasomal degradation.[60] Upon stress-activation, JNK phosphorylates p53, thus reducing its degradation.[61] The subsequent increased stability and accumulation of p53 may contribute to HS-induced cell-cycle arrest and/or apoptosis. Apart from its apoptosis promoting activities, JNK can also causes the phosphorylation of HSF1. Two opposing effects have been reported. On one hand, the phosphorylation of serine 363 of HSF1 by JNK suppresses its transcriptional activity, in a manner similar to ERK phosphorylation.[62] Under more severe HS conditions, JNK phosphorylation at an unknown site seems to enhance HSF1 activity and stabilization.[63]

In some cells, blocking JNK activation during HS is sufficient for protection and accordingly, the state of thermotolerance that develops after a priming HS closely correlates with the downregulation of JNK.[64,65] As in the case of ERK, Hsp70 negatively regulates JNK activation. This inhibitory action arises either through the direct binding of Hsp70 to JNK, which prevents its activation, or by the Hsp70-mediated protection of the heat-denatured JNK phosphatase.[47,66,67] The mechanism involved in this protection is unclear since the ATPase activity (hence the refolding activity) of Hsp70 is not required for blocking the JNK pathway.[68] This might explain the surprising finding that in some cells, a chaperone-defective mutant of Hsp70 can still give protection against HS.[69,70] It is intriguing that the regulation of the ERK and JNK pathways following heat stress are quite similar, whereas their respective function in the fate of heat-shocked cells is antagonistic.

Activation of the p38 Pathway (Fig. 2)

HS activates the p38 pathway by a mechanism that is both distinct from ERK and JNK and very specific to HS. In contrast to ERK, p38 activation is not antagonized by a dominant negative mutant of EGFR nor a specific EGFR inhibitor, tyrphostin AG1478.[27] Moreover, suramin does not block its activation whereas it completely prevents the activation of ERK by HS and the activation of JNK by UV light and hyperosmotic shock.[27,28,30,53] Unlike ERK and JNK, HS-induced p38 is not due to the inactivation of a phosphatase.[46] The p38 activation by

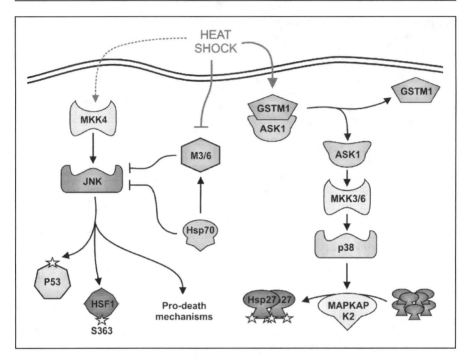

Figure 2. Major players in the HS-activation of the p38 and JNK pathways. See text for details.

HS rather needs the successive induction of a specific set of kinases, namely, the MAP3K apoptosis signal-regulating kinase-1 (ASK1) followed by the MAP2K MKK3/6.[53,71] ASK1 is the upstream MAP3K also responsible for relaying the oxidative stress signal to JNK and p38 via the dissociation from Ask1 of the redox sensible inhibitor thioredoxin.[72,73] HS-activation of ASK1 instead involves the dissociation of another repressor, the glutathione S-transferase-μ1 (GSTM1).[71] Accordingly, the overexpression of GSTM1 inhibits the activation of p38 by HS.[71] However, the specific factors that lead to the dissociation of GSTM1 from ASK1 upon HS are still unknown. One hypothesis is that GSTM1, which is known to bind hydrophobic molecules, could be titrated by the production of sphingosine and ceramide during HS, thus releasing the basal repression on ASK1.[74-77]

The importance of p38 activation during HS is underscored by the fact that one downstream target of the pathway is a major thermoprotective Hsp, Hsp27/HspB1.[78,79] Hsp27 is phosphorylated within minutes of HS by MAPKAP kinase 2 (MAPKAPK-2), a direct target of p38.[18,80] Hsp27 phosphorylation is transient, last typically 2 to 3 h and vanishes before the protein starts to increase its concentration as a result of the transcriptional activation of the HS genes. Phosphorylation induces major structural changes in the supramolecular organization of Hsp27, shifting it from large oligomers towards dimers, and appears to have a thermoprotective role since nonphosphorylatable mutants of Hsp27 are much less active in thermoprotection.[81,82] Whereas induction of p38 may be very useful in the first minutes of HS to rapidly promote, through Hsp27, intrinsic protective activity, its inappropriate activation can also be very harmful. In fact, numerous studies have characterized pro-apoptotic functions for p38 that are very similar to those of JNK.[83-87] Paradoxically, even Hsp27 phosphorylation can become harmful in some circumstances, particularly when Hsp27 is present at high concentration or when the ERK pathway is not activated at the same time.[88] Hsp27 is an actin polymerization factor and its phosphorylation enhances the stability and even promotes the polymerization of actin filaments under stressful conditions, such as HS and oxidative stress.[81,89] Overstimulated or badly timed phosphorylation of Hsp27 can lead to extensive cell blebbing and apoptosis by mediating inappropriate actin polymerization.[83,88] For this reason, activation of p38 is tightly regulated. Like JNK, p38 is not reactivated during the induced-state of thermotolerance and its capacity to induce Hsp27 phosphorylation inversely correlates with cell thermoresistance. Indeed, a first HS treatment totally desensitizes cell to reinduction of Hsp27 phosphorylation by any subsequent HS in the time frame corresponding to the thermotolerant state.[79] During this period, all the abovementioned elements of the p38 pathway become completely insensitive to another heat treatment.[71,90] In contrast to ERK and JNK, p38 desensitization is not a Hsp70-related effect.[90] Observations inferred from classical homologous and heterologous desensitization experiments strongly suggest that the mechanism of activation of p38 by HS is highly specific. Indeed, the HS-desensitized p38 pathway remains fully activable by other stresses such as hyperosmolarity, arsenite and H_2O_2, and by many agonists such as EGF, platelet-derived growth factor (PDGF), tumor necrosis factor (TNF), fresh serum, and tetradecanoylphorbol-13 acetate (TPA; unpublished results).[90] The mechanisms responsible for this desensitization are unknown. Indirect evidence suggested the existence of an upstream element of the p38 pathway, which is rapidly inactivated by heat stress.[90]

Activation of the PKB Pathway (Fig. 1)

Concomitantly to MAPK stimulation, thermal stress also induces the well-known pro-survival pathway composed of phosphatidylinositol 3-kinase (PI3K) and PKB. Probably by the action of a tyrosine kinase, HS causes the rapid activation of PI3K that generates the formation of the lipid-bound phosphatidylinositol 3,4-diphosphate (PIP2) and phosphatidylinositol 3,4,5-triphosphate (PIP3).[27] The accumulation of PIP3 favors the membrane translocation of PKB that is subsequently activated by phosphoinositide-dependent kinase 1 and 2 (PDK1/2).[91] The activation of PKB during HS needs the prior activation of PI3K since the pretreatment with any of the PI3K inhibitors, namely wortmannin, LY294002 and caffeine (unpublished results) completely block this induction.[92,93] As for the ERK pathway, PKB activation mediates a primordial survival response in heat-stressed cells. Thus, the inhibition of PKB augments drastically the apoptosis of heat-shocked cells and increases significantly their susceptibility to cell death even under mild HS conditions.[64,94,95] Unpublished data from our lab suggest that, in contrast to the MAPK, the activation of PKB is not desensitized by a prior HS treatment. This suggests that activation of PKB is important for the survival of heat-shocked cells even in the course of thermotolerance when Hsp are fully expressed. However, how this survival function is mediated during HS has not been investigated.

Since PKB is known to inhibit many proteins closely implicated in apoptosis, its rapid activation during HS likely dampers many coinduced pro-death mechanisms. For example, PKB activation can restrict the induction of MKK4/JNK pathway, inactivates the pro-apoptotic protein Bad, influence the stability of p53 and affects the transcriptional response to apoptotic stimuli by acting on Forkhead factors.[96-100] Inhibition of glycogen synthase kinase-3 β (GSK3β) by PKB phosphorylation is another event that could play an important role in the regulation of the HS response. Indeed, GSK3β overexpression impairs HSF1 activation by HS and results in reduction of Hsp70 production.[93] This effect seems to be mediated by the GSK3β directed-phosphorylation of HSF1 on serine 303.[41] This phosphorylation, in conjunction to the ERK-mediated phosphorylation of serine 307, is another essential step for the binding of 14-3-3ε to HSF1.[41,44] Thus, by inhibiting GSK3β, PKB might indirectly modulates HSF1 stimulation.

Like the MAPK, the Hsp also seem to modulate the activity of PKB. Hsp90 has a major impact on PKB regulation since both PDK1 and PKB are clients of Hsp90 and are rapidly degraded in the presence of Hsp90 inhibitors.[101,102] PKB also interact with Hsp27, however, the significance of this interaction is not clear, neither for the protective activity of Hsp27 nor for that of PKB.[103] It was shown that under oxidative stress, PKB can also negatively regulate

ASK1, thereby attenuating the activation of the p38 pathway.[104] These data suggest that PKB might play a role in the activation/desensitization of the p38 pathway by HS. However, the HS activation of the p38 pathway is not impaired by the inhibition of PKB, nor does the chemical inhibition of p38 by SB203580 interferes with the activation of PKB by HS.[92,93] Thus, despite the possible links between the two pathways, current evidence indicates that there is no direct cross-talk involved during heat stress.

Molecular Origin of the Heat Shock Signal

What is (are) the initial event(s) that trigger(s) the activation of these signaling pathways? As a global physical stress, HS profoundly alters all cellular structures and metabolic processes. Heat rapidly induces structural changes in cytoskeletal integrity: intermediate filaments are destabilized and form aggregates around the nucleus, the actin network is reorganized and microtubules are disintegrated.[105-107] In some cell types, the Golgi apparatus is fragmented and mitochondria become swollen.[107] The respiratory chain reaction of mitochondria is greatly reduced by the decoupling of oxidative phosphorylation with the electron chain transport.[108] Perturbations at the initiation and elongation steps impaired strongly DNA replication, whereas splicing of RNA is inhibited and protein synthesis is strongly reduced and redirected toward the preferential translation of Hsp messenger RNA.[109,110] In theory, activation of kinase cascades may occur as a feedback reaction to the accumulation of any of these damages. Another possibility, which intuitively would appear better for cell homeostasis, is that the signal might be generated before any damage occurs,. For this, one needs to postulate the existence of an efficient HS-sensor.

HS is a potent proteotoxic stress and as such, a compelling hypothesis would be that denatured proteins provide the initial signal for the activation of the MAPK and PKB pathways. As mentioned before, accumulation of denatured proteins, by titrating the basally expressed Hsp, are responsible in a large part for the activation of HSF. The role of protein denaturation is, however, much less clear in the case of kinase activation. The finding that Hsp70 overexpression can block the HS-activation of these pathways could support this view.[34,111] However, the results are difficult to interpret since chaperone dead mutants of Hsp70 work as well as wild type.[68] It was also demonstrated that p38 and JNK are induced upon the accumulation and aggregation of polyglutamine-containing proteins such as huntingtin and androgen receptor, which are responsible for the Huntington and Kennedy's diseases, respectively.[47,112] However, in contrast to HS where denatured proteins accumulate in minutes, the actual aggregation of polyglutamine is a slow process that takes days and thus it is not possible to determine whether the activation process is direct. Finally, we have found that some pathways such as the PKB pathway are activated in thermotolerant cells (HS-primed cells that have accumulated the full complement of Hsp) as well as in control cells, suggesting that the bulk denaturation of proteins, which is expected to occur at a much reduced rate in thermotolerant cells, is not the early signaling event for activation of these pathways. More subtle changes in the conformation (activation) of specific molecules are more likely to be at the origin of the signal for at least some of these pathways.

Many pieces of indirect evidence suggest a model in which the initial HS signal would originate from the cellular membrane and involve receptors or receptor-like molecules.[113] This concept is not unique to thermal stress since activation of the ERK and JNK pathways by UV light and osmotic shock also involves agonist-independent receptor activation.[28-30] It has been shown repeatedly that HS causes an overall gradual increase in the fluidity of the cell membrane as the temperature increases.[114,115] Interestedly, the hyperfluidization of cell membranes induces an Hsp response very similar to that induced by HS.[116-118] (see also the chapter by Laszlo Vigh in this volume for more details). HS-induced increase in membrane fluidity might also accelerate the clustering of receptors into lipid rafts, thereby changing the activation threshold of certain receptors leading to agonist-independent activation. The ERK pathway is already known to be heat-activated by the agonist-independent activation of a receptor, probably EGFR.[27] PKB is commonly activated via PI3K downstream of membrane receptors and may be similarly activated by HS.[119] Furthermore, the HS desensitization of the p38 pathway is reminiscent of the growth factor receptor desensitization that follows their stimulation. Interestingly, clathrin, one of the major components of the machinery involved in the internalization and desensitization of these receptors, is tyrosine-phosphorylated in response to HS (Dorion, S and Landry J, unpublished results) and H_2O_2. Under oxidative stress, this phosphorylation seems to affect positively the endocytosis of certain molecules, such as transferrin.[120] Moreover, the phosphorylation of clathrin by the tyrosine kinase Src has already been shown to favor the uptake of the agonist-stimulated EGFR.[121] The possibility that heat might activate some specific membrane receptor is further accredited by the discovery of a receptor for noxious heat stimuli in nociceptive neurons.[122] Curiously, this receptor was first described as a cation channel activated by capsaicin, the irritating ingredient of chili peppers. In the case of HS, a modification in the structure or the gating potential of the receptor is responsible for activation.[123,124] Whether such HS receptor exists in all cell types is not clear. However, capsaicin does activate p38 in many cells types (Dorion, S and Landry J, unpublished results).

In addition to facilitating receptor activation, HS-induced changes in the fluidity of the bilayer lipid membrane might alter membrane tension and stretching. Membrane stretching leads to the activation of the MAPK pathways in a variety of cell types.[125,126] Furthermore, in yeast, the cell wall mechanosensor Wsc1 is required for HS-activation of a specific stress kinase pathway composed of PKC1 and Mpk1 and is a critical determinant for the induction of thermotolerance.[127,128] Yeast strains defective in Wsc1 or any of the proteins of the downstream HS-activated pathway are characterized by increased cell lysis at high temperatures, a phenotype very similar to that observed in HSF1 mutant strains.[128-130] The plasma membrane may still contribute in another way to the early HS-induced cell responses. HS induces the production of certain lipid signaling molecules derived from sphingolipids, such as ceramide.[75-77] These lipids play pivotal roles in the stimulation of signaling pathways under a variety of stress conditions.[131] For example, the formation of ceramide following exposure to environmental stresses can lead to the activation of the JNK pathway and promotes apoptosis.[132] An earlier study described that ceramide could induce the enhancement of gene transcription for αB-crystallin, one of the small Hsp.[75] Interestedly, another product of ceramide metabolism, sphingosine 1-phosphate, might contribute to the induction of the p38 pathway.[133]

A final possibility is that the chromatin acts as a thermosensor and is at the origin of the early HS signal. This hypothesis is suggested from the observation that ataxia-telangiectasia mutated (ATM) is activated during HS and that two probable targets of this kinase, the histone H2AX and p53 are rapidly phosphorylated in the course of HS (our unpublished observation).[134-137] ATM is a member of the phosphoinositide-3-kinase-related protein kinase (PIKK) family considered as major players in the initiation of the genotoxic stress response and phosphorylation of H2AX is generally considered as a marker of DNA damage. It is unlikely that HS induces significant DNA damage.[138] It appears more likely that, as shown for hyperosmotic stress, induction of H2AX and ATM occur as a response to HS-induced perturbations in the structure of chromatin.[139] A large molecular organization such as the chromatin is in all likeliness a very sensitive sensor for small changes in temperature. This is particularly interesting considering that a molecular connection has already been suggested and partly characterized between PIKK and MAPK and PKB activation.[38,140-142]

Conclusion

The rapidity of activation, specificity and tight regulation of the signaling pathways activated by HS suggest an important role in the early cell response. One can envision that these pathways not only connect to cell death and survival pathways but also to numerous cellular processes allowing cells to rapidly react and thereby keeping an optimal functional state under mild variation in temperature. An important aspect in future work will be to integrate all the information obtained on individual pathways, in individual cell lines and at different temperatures into a global picture of the HS response.

Acknowledgements

The research in the authors' laboratory was supported by the Canadian Institutes of Health Research and the Canada Research Chair in Stress Signal Transduction. S.I. Nadeau was recipient of a studentship from the Fonds de recherche en santé du Québec.

References

1. Gerner EW, Schneider MJ. Induced thermal resistance in HeLa cells. Nature 1975; 256:500-502.
2. Landry J, Bernier D, Chretien P et al. Synthesis and degradation of heat shock proteins during development and decay of thermotolerance. Cancer Res 1982; 42:2457-2461.
3. Li GC, Werb Z. Correlation between synthesis of heat shock proteins and development of thermotolerance in Chinese hamster fibroblasts. Proc Natl Acad Sci USA 1982; 79:3218-3222.
4. Subjeck JR, Sciandra JJ, Chao CF et al. Heat shock proteins and biological response to hyperthermia. Br J Cancer Suppl 1982; 45:127-131.
5. Lindquist S. The heat-shock response. Annu Rev Biochem 1986; 55:1151-1191.
6. Hohfeld J, Cyr DM, Patterson C. From the cradle to the grave: Molecular chaperones that may choose between folding and degradation. EMBO Rep 2001; 2:885-890.
7. Friant S, Meier KD, Riezman H. Increased ubiquitin-dependent degradation can replace the essential requirement for heat shock protein induction. EMBO J 2003; 22:3783-3791.
8. Ananthan J, Goldberg AL, Voellmy R. Abnormal proteins serve as eukaryotic stress signals and trigger the activation of heat shock genes. Science 1986; 232:522-524.
9. Voellmy R. On mechanisms that control heat shock transcription factor activity in metazoan cells. Cell Stress Chaperones 2004; 9:122-133.
10. Sarge KD, Murphy SP, Morimoto RI. Activation of heat shock gene transcription by heat shock factor 1 involves oligomerization, acquisition of DNA-binding activity, and nuclear localization and can occur in the absence of stress. Mol Cell Biol 1993; 13:1392-1407.
11. Marchler G, Wu C. Modulation of Drosophila heat shock transcription factor activity by the molecular chaperone DROJ1. EMBO J 2001; 20:499-509.
12. Morimoto RI. Cells in stress: Transcriptional activation of heat shock genes. Science 1993; 259:1409-1410.
13. Zou J, Guo Y, Guettouche T et al. Repression of heat shock transcription factor HSF1 activation by HSP90 (HSP90 complex) that forms a stress-sensitive complex with HSF1. Cell 1998; 94:471-480.
14. Huot J, Lambert H, Lavoie JN et al. Characterization of 45-kDa/54-kDa HSP27 kinase, a stress-sensitive kinase which may activate the phosphorylation-dependent protective function of mammalian 27-kDa heat-shock protein HSP27. Eur J Biochem 1995; 227:416-427.
15. Guay J, Lambert H, Gingras-Breton G et al. Regulation of actin filament dynamics by p38 map kinase-mediated phosphorylation of heat shock protein 27. J Cell Sci 1997; 110:357-368.
16. Dubois MF, Bensaude O. MAP kinase activation during heat shock in quiescent and exponentially growing mammalian cells. FEBS Lett 1993; 324:191-195.
17. Adler V, Schaffer A, Kim J et al. UV irradiation and heat shock mediate JNK activation via alternate pathways. J Biol Chem 1995; 270:26071-26077.
18. Rouse J, Cohen P, Trigon S et al. A novel kinase cascade triggered by stress and heat shock that stimulates MAPKAP kinase-2 and phosphorylation of the small heat shock proteins. Cell 1994; 78:1027-1037.
19. Widmann C, Gibson S, Jarpe MB et al. Mitogen-activated protein kinase: Conservation of a three-kinase module from yeast to human. Physiol Rev 1999; 79:143-180.
20. Garrington TP, Johnson GL. Organization and regulation of mitogen-activated protein kinase signaling pathways. Curr Opin Cell Biol 1999; 11:211-218.
21. Chang L, Karin M. Mammalian MAP kinase signalling cascades. Nature 2001; 410:37-40.
22. Konishi H, Matsuzaki H, Tanaka M et al. Activation of RAC-protein kinase by heat shock and hyperosmolarity stress through a pathway independent of phosphatidylinositol 3-kinase. Proc Natl Acad Sci USA 1996; 93:7639-7643.
23. Kyriakis JM. Making the connection: Coupling of stress-activated ERK/MAPK (extracellular-signal-regulated kinase/mitogen-activated protein kinase) core signalling modules to extracellular stimuli and biological responses. Biochem Soc Symp 1999; 64:29-48.
24. Chen F, Torres M, Duncan RF. Activation of mitogen-activated protein kinase by heat shock treatment in Drosophila. Biochem J 1995; 312:341-349.
25. Aikawa R, Komuro I, Yamazaki T et al. Oxidative stress activates extracellular signal-regulated kinases through Src and Ras in cultured cardiac myocytes of neonatal rats. J Clin Invest 1997; 100:1813-1821.

26. Chen W, Martindale JL, Holbrook NJ et al. Tumor promoter arsenite activates extracellular signal-regulated kinase through a signaling pathway mediated by epidermal growth factor receptor and Shc. Mol Cell Biol 1998; 18:5178-5188.

27. Lin RZ, Hu ZW, Chin JH et al. Heat shock activates c-Src tyrosine kinases and phosphatidylinositol 3-kinase in NIH3T3 fibroblasts. J Biol Chem 1997; 272:31196-31202.

28. Sachsenmaier C, Radler-Pohl A, Zinck R et al. Involvement of growth factor receptors in the mammalian UVC response. Cell 1994; 78:963-972.

29. Coffer PJ, Burgering BM, Peppelenbosch MP et al. UV activation of receptor tyrosine kinase activity. Oncogene 1995; 11:561-569.

30. Rosette C, Karin M. Ultraviolet light and osmotic stress: Activation of the JNK cascade through multiple growth factor and cytokine receptors. Science 1996; 274:1194-1197.

31. Rao GN. Hydrogen peroxide induces complex formation of SHC-Grb2-SOS with receptor tyrosine kinase and activates Ras and extracellular signal-regulated protein kinases group of mitogen-activated protein kinases. Oncogene 1996; 13:713-719.

32. Kataoka K, Miura M. Insulin-like growth factor I receptor does not contribute to heat shock-induced Activation of Akt and extracellular signal-regulated kinase (ERK) in mouse embryo fibroblasts. J Radiat Res (Tokyo) 2004; 45:141-144.

33. Ng DC, Bogoyevitch MA. The mechanism of heat shock activation of ERK mitogen-activated protein kinases in the interleukin 3-dependent ProB cell line BaF3. J Biol Chem 2000; 275:40856-40866.

34. Yaglom J, O'Callaghan-Sunol C, Gabai V et al. Inactivation of dual-specificity phosphatases is involved in the regulation of extracellular signal-regulated kinases by heat shock and hsp72. Mol Cell Biol 2003; 23:3813-3824.

35. Song J, Takeda M, Morimoto RI. Bag1-Hsp70 mediates a physiological stress signalling pathway that regulates Raf-1/ERK and cell growth. Nat Cell Biol 2001; 3:276-282.

36. Wang X, Martindale JL, Holbrook NJ. Requirement for ERK activation in cisplatin-induced apoptosis. J Biol Chem 2000; 275:39435-39443.

37. Bacus SS, Gudkov AV, Lowe M et al. Taxol-induced apoptosis depends on MAP kinase pathways (ERK and p38) and is independent of p53. Oncogene 2001; 20:147-155.

38. Tang D, Wu D, Hirao A et al. ERK activation mediates cell cycle arrest and apoptosis after DNA damage independently of p53. J Biol Chem 2002; 277:12710-12717.

39. Ballif BA, Blenis J. Molecular mechanisms mediating mammalian mitogen-activated protein kinase (MAPK) kinase (MEK)-MAPK cell survival signals. Cell Growth Differ 2001; 12:397-408.

40. Woessmann W, Meng YH, Mivechi NF. An essential role for mitogen-activated protein kinases, ERKs, in preventing heat-induced cell death. J Cell Biochem 1999; 74:648-662.

41. Chu B, Soncin F, Price BD et al. Sequential phosphorylation by mitogen-activated protein kinase and glycogen synthase kinase 3 represses transcriptional activation by heat shock factor-1. J Biol Chem 1996; 271:30847-30857.

42. Holmberg CI, Tran SE, Eriksson JE et al. Multisite phosphorylation provides sophisticated regulation of transcription factors. Trends Biochem Sci 2002; 27:619-627.

43. He B, Meng YH, Mivechi NF. Glycogen synthase kinase 3beta and extracellular signal-regulated kinase inactivate heat shock transcription factor 1 by facilitating the disappearance of transcriptionally active granules after heat shock. Mol Cell Biol 1998; 18:6624-6633.

44. Wang X, Grammatikakis N, Siganou A et al. Regulation of molecular chaperone gene transcription involves the serine phosphorylation, 14-3-3 epsilon binding, and cytoplasmic sequestration of heat shock factor 1. Mol Cell Biol 2003; 23:6013-6026.

45. Wang X, Grammatikakis N, Siganou A et al. Interactions between extracellular signal-regulated protein kinase 1, 14-3-3epsilon, and heat shock factor 1 during stress. J Biol Chem 2004; 279:49460-49469.

46. Meriin AB, Yaglom JA, Gabai VL et al. Protein-damaging stresses activate c-Jun N-terminal kinase via inhibition of its dephosphorylation: A novel pathway controlled by HSP72. Mol Cell Biol 1999; 19:2547-2555.

47. Merienne K, Helmlinger D, Perkin GR et al. Polyglutamine expansion induces a protein-damaging stress connecting heat shock protein 70 to the JNK pathway. J Biol Chem 2003; 278:16957-16967.

48. Palacios C, Collins MK, Perkins GR. The JNK phosphatase M3/6 is inhibited by protein-damaging stress. Curr Biol 2001; 11:1439-1443.

49. Meriin AB, Gabai VL, Yaglom J et al. Proteasome inhibitors activate stress kinases and induce Hsp72. Diverse effects on apoptosis. J Biol Chem 1998; 273:6373-6379.

50. Zanke BW, Boudreau K, Rubie E et al. The stress-activated protein kinase pathway mediates cell death following injury induced by cis-platinum, UV irradiation or heat. Curr Biol 1996; 6:606-613.

51. Nishina H, Fischer KD, Radvanyi L et al. Stress-signalling kinase Sek1 protects thymocytes from apoptosis mediated by CD95 and CD3. Nature 1997; 385:350-353.
52. Yang D, Tournier C, Wysk M et al. Targeted disruption of the MKK4 gene causes embryonic death, inhibition of c-Jun NH2-terminal kinase activation, and defects in AP-1 transcriptional activity. Proc Natl Acad Sci USA 1997; 94:3004-3009.
53. Maroni P, Bendinelli P, Zuccorononno C et al. Cellular signalling after in vivo heat shock in the liver. Cell Biol Int 2000; 24:145-152.
54. Han SI, Oh SY, Woo SH et al. Implication of a small GTPase Rac1 in the activation of c-Jun N-terminal kinase and heat shock factor in response to heat shock. J Biol Chem 2001; 276:1889-1895.
55. Tournier C, Hess P, Yang DD et al. Requirement of JNK for stress-induced activation of the cytochrome c-mediated death pathway. Science 2000; 288:870-874.
56. Deng Y, Ren X, Yang L et al. A JNK-dependent pathway is required for TNFalpha-induced apoptosis. Cell 2003; 115:61-70.
57. Enomoto A, Suzuki N, Liu C et al. Involvement of c-Jun NH2-terminal kinase-1 in heat-induced apoptotic cell death of human monoblastic leukaemia U937 cells. Int J Radiat Biol 2001; 77:867-874.
58. Yamamoto K, Ichijo H, Korsmeyer SJ. BCL-2 is phosphorylated and inactivated by an ASK1/Jun N-terminal protein kinase pathway normally activated at G(2)/M. Mol Cell Biol 1999; 19:8469-8478.
59. Kharbanda S, Saxena S, Yoshida K et al. Translocation of SAPK/JNK to mitochondria and interaction with Bcl-x(L) in response to DNA damage. J Biol Chem 2000; 275:322-327.
60. Fuchs SY, Adler V, Buschmann T et al. JNK targets p53 ubiquitination and degradation in nonstressed cells. Genes Dev 1998; 12:2658-2663.
61. Fuchs SY, Adler V, Pincus MR et al. MEKK1/JNK signaling stabilizes and activates p53. Proc Natl Acad Sci USA 1998; 95:10541-10546.
62. Dai R, Frejtag W, He B et al. c-Jun NH2-terminal kinase targeting and phosphorylation of heat shock factor-1 suppress its transcriptional activity. J Biol Chem 2000; 275:18210-18218.
63. Park J, Liu AY. JNK phosphorylates the HSF1 transcriptional activation domain: Role of JNK in the regulation of the heat shock response. J Cell Biochem 2001; 82:326-338.
64. Gabai VL, Yaglom JA, Volloch V et al. Hsp72-mediated suppression of c-Jun N-terminal kinase is implicated in development of tolerance to caspase-independent cell death. Mol Cell Biol 2000; 20:6826-6836.
65. Gabai VL, Meriin AB, Mosser DD et al. Hsp70 prevents activation of stress kinases. A novel pathway of cellular thermotolerance. J Biol Chem 1997; 272:18033-18037.
66. Park HS, Lee JS, Huh SH et al. Hsp72 functions as a natural inhibitory protein of c-Jun N-terminal kinase. EMBO J 2001; 20:446-456.
67. Volloch V, Gabai VL, Rits S et al. HSP72 can protect cells from heat-induced apoptosis by accelerating the inactivation of stress kinase JNK. Cell Stress Chaperones 2000; 5:139-147.
68. Yaglom JA, Gabai VL, Meriin AB et al. The function of HSP72 in suppression of c-Jun N-terminal kinase activation can be dissociated from its role in prevention of protein damage. J Biol Chem 1999; 274:20223-20228.
69. Li GC, Li L, Liu RY et al. Heat shock protein hsp70 protects cells from thermal stress even after deletion of its ATP-binding domain. Proc Natl Acad Sci USA 1992; 89:2036-2040.
70. Volloch V, Gabai VL, Rits S et al. ATPase activity of the heat shock protein hsp72 is dispensable for its effects on dephosphorylation of stress kinase JNK and on heat-induced apoptosis. FEBS Lett 1999; 461:73-76.
71. Dorion S, Lambert H, Landry J. Activation of the p38 signaling pathway by heat shock involves the dissociation of glutathione S-transferase Mu from Ask1. J Biol Chem 2002; 277:30792-30797.
72. Ichijo H, Nishida E, Irie K et al. Induction of apoptosis by ASK1, a mammalian MAPKKK that activates SAPK/JNK and p38 signaling pathways. Science 1997; 275:90-94.
73. Saitoh M, Nishitoh H, Fujii M et al. Mammalian thioredoxin is a direct inhibitor of apoptosis signal-regulating kinase (ASK) 1. EMBO J 1998; 17:2596-2606.
74. Hayes JD, Pulford DJ. The glutathione S-transferase supergene family: Regulation of GST and the contribution of the isoenzymes to cancer chemoprotection and drug resistance. Crit Rev Biochem Mol Biol 1995; 30:445-600.
75. Chang Y, Abe A, Shayman JA. Ceramide formation during heat shock: A potential mediator of alpha B-crystallin transcription. Proc Natl Acad Sci USA 1995; 92:12275-12279.
76. Chung N, Jenkins G, Hannun YA et al. Sphingolipids signal heat stress-induced ubiquitin-dependent proteolysis. J Biol Chem 2000; 275:17229-17232.

77. Dickson RC, Nagiec EE, Skrzypek M et al. Sphingolipids are potential heat stress signals in Saccharomyces. J Biol Chem 1997; 272:30196-30200.

78. Chrétien P, Landry J. Enhanced constitutive expression of the 27-kDa heat shock proteins in heat-resistant variants from Chinese hamster cells. J Cell Physiol 1988; 137:157-166.

79. Landry J, Chretien P, Laszlo A et al. Phosphorylation of HSP27 during development and decay of thermotolerance in Chinese hamster cells. J Cell Physiol 1991; 147:93-101.

80. Huot J, Lambert H, Lavoie JN et al. Characterization of 45-kDa/54-kDa HSP27 kinase, a stress-sensitive kinase which may activate the phosphorylation-dependent protective function of mammalian 27-kDa heat-shock protein HSP27. Eur J Biochem 1995; 227:416-427.

81. Lavoie JN, Lambert H, Hickey E et al. Modulation of cellular thermoresistance and actin filament stability accompanies phosphorylation-induced changes in the oligomeric structure of heat shock protein 27. Mol Cell Biol 1995; 15:505-516.

82. Lambert H, Charette SJ, Bernier AF et al. HSP27 multimerization mediated by phosphorylation-sensitive intermolecular interactions at the amino terminus. J Biol Chem 1999; 274:9378-9385.

83. Deschesnes RG, Huot J, Valerie K et al. Involvement of p38 in apoptosis-associated membrane blebbing and nuclear condensation. Mol Biol Cell 2001; 12:1569-1582.

84. Brenner B, Koppenhoefer U, Weinstock C et al. Fas- or ceramide-induced apoptosis is mediated by a Rac1-regulated activation of Jun N-terminal kinase/p38 kinases and GADD153. J Biol Chem 1997; 272:22173-22181.

85. Toyoshima F, Moriguchi T, Nishida E. Fas induces cytoplasmic apoptotic responses and activation of the MKK7-JNK/SAPK and MKK6-p38 pathways independent of CPP32-like proteases. J Cell Biol 1997; 139:1005-1015.

86. Frasch SC, Nick JA, Fadok VA et al. p38 mitogen-activated protein kinase-dependent and -independent intracellular signal transduction pathways leading to apoptosis in human neutrophils. J Biol Chem 1998; 273:8389-8397.

87. Desbiens KM, Deschesnes RG, Labrie MM et al. c-Myc potentiates the mitochondrial pathway of apoptosis by acting upstream of apoptosis signal-regulating kinase 1 (Ask1) in the p38 signalling cascade. Biochem J 2003; 372:631-641.

88. Huot J, Houle F, Rousseau S et al. SAPK2/p38-dependent F-actin reorganization regulates early membrane blebbing during stress-induced apoptosis. J Cell Biol 1998; 143:1361-1373.

89. Guay J, Lambert H, Gingras-Breton G et al. Regulation of actin filament dynamics by p38 map kinase-mediated phosphorylation of heat shock protein 27. J Cell Sci 1997; 110:357-368.

90. Dorion S, Berube J, Huot J et al. A short lived protein involved in the heat shock sensing mechanism responsible for stress-activated protein kinase 2 (SAPK2/p38) activation. J Biol Chem 1999; 274:37591-37597.

91. Vanhaesebroeck B, Alessi DR. The PI3K-PDK1 connection: More than just a road to PKB. Biochem J 2000; 346:561-576.

92. Shaw M, Cohen P, Alessi DR. The activation of protein kinase B by H2O2 or heat shock is mediated by phosphoinositide 3-kinase and not by mitogen-activated protein kinase-activated protein kinase-2. Biochem J 1998; 336:241-246.

93. Bijur GN, Jope RS. Opposing actions of phosphatidylinositol 3-kinase and glycogen synthase kinase-3beta in the regulation of HSF-1 activity. J Neurochem 2000; 75:2401-2408.

94. Bang OS, Ha BG, Park EK et al. Activation of Akt is induced by heat shock and involved in suppression of heat-shock-induced apoptosis of NIH3T3 cells. Biochem Biophys Res Commun 2000; 278:306-311.

95. Ma N, Jin J, Lu F et al. The role of protein kinase B (PKB) in modulating heat sensitivity in a human breast cancer cell line. Int J Radiat Oncol Biol Phys 2001; 50:1041-1050.

96. Park HS, Kim MS, Huh SH et al. Akt (protein kinase B) negatively regulates SEK1 by means of protein phosphorylation. J Biol Chem 2002; 277:2573-2578.

97. Levresse V, Butterfield L, Zentrich E et al. Akt negatively regulates the cJun N-terminal kinase pathway in PC12 cells. J Neurosci Res 2000; 62:799-808.

98. Kennedy SG, Kandel ES, Cross TK et al. Akt/Protein kinase B inhibits cell death by preventing the release of cytochrome c from mitochondria. Mol Cell Biol 1999; 19:5800-5810.

99. Ogawara Y, Kishishita S, Obata T et al. Akt enhances Mdm2-mediated ubiquitination and degradation of p53. J Biol Chem 2002; 277:21843-21850.

100. Rena G, Guo S, Cichy SC et al. Phosphorylation of the transcription factor forkhead family member FKHR by protein kinase B. J Biol Chem 1999; 274:17179-17183.

101. Sato S, Fujita N, Tsuruo T. Modulation of Akt kinase activity by binding to Hsp90. Proc Natl Acad Sci USA 2000; 97:10832-10837.

102. Fujita N, Sato S, Ishida A et al. Involvement of Hsp90 in signaling and stability of 3-phosphoinositide-dependent kinase-1. J Biol Chem 2002; 277:10346-10353.

103. Konishi H, Matsuzaki H, Tanaka M et al. Activation of protein kinase B (Akt/RAC-protein kinase) by cellular stress and its association with heat shock protein Hsp27. FEBS Lett 1997; 410:493-498.

104. Zhang R, Luo D, Miao R et al. Hsp90-Akt phosphorylates ASK1 and inhibits ASK1-mediated apoptosis. Oncogene 2005; 24:3954-3963.

105. Glass JR, DeWitt RG, Cress AE. Rapid loss of stress fibers in Chinese hamster ovary cells after hyperthermia. Cancer Res 1985; 45:258-262.

106. Iida K, Iida H, Yahara I. Heat shock induction of intranuclear actin rods in cultured mammalian cells. Exp Cell Res 1986; 165:207-215.

107. Welch WJ, Suhan JP. Morphological study of the mammalian stress response: Characterization of changes in cytoplasmic organelles, cytoskeleton, and nucleoli, and appearance of intranuclear actin filaments in rat fibroblasts after heat-shock treatment. J Cell Biol 1985; 101:1198-1211.

108. Patriarca EJ, Maresca B. Acquired thermotolerance following heat shock protein synthesis prevents impairment of mitochondrial ATPase activity at elevated temperatures in Saccharomyces cerevisiae. Exp Cell Res 1990; 190:57-64.

109. Warters RL, Stone OL. The effects of hyperthermia on DNA replication in HeLa cells. Radiat Res 1983; 93:71-84.

110. Yost HJ, Petersen RB, Lindquist S. RNA metabolism: Strategies for regulation in the heat shock response. Trends Genet 1990; 6:223-227.

111. Mosser DD, Caron AW, Bourget L et al. Role of the human heat shock protein hsp70 in protection against stress-induced apoptosis. Mol Cell Biol 1997; 17:5317-5327.

112. LaFevre-Bernt MA, Ellerby LM. Kennedy's disease. Phosphorylation of the polyglutamine-expanded form of androgen receptor regulates its cleavage by caspase-3 and enhances cell death. J Biol Chem 2003; 278:34918-34924.

113. Vigh L, Escriba PV, Sonnleitner A et al. The significance of lipid composition for membrane activity: New concepts and ways of assessing function. Prog Lipid Res 2005; 44:303-344.

114. Dynlacht JR, Fox MH. The effect of 45 degrees C hyperthermia on the membrane fluidity of cells of several lines. Radiat Res 1992; 130:55-60.

115. Mejia R, Gomez-Eichelmann MC, Fernandez MS. Membrane fluidity of Escherichia coli during heat-shock. Biochim Biophys Acta 1995; 1239:195-200.

116. Carratu L, Franceschelli S, Pardini CL et al. Membrane lipid perturbation modifies the set point of the temperature of heat shock response in yeast. Proc Natl Acad Sci USA 1996; 93:3870-3875.

117. Horvath I, Glatz A, Varvasovszki V et al. Membrane physical state controls the signaling mechanism of the heat shock response in Synechocystis PCC 6803: Identification of hsp17 as a "fluidity gene". Proc Natl Acad Sci USA 1998; 95:3513-3518.

118. Balogh G, Horvath I, Nagy E et al. The hyperfluidization of mammalian cell membranes acts as a signal to initiate the heat shock protein response. FEBS J 2005; 272:6077-6086.

119. Burgering BM, Coffer PJ. Protein kinase B (c-Akt) in phosphatidylinositol-3-OH kinase signal transduction. Nature 1995; 376:599-602.

120. Ihara Y, Yasuoka C, Kageyama K et al. Tyrosine phosphorylation of clathrin heavy chain under oxidative stress. Biochem Biophys Res Commun 2002; 297:353-360.

121. Wilde A, Beattie EC, Lem L et al. EGF receptor signaling stimulates SRC kinase phosphorylation of clathrin, influencing clathrin redistribution and EGF uptake. Cell 1999; 96:677-687.

122. Caterina MJ, Schumacher MA, Tominaga M et al. The capsaicin receptor: A heat-activated ion channel in the pain pathway. Nature 1997; 389:816-824.

123. Clapham DE. TRP channels as cellular sensors. Nature 2003; 426:517-524.

124. Voets T, Droogmans G, Wissenbach U et al. The principle of temperature-dependent gating in cold- and heat-sensitive TRP channels. Nature 2004; 430:748-754.

125. Li C, Xu Q. Mechanical stress-initiated signal transductions in vascular smooth muscle cells. Cell Signal 2000; 12:435-445.

126. Sadoshima J, Izumo S. Mechanical stretch rapidly activates multiple signal transduction pathways in cardiac myocytes: Potential involvement of an autocrine/paracrine mechanism. EMBO J 1993; 12:1681-1692.

127. Gray JV, Ogas JP, Kamada Y et al. A role for the Pkc1 MAP kinase pathway of Saccharomyces cerevisiae in bud emergence and identification of a putative upstream regulator. EMBO J 1997; 16:4924-4937.

128. Verna J, Lodder A, Lee K et al. A family of genes required for maintenance of cell wall integrity and for the stress response in Saccharomyces cerevisiae. Proc Natl Acad Sci USA 1997; 94:13804-13809.

129. Kamada Y, Jung US, Piotrowski J et al. The protein kinase C-activated MAP kinase pathway of Saccharomyces cerevisiae mediates a novel aspect of the heat shock response. Genes Dev 1995; 9:1559-1571.
130. Imazu H, Sakurai H. Saccharomyces cerevisiae heat shock transcription factor regulates cell wall remodeling in response to heat shock. Eukaryot Cell 2005; 4:1050-1056.
131. Jenkins GM. The emerging role for sphingolipids in the eukaryotic heat shock response. Cell Mol Life Sci 2003; 60:701-710.
132. Verheij M, Bose R, Lin XH et al. Requirement for ceramide-initiated SAPK/JNK signalling in stress-induced apoptosis. Nature 1996; 380:75-79.
133. Kozawa O, Tanabe K, Ito H et al. Sphingosine 1-phosphate regulates heat shock protein 27 induction by a p38 MAP kinase-dependent mechanism in aortic smooth muscle cells. Exp Cell Res 1999; 250:376-380.
134. Kaneko H, Igarashi K, Kataoka K et al. Heat shock induces phosphorylation of histone H2AX in mammalian cells. Biochem Biophys Res Commun 2005; 328:1101-1106.
135. Takahashi A, Matsumoto H, Nagayama K et al. Evidence for the involvement of double-strand breaks in heat-induced cell killing. Cancer Res 2004; 64:8839-8845.
136. Miyakoda M, Suzuki K, Kodama S et al. Activation of ATM and phosphorylation of p53 by heat shock. Oncogene 2002; 21:1090-1096.
137. Guan J, Stavridi E, Leeper DB et al. Effects of hyperthermia on p53 protein expression and activity. J Cell Physiol 2002; 190:365-374.
138. Kampinga HH, Laszlo A. DNA double strand breaks do not play a role in heat-induced cell killing. Cancer Res 2005; 65:10632-10633.
139. Bakkenist CJ, Kastan MB. DNA damage activates ATM through intermolecular autophosphorylation and dimer dissociation. Nature 2003; 421:499-506.
140. Fritz G, Kaina B. Late activation of stress kinases (SAPK/JNK) by genotoxins requires the DNA repair proteins DNA-PKcs and CSB. Mol Biol Cell 2006; 17:851-861.
141. Viniegra JG, Martinez N, Modirassari P et al. Full activation of PKB/Akt in response to insulin or ionizing radiation is mediated through ATM. J Biol Chem 2005; 280:4029-4036.
142. Wang X, McGowan CH, Zhao M et al. Involvement of the MKK6-p38gamma cascade in gamma-radiation-induced cell cycle arrest. Mol Cell Biol 2000; 20:4543-4552.

Chapter 11

Membrane-Regulated Stress Response:
A Theoretical and Practical Approach

László Vígh,* Zsolt Török, Gábor Balogh, Attila Glatz, Stefano Piotto and Ibolya Horváth

Abstract

A number of observations have lent support to a model in which thermal stress is trans-duced into a signal at the level of the cellular membranes. Our alternative, but not exclusive, approach is based on the concept that the initial stress-sensing events are associated with the physical state and lipid composition of cellular membranes, i.e., the subtle alteration(s) of membrane fluidity, phase state, and/or microheterogeneity may operate as a cellular thermometer. In fact, various pathological states and aging are associated with typical "membrane defects" and simultaneous dysregulation of heat shock protein synthesis. The dis-covery of nonproteotoxic membrane-lipid interacting compounds, capable of modulating mem-brane microdomains engaged in primary stress sensing may be of paramount importance for the design of new drugs with the ability to induce or attenuate the level of particular heat shock proteins.

Introduction

During recent decades, a mounting volume of evidence has demonstrated that, when sub-jected to a wide variety of environmental assaults, ranging from extreme temperatures to meta-bolic poisons, cells of widely differing origins respond in a highly conserved and universal man-ner, producing stress proteins. Functional analysis of the minimal stress proteome of all cellular organisms has revealed at least 6 distinct categories, providing information concerning key as-pects of the cellular stress response.[1] Seemingly different stress stimuli always led to an increased expression of the same group of proteins, and many of the agents that induced the stress protein response were protein denaturants. As a landmark result of these studies, the injection of dena-tured proteins into living cells was sufficient to induce a stress protein response.[2] These observa-tions gave rise to the conclusion, that the stress response is universally initiated by the accumu-lation of denatured or misfolded proteins.[3] Following the establishment of a novel role for certain stress proteins as molecular chaperones in unstressed cells, from the early 1990s most of the investigators in this field turned their attention to chaperone research. Scientists began to realize the enormous impact to be expected from the pharmacological manipulation of the expression of stress proteins. Paradoxically, in parallel with efforts aimed at the development of heat shock protein (HSP)-inducing or inhibiting agents, the basic question of the molecular mechanism of stress sensing and signaling again became timely and of enormous interest.[4,5]

*Corresponding Author: László Vígh—Institute of Biochemistry, Biological Research Center of the Hungarian Academy of Sciences, H-6726, Szeged, Temesvári Krt. 62, Hungary. Email: Vigh@brc.hu

Molecular Aspects of the Stress Response: Chaperones, Membranes and Networks, edited by Peter Csermely and László Vígh. ©2007 Landes Bioscience and Springer Science+Business Media.

The discovery of compounds such as bimoclomol and derivatives that can upregulate HSPs without evidence of proteotoxicity[6,7] stimulated us to reinvestigate this question. We discuss here a model in which the initial stress signal may originate from the cellular membranes, related to the fluidity and microdomain organization and may involve lipases, receptors, receptor-like molecules. Theoretically, pharmacologic methods are applicable to correct membrane defects and normalize the dysregulated stress protein response in disease states.[8] The study of "membrane-based" cellular stress poses a number of unique intellectual and technical questions, which will also be discussed. The discovery of new drugs is a process generally based on the knowledge of an exactly defined target molecule. In the literature, drug discovery is often referred to as involving high-throughput methods that utilize combinatorial chemistry, genomics and proteomics information as starting point. Changes in the lipid organization of the plasma membrane can modify cell signaling events, but lipid membranes are not a conventional target for the design of new drugs.[9] Consequently, computer simulations are generally confined to the investigation of drug-substrate interactions. In this context, we set out to explore some recent advances in the computer simulation of lipid membranes.

The Evolution of the "Membrane Sensor" Hypothesis with the Aid of Unicellular Stress Models: The Beauty of Simplicity

It has long been well known that, during abrupt temperature fluctuations, membranes prove to be among the most thermally sensitive macromolecular structures within cells. Not surprisingly, therefore, membranes are one of the main cellular targets of temperature adaptation.[10,11] Many organisms are known to adapt via the extent of unsaturation of their constituent membrane lipids.[12] The readily transformable unicellular cyanobacteria have proved to be particularly suitable for studies of temperature stress response at the molecular level. The general features of their plasma and thylakoid membranes resemble those of the chloroplasts of higher plants in terms of both lipid composition and the assembly of membranes.[13]

As it is totally nonspecific in its effects on the cellular metabolism, temperature is a blunt instrument with which to dissect the mechanisms involved in this response. Nevertheless, application of the technique of homogeneous catalytic hydrogenation[14-18] allowed the controlled in vivo in situ hydrogenation of unsaturated membrane lipids with apparently no effects other than ordering of the membrane interior.

Catalytic hydrogenation has been demonstrated to be a useful method for investigation of the molecular basis of cellular chilling sensitivity in various in vivo unicellular model systems, such as cyanobacteria,[14,15] the green alga *Dunaliella salina*[19] or the ciliated protozoan *Tetrahymena mimbres*,[20] and also in vitro, using isolated membranes of very different origins.[18] The phase behavior of membrane lipids of the cyanobacterium *Anacystis nidulans* was believed to be the primary factor underlying the chilling sensitivity of this organism, and catalytic hydrogenation studies were undertaken to examine this hypothesis. With appropriate selection of the reaction conditions, the initial phase of in vivo membrane lipid hydrogenation was confined to the outer cell surface membrane. Such a cytoplasmic membrane-selective in vivo hydrogenation was associated with an increased leakiness to solutes, and fluid-to-gel-phase separation of the membrane components occurred at relatively high temperatures.[15] As judged by the K^+ release, (reflecting the leakiness of the cytoplasmic membrane), and the rate of O_2 evolution (an index of the integrity of the inner, thylakoid membrane), 28°C-grown cells subjected to surface membrane-specific hydrogenation displayed a susceptibility to chilling injury similar to that of their 38°C-grown counterparts. This finding furnished the first support for the hypothesis that the phase behavior of the cytoplasmic membrane, as modulated by the degree of unsaturation of the constituent membrane lipids, is directly related to the susceptibility of the cells to chilling damage.

The surface membrane-selective hydrogenation of another blue-green alga, *Synechocystis*, stimulated the transcription of the desaturase *desA* gene in the same way[21] as cooling.[22] Once again, only the plasma membrane was hydrogenated, and it was therefore proposed that this

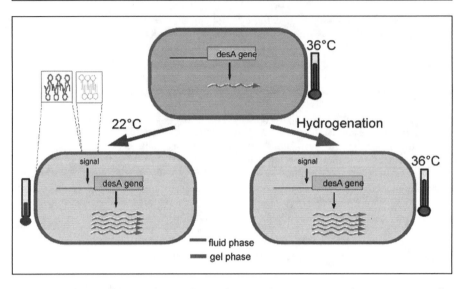

Figure 1. Activation of the desaturase *desA* in the cyanobacterium *Synechocystis* PCC 6803 by low-temperature stress and by isothermal catalytic hydrogenation of the unsaturated membrane lipids.[21]

membrane must act as a sensor of altered saturation, or "fluidity", thereby initiating a control response, the ultimate effect of which is increased transcription of a desaturase gene, *desA* (Fig. 1). The desaturase proteins are then synthesized de novo and targeted to both the plasma and thylakoid membrane in order to catalyze the desaturation of lipid fatty acyl chains and compensate for the decrease in membrane fluidity caused by hydrogenation or exposure to low temperature. Isothermally hydrogenated cells displayed a heightened sensitivity to cooling in the activation of *desA* transcription, implying that membrane hydrogenation and cooling have additive effects, and that cooling exerts its effect through changes in membrane fluidity.[21] It later emerged that there is a very tight coordination between the physical state of the membranes and the expression of cold-inducible genes, other than *desA*. The operation of a thermosensor, histidine kinase Hik33, which perceives the cold-signal via the rigidification of membrane lipids and ultimately conveys signals to desaturase genes, was later elucidated in detail.[23] The sensory kinase Hik33 includes a type-P linker, a leucine zipper, and a PAS domain. The type-P linker contains two helical regions in tandem, which are assumed to transduce stress signals via intramolecular structural changes that result from interactions between the two helical regions and lead to the intermolecular dimerization of Hik33 monomers. Through the application of either the exposure of cells to low temperature or catalytic membrane hydrogenation, it was documented that, whereas the more ordered (less "fluid") lipid molecules move closer, this membrane rearrangement pushes the linker domains of Hik33 closer to each other, with resultant autophosphorylation of the histidine kinase domains (P). Hik33 is also involved in the perception of hyperosmotic stress and salt stress.[24] The activated Hik33 dimers phosphorylate the response regulator (Fig. 2A). The existence of similar two-component systems, consisting of a membrane-associated Hik, and a cognate response regulator (Rre), has likewise been well established in *E. coli* and *B. subtilis*. Their regulation in response to stress can be either positive or negative.[24]

In conclusion, the discovery of a feedback between the membrane fluidity and the expression of genes for desaturases on cyanobacterial models suggested the presence of a surface-membrane-embedded temperature sensor. In a wider sense, these results yielded the first direct evidence that the dynamic properties and structure of membranes affect the gene

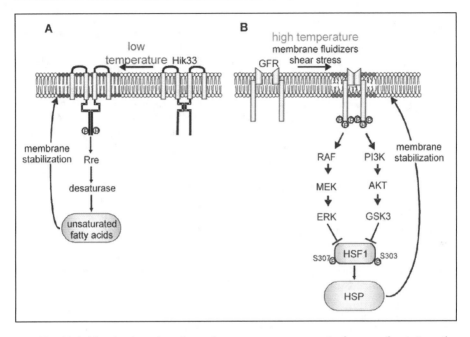

Figure 2. A) Cold activation of a primary low-temperature sensor in the cyanobacterium, the membrane-associated histidine kinase, Hik33. Via response regulator(s), Rre(s), Hik33 is ultimately linked to fatty acid desaturases and can be stimulated by membrane rigidification and simultaneous formation of the gel lipid phase, attained either by low- temperature shift or by catalytic membrane lipid hydrogenation.[21] B) A heat shock (shear stress, membrane fluidizers, etc.) induced increase in membrane fluidity can activate growth factor receptors (GFRs) even in a ligand-independent manner. Membrane hyperfluidization may accelerate the clustering of GFRs into lipid microdomains and their autophosphorylation. This leads to a down-stream signal cascade through RAF and PI3K, leading to ERK and GSK3 phosphorylation, respectively, with concomitant upregulation of HSP synthesis (typically as described by Khalegue et al[62] for ErbB activation in tumor cells). Note the striking similarity of the low-temperature induced (probably fluid-to-gel phase transition) membrane changes on the activation of HIK33, leading to upregulated desaturases, and, the high-temperature-induced upregulation of HS genes via GFR dimerization. Both end-products, i.e., lipids with more unsaturated acyl chains and certain subpopulations of HSPs, are capable of stabilizing membranes under extreme temperature stress.

transcriptional activity . On the other hand, this observation served as a landmark in studies of the potential roles of the lipid phase and fluidity changes of membranes in high-temperature sensing and concomitant activation of heat shock (HS) genes.[25]

In fact, the genetic modification of the lipid unsaturation and thereby the membrane physical state in the yeast *Saccharomyces cerevisiae* obtained by over-expression of a desaturase gene, was shown to reset the optimal temperature of heat shock response (HSR) and altered the expression of the h*sp70* and h*sp82* genes.[26] In agreement with the above findings, in a series of studies, Curran and coworkers demonstrated that the heat sensitivity of the HSR[27] and the general stress response[28] (GSR) pathways depend critically on the fatty acid composition of the membrane lipids present in the yeast cells. Furthermore, these studies excluded direct thermal denaturation of the cellular proteins as the trigger for the activation of these major stress-signaling pathways since the addition of different fatty acids to the growth media led to different thresholds of the HSR. It was suggested that HS is detected by a membrane-linked thermostat(s), whose activation is a consequence not only of the elevated temperature, but also of the specific composition and physical state of the membrane lipids.

HS promotes at least two MAPK-driven signal transduction pathways in *Schizosaccharomyces pombe*. Interestingly, their high-temperature-induced phosphorylations display striking differences. The high-temperature-induced phosphorylation of MAPK Sty1/Spc1 appears to depend on the inhibitory phosphatases, Pyp1,2, rather than the downstream MAPKK Wis1,[29] similarly as for mammalian ERK kinase (for details, see the chapter by Nadeau and Landry). In contrast, the HS- (and other)-induced phosphorylation of the "cell-integrity" responsive MAPK, Pmk1 (p42/44), seems to be completely dependent on the donwnstream MAPK kinases. Furthermore, the time course of phosphorylation and the cellular localization of the two MAPKs display intriguing differences,[30] holding out the promise of an understanding of their distinct, stress-stimulated operation in real-time measurements. Besides the above-mentioned altered features, both sty1[+] and pmk1[+] proved to be dispensable under normal conditions, but the mutants exhibit a heat-sensitive phenotype.[30,31] Deletion of any of the two MAPKs caused osmosensitivity,[32] whereas pmk1 is involved in the "cell-integrity" pathway.[33] In contrast with these MAPKs, the HS factor (HSF) is required for growth even in the absence of HS. The *Drosophila melanogaster* HSF can rescue the lethality, and the *S. pombe* HSF is able to bind to the HS element (HSE) of the human *hsp70* gene.[34] It has also been demonstrated that strains bearing different *hsf1* deletions exhibit an altered growth capability when exposed to different kinds of stress (e.g., heat or cadmium treatment).[35] The striking similarities of the human and *S. pombe* signal transduction pathways suggest that *S. pombe* might serve as a powerful and complementary model with which to explore membrane-associated stress sensors and stress signaling in mammalian cells.

A close correlation has also been observed between the physical state of the thylakoid membrane and the threshold temperatures required for maximal activation of HS-inducible genes (i.e., *dnaK, groESL, cpn60* and *hsp17*)[36,13] in *Synechocystis* cells. Overproduction of a *Synechocystis* Δ^{12} desaturase in *Salmonella typhi*, which is inactive enzymatically under the experimental conditions, but which inserts into the membranes, was also able to cause resetting of the HSR in parallel with a considerable effect on the virulence of this pathogen. A substantially higher membrane protein content (i.e., an unbalanced protein/phospholipid ratio) was found in the membranes of the transformed cells. As evidence that the desaturase-transformed cells are unable to accommodate the extra membrane protein properly, they displayed an elevated permeability in their outer membrane even under nonstressed conditions (Colonna-Romano S, Eletto AM, Török Zs et al unpublished). An unbalanced membrane phospholipid composition was shown to affect the expression of several regulatory genes in *E. coli*.[37] Overproduction of the membrane-bound sn-glycerol-acyltransferse in *E. coli* triggered a HSR.[38] Treatment of *E. coli* with nonlethal doses of heat or of the membrane-perturbing agent benzyl alcohol caused transient membrane fluidization and permeabilization, and induced the transcription of the entire array of heat shock genes in a sigma-32 (σ^{32})-dependent manner. This early response was followed by a rapid adaptation (priming) of the cells to an otherwise lethal elevated temperature, in strong correlation with an observed remodeling of the composition and alkyl chain unsaturation of the membrane lipids. Just like the activation of *hs* genes, however, the acquisition of cellular thermotolerance in benzyl alcohol-primed *E. coli* cells was unrelated to protein denaturation.[39] The benzyl alcohol-mediated induction of chaperones in *E. coli* improves the native folding of aggregation-prone recombinant proteins, thereby substituting the more demanding approach of chaperone coexpression.[40] When heat treatment is to be avoided, as in the case of heterologously expressing heat-sensitive proteins, the membrane fluidizer-induced expression of recombinant proteins by a conditional HS promoter may serve as a novel and efficient biotechnological tool.[41]

We assume that, both in *E. coli* and in *Salmonella*, a high-temperature signal is transduced, in part, via the CpxA-CpxR phospho-relay system. CpxA is a histidine kinase that contains two transmembrane regions, and CpxR is a response regulator that functions as a transcription factor to regulate the expression of heat-inducible genes.[42] Since the activity of CpxA is greatly influenced by the composition of membrane lipids,[43] it is tempting to suggest that CpxA might also sense changes in the physical state of the membrane lipids of *E. coli* and *Salmonella* cells exposed to high-temperature stress.[39]

A reporter system based on the highly thermostable enzyme lichenase was recently designed for the analysis of *Synechocystis* HS promoters when heterologously induced in *E. coli*. Suprisingly, the cyanobacterial *groESL* promoter was not only recognized by the *E. coli* transcriptional apparatus, but was regulated by changes in the membrane fluidity of the host attained by benzyl alcohol administration. In *E. coli*, most HS genes are controlled by σ,[32] the product of the *rpoH* gene, which is completely absent in cyanobacteria. Instead, a CIRCE (Controling Inverted Repeat of Chaperone Expression) element is present upstream of the *Synechocystis groESL operon*, which is controled by the repressor *hrcA* (heat regulation at CIRCE).[13,44] Whereas its mechanism is unknown, the data demonstrate that the membrane physical properties of a Gram-negative bacterium, such as *E. coli* are capable of regulating the HS promoter from a distantly related photosynthetic cyanobacterium, apparently in a σ^{32}- independent manner (Shigapova N, Balogi Zs, Fodor E et al, unpublished). Overall, a common denominator for the effects of cold stress and HS, i.e., a change in membrane fluidity, clearly appears to play the primary role in the perception of temperature changes causing the transcriptional induction of desaturases at cold and a reset of HS gene transcription at high temperature in these unicellular models.

Evidence Concerning the Operation of Membrane-Associated Stress Sensing and Signaling Mechanisms in Mammalian Cells. Membrane Lipids May Provide the Molecular Switch for Stress Sensing and Signaling

There is a large body of evidence, much of which was earlier ignored or simply overlooked, that membrane-associated stress sensing and signaling mechanisms operate in mammalian cells. The formation of isofluid membrane states by the administration of benzyl alcohol and heptanol, which corresponded to mild HS-induced hyperfluidization, resulted in almost identical downshifts in the temperature thresholds of the HSR, accompanied by the synthesis of stress proteins at the growth temperature in human erythroleukemic (K562) cells.[45] Like thermal stress, membrane fluidizers elicited nearly identical rises of cytosolic Ca^{2+}, in both Ca^{2+}-containing and Ca^{2+}-free media and closely similar extents of increase in mitochondrial hyperpolarization. Most importantly, neither of the membrane fluidizers caused any detectable protein denaturation at concentrations that induced the HSP response. Again, the induction of HSP70, HSP40 and HSP27 by paeoniflorin, a major active constituent of a herbal medicine, was shown to be mediated by HSF1, in the complete absence of proteotoxicity.[46] The coinducing effect of the hydroxylamine derivative bimoclomol on HSP expression[6] was also demonstrated to be mediated via HSF1,[47] and the presence of bimoclomol did not affect protein denaturation in various mammalian cells. Instead, the compound (and its analogs) specifically interacts with and significantly increases the fluidity of negatively-charged membrane lipids. In accordance with this, the HSP-coinducing activity of bimoclomol appeared susceptible to the fatty acid composition and membrane physical state of the target cells (Vigh L., unpublished). Additionally, bimoclomol was revealed to be an efficient inhibitor of bilayer-to-nonbilayer lipid-phase transitions.[7] It has been suggested that, while sensitizing the cellular membranes under mild HS conditions, the drug and its analogs ensure simultaneous protection against irreversible membrane damage at higher temperatures.[7] Whereas bimoclomol was observed to have potential therapeutic value in the treatment of diabetes, a cardiac dysfunction and cerebrovascular disorders,[5,7] another, closely-related bimoclomol analog, arimoclomol, delayed the disease progression in amyotrophic lateral sclerosis.[48] As reviewed elsewhere,[5,8] pharmacological activation of the HSP response with lipid-interacting HSP coinducers may be a successful therapeutic approach for the treatment of most varied diseases.

It was recently suggested that the mode of action of non proteotoxic membrane perturbants (e.g., benzyl alcohol) resembles mild, "fever-like" HS[49] and in its underlying mechanism can be analogous to shear stress. The fact that hemodynamic shear stress, an important determinant of vascular remodeling and atherogenesis, is able to increase HSP expression in cells under isothermal conditions, is well established and has significant clinical consequences. HSP70 induction has been demonstrated in the arterial wall in response to acute hypertension, balloon angioplasty and advanced lesions of atherosclerosis.[50] The mechanical stress-induced Hsp70 expression in arterial smooth muscle cells was shown to be regulated by the small G proteins Ras and Rac via PI3K.[51] The key shear stress responses, including the HSP response, are apparently linked within a single, integrin-mediated pathway. A fluorescence resonance energy transfer (FRET) study localized activated Rac1 in the direction of flow during shear stress. It was shown that, in parallel with increased membrane fluidity in the upstream side of the cells, shear stress results in the remodeling of focal adhesions, with new adhesions forming preferentially toward the downstream edge of the cells. Active Rac1 binds preferentially to low-density, cholesterol-rich membranes (rafts) and this binding step is specifically determined, at least in part, by the composition and physical state of the membrane lipids.[52]

Analogously to shear stress or oxidative stress, moderate HS has been observed to induce membrane translocation of Rac1 and membrane ruffling in a Rac1-dependent manner, whereas increased HSP expression is paralleled by the activation of HSF1.[53] In favor of a membrane stress sensor model, it has been suggested that the potential sensing mechanism of mild HS is based on membrane fluidization and/or rearrangement,[1] closely resembling the HSP induction attained by nonproteotoxic membrane fluidizers. Membrane rearrangement by mild HS or membrane hyperfluidization may activate growth factor receptor tyrosine kinases by causing their nonspecific clustering.[8] As described earlier, the activation of such cell surface receptors has other potential consequences, including the activation of PI3K, which in turn activates Rac1. Rac1 may stimulate NADPH oxidase which produces H_2O_2 and therefore the stress-stimulated nonspecific clustering of cell surface receptors, providing a possible avenue for the oxidative burst and activation of the H_2O_2-induced stress signaling mechanism.[1] A Rac1-dependent HS signal pathway very probably plays an important role in physiological thermal stress responses such as fever. Rather than acting as a proteotoxic stress, fever may function as a key, membrane-mediated signal required for resetting the body conditions. In contrast with the poikilotherms, mammalian acclimation is limited to a very narrow range of changes in body temperature; accordingly, little is known about membrane and lipid remodeling under such acclimation conditions. The marked changes in the affinity of various G-protein-coupled receptors and Na-K-ATPase activity during the course of heat acclimation in mammals imply the possibility of significant and specific changes in membrane lipid composition,[54] since alterations in the functions of these proteins have been documented to be strongly membrane lipid-dependent.[9]

In conclusion, while many inducers of the HSR may function through a protein unfolding mechanism, the above observations clearly suggest that some inducers may work through a distinct mechanism.[55-57] Moreover, a membrane sensor model could explain specific HSP expression patterns with different sensors. This model predicts that the plasma membrane, which is the barrier to the external environment and well suited for sensing thermal stress, acts as an important regulatory interface. Hence, even subtle alterations in the lipid phase of surface membranes ("membrane defects") may influence membrane-initiated stress-signaling processes by causing changes in fluidity, membrane thickness or the clustering of receptors or other proteins localized in the surface membranes.

It was suggested recently that in such prominent disease states as insulin-resistant diabetes and cancer, where the directions of HSP dysregulation and membrane fluidity run in parallel, but strikingly opposite manners, there must exist a conserved signaling cascade which is uniformly controlled by membrane hyperstructures and ultimately affects the level of HSP expression as well (Vigh L, Horvath I, Maresca B et al, unpublished). We believe that the signaling from the transmembrane growth factor receptors to HS genes fulfills such a criterion. PI3K, Akt and GSK3, a negative regulator of HSF1 activation,[58] are central components of this signaling cascade . Diabetes is associated with a low HSP level, and with decreased PI3K and enhanced GSK3 activities.[59] Correction of a low HSP state improves insulin resistance. A close chemical relative of bimoclomol, BRX-220, has been reported to increase the insulin sensitivity of both Zucker diabetic fatty rats and streptozotocin-diabetic rats.[60] Another bimoclomol analog, the HSP coinducer BGP-15, has passed successfully a Phase IIa human clinical trial as an insulin-sensitizer compound. Similarly to several other insulin sensitizers,[61] BGP-15 upregulated Akt kinase and acted as an GSK3 inhibitor. However, the operation of such a signaling cascade also provides a potential mechanism for the widespread elevation in HSP expression in cancer (Fig. 2B). As recently documented by Khaleque et al,[62] an HSP elevation in tumor cells can be induced by the highly malignant growth factor heregulin β1 (HRGβ1), which causes homo- and heterodimerization between each member of the four ErbB receptor molecules via "horizontal signaling" in the plane of the plasma membrane. Just like the intermolecular dimerization of membrane-associated Hik33, which serves as the first key step in low-temperature sensing and signaling, with resultant upregulation of desaturases in prokaryotes (Fig. 2A), the formation of raft-associated GFR dimers is followed by tyrosine phosphorylation of their intracellular domains in mammalian cells (Fig. 2B). The major downstream signaling pathways include the Ras-Raf1-Mek-ERK and PI3K-PDK1-Akt pathways.[62] HRGβ1 appears to be linked to HSP expression by its activation of HSF1 through inhibition of the constitutive kinase GSK3.

It may be noted that the transmission of a signal from the cell surface to HS genes following mild HS, shear stress or growth factors, with the involvement of the activation of receptors or receptor-like molecules and their clustering into lipid rafts, is uniformly dependent on precise regulation by the lipid composition of the membranes. The localization of growth factor receptors to distinct microdomains appears to modulate both their ligand binding and tyrosine kinase activities.[63] Moreover, the major signal termination mechanism, i.e., the lateral movement of dimerized receptors in the plane of the surface membrane and the delivery of activated receptor-containing endosomes, is totally lipid-dependent.[64,65]

The precise role played by lipid microdomains in membrane-directed stress sensing and signaling is far from clear. Their studies have been hampered by the lack of suitable physical methods for the visualization of membrane microdomains in intact cells.[8,66] As compared with unicellular model organisms, the membrane structures in mammalian cells display enormous lipid variety, multiple functions, locations, associations and intimate links with neighbouring membranes, and this makes their study extremely difficult. It is tempting to speculate that one of the major roles of the more than one thousand lipid molecular species in mammalian membranes is to provide an on-off switch for signaling events at the membrane level.[66,67] Membrane lipids are among the molecules that adapt best in response to various perturbations. Even subtle changes in the compositions of acyl chains or head groups can alter the packing arrangements of lipids within a bilayer. As a chain reaction, altered lipid packing properties change the balance between bilayer and nonbilayer lipids, affect the bilayer stability and fluidity, and also the lipid-protein interactions and microdomain organizations. External factors, including temperature, chemicals, ions, radiation, pressure, nutrients, the growth phase of the culture, etc., are all capable of changing the membrane packing order and lipid composition.[8] Finally, it should be born in mind that the lipid-selective association of certain HSPs may also lead to an increased membrane molecular order and may result in downregulation of the HS gene expression.[68-70]

In recent years, our understanding of the plasma membrane has changed considerably as our knowledge of lipid microdomains has expanded. These include structures known as lipid rafts and caveolae, which are readily identified by their unique lipid constituents. Cholesterol, sphingolipids and specific phospholipids with more saturated fatty acyl chain moieties are typically highly enriched in these lipid microdomains. Since lipid microdomains have been widely shown to play important roles in the compartmentalization, modulation and integration of cell signaling, we suggest that these microdomains may additionally have an influential role in stress sensing and signaling. Favorable interactions between cholesterol and the saturated lipids result in patches of liquid-ordered (L_o) or raft domains from the remaining liquid-disordered (L_d) membranes. Specific signaling proteins are targeted to or concentrated in rafts in consequence of the greater solubility of their lipid anchors in L_o than in L_d compartments.[71] Since the raft structure is also dependent on the thermally controlled lipid-phase behavior, even mild changes in temperature could result in a fundamentally altered solubility and consequently redistribution of these proteins in the rafts. More severe heat may result in complete raft dissolution and disruption of the signaling activities.[8] It was recently documented by means of single-molecule microscopy that diffusional trapping through protein-protein interactions does indeed create plasma membrane microdomains that concentrate or exclude cell surface proteins, facilitating cell signalling.[72] Different fluorescence imaging techniques, adding the beauty of visualization to the scientific information,[73] in conjunction with computational modeling techniques (see later), can promote a better understanding of the network of complex molecular interactions taking place within membranes under stress conditions.

Stress Response Profiling: Can We "Zoom In" on Membrane Hyperstructures Engaged in the Generation of Stress Signal?

Most of the experiments targeted toward an understanding of stress signaling are performed on one or just a few cells, or on many cells, but without a knowledge of the prehistory of each individual cell. It is critical that cells in culture can be quite heterogeneous and present a wide variety of phenotypes. This heterogeneity can result from the cells being in different stages of the cell cycle, or from other intrinsic differences between the cells. A recent study of transcription activation in a serum starvation assay highlighted the wide variety of responses among individual cells, even though the large differences between the control and treatment groups were scored in a population assay such as an immunoblot.[74] For this reason, it is necessary to record data from many individual cells so as to aquire a statistically significant sampling of cellular behavior and dynamics. This quest depends heavily on molecular imaging, which shows when and where genetically or biochemically defined molecules, signals or processes appear, interact and disappear, in time and space. With high-content cellular analysis, it is necessary to perform high-throughput phenotype profiling, linking gene expression to biochemical signaling pathways in the cell and, ultimately, to cell behavior.[74] During or after the imposition of stress on cells, it becomes possible to zoom in on individual cells or individual fluorescently tagged molecules, using an ultrasensitive, high-speed camera to observe what happens as each cell reacts to the particular treatment.[75] A change may occur in the topology of a particular membrane domain, a candidate from which an initial HS signal could originate, or a membrane receptor may be activated, allowing the monitoring in real time as a receptor complex responds and activates signaling pathways. An intracellular signaling molecule tagged with a fluorescent protein could light up as it interacts with another component of the signaling pathway, perhaps relaying instructions to the nucleus to activate or deactivate a particular stress defensive gene. The particularly valuable aspect of this methodology is not the astounding visual images it produces, but rather the abundant and diverse data that can be extracted from those images—data that afford a better understanding of what is happening in the cell in response to stress. Image analysis of this nature requires efficient imaging algorithms, such as the open source Cell Profiler (www.cellprofiler.org), which provides a selection of tools for automatic analysis of the cell morphology, and the localization and intensity distribution of fluorescent markers (Carpenter AE, Jones TR, Lamprecht M et al, unpublished). Investigators can exploit the flexibility of the data analysis software to design their own algorithms from the building blocks provided, customizing the data-mining tools to optimize the data output. The potential applications and ultimate value of high content screening and cellular image analysis are limited only by the imagination and expertise of the investigator who is using them to probe stress sensing and cell behavior.

Can We Point to Lipid Molecular Species Engaged in Stress Sensing and Signaling?

The diversity of the membrane constituent phospholipids originates from the differences in head groups, acyl chains, and the degree of unsaturation, resulting in several hundred coexisting lipid molecular species within a given cell. This polymorphism permits the regulation of membrane fluidity in poikilothermal organisms. The organized self-assembly of functional signal-generating protein-lipid complexes leads to a wide range of chemical and physical membrane properties, through which the membrane-associated events are regulated and coordinated.

By means of a combination of thin-layer separation techniques with MS in the cyanobacterium *Synechocystis,* a highly saturated monoglucosyldiacylglycerol (MGlcDG) molecular species ("heat shock lipid") was identified; this has the ability to cause membrane rigidification and to counteract the formation of nonbilayer structures during heat/light stress. Moreover, MGlcDG expressed the strongest interaction with the membrane-stabilizing small HSP, HSP17.[76] Specific HSP-lipid interactions may be unrecognized means for the spatial separation and distinct compartmentalization of HSPs to lipid domains, which are thought to be involved in various signaling pathways.[8] Another HS lipid, cholesteryl glucoside (CGl), was shown to accumulate rapidly in cells from molds to humans following exposure to environmental stress. CGl production is followed by the activation of certain PKCs and the induction of HSPs.[77,78]

Lipids and lipid-derived metabolites may also contribute to the early HS-induced cell responses. Among others, arachidonate, certain prostaglandins, IP3 and ceramides have been proved to be second messengers of the HSR.[8] The source lipids of signaling molecules are localized in different compartments of the cell. The mapping of time and stimulus-dependent changes in the lipidome of plasmamembrane, rafts, mitochondria ER, Golgi or nuclear membranes requires appropriate subcellular fractionation and lipid isolation. In principle, complete characterization of the lipid components of these membranes or membrane microdomains is essential for an understanding of how the membrane senses and copes with environmental stress.

The cross-talk between the raft proteins and lipids would trigger a remodeling of the newly-formed raft to optimize the stability of the new structure. This might be accomplished by excluding some proteins and lipids from the raft, while recruiting others. It is tempting to speculate that, in response to stress conditions, chemical (e.g., oxidation) and physical (e.g., fluidity) changes in these lipids might result in the recruitment/exclusion of other proteins and/or lipids, leading to a unique remodeled composition, thereby provoking the generation of stress signals.

However, it should be pointed out that no two rafts or membrane preparations yield the same analytical result, and there is no isolation procedure that does not involve loss of several components during "disassembly" of the cells. These caveats provide the impulse for the development of whole-cell or organ analysis with the aim of the creation of overall lipid maps after different perturbations. The ESI/MS of lipids is the most sensitive, discriminating and direct method for identification of the complete set of lipid molecular species.[79-81]

Furthermore, lipidomics today is extended toward the more polar extraction procedures and the inspection of unknown peaks of possible new moieties. In this manner, the entire spectrum of new metabolites can be identified and the potential signaling functions of these molecules can be determined. The question arises as to how the observed lipidomic and metabolic fingerprints will help in the elucidation of the sensing and signaling of stress. The plethora of new data requires different statistical approaches, which are not limited to simple statistical metrics. Supervised and unsupervised classification methods, such as discriminant, cluster and principal component analysis, multidimensional scaling or other pattern recognition techniques, will be important[82,83] for the identification of significant and reproducible changes in the patterns of lipid species changes in response to stress or other stimuli. An understanding of the potential signaling function of these metabolites or the importance of

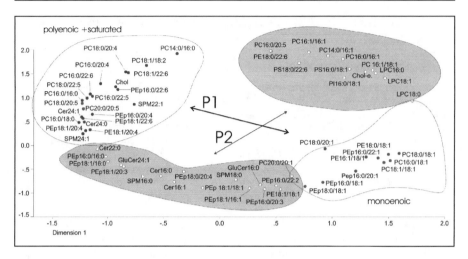

Figure 3. The lipidome of the cells was determined by direct-injection ESI tandem mass spectrometry (Balogh and Liebisch, unpublished) and the molecular species compositions of PC (phosphatidylcholine), PE (phosphatidylethanolamine), PEP (Phoshaptidyl- ethanolamine plasmalogen), PS (Phosphatidylserine), PI (Phosphatidylinositol) , SPM (Sphingomyelin), LPC (Lysophosphatydilcholine), Cer (Ceramide), Glucer (Glucosyl ceramide) , Chol (Cholesterol) and Chol-e (Cholesterol esther) were calculated. First, the meaningful components (based on measurement accuracy and the relative quantity) were selected (approximately 100 of 300 species) and the values were transformed to Z-scores. For multidimensional scaling, the Euclidean distance was used. Two principal components (P1 and P2) were extracted. The groups represent the positive or negative participation associated with a given component.

a specific lipid composition of a membrane domain requires the correlation of the alterations with physiological data with the aid of system-biological methods.

In order to illustrate the power of these statistical procedures, the lipidome changes of B16 mouse melanoma cells under different circumstances are presented in Figure 3. B16 cells were subjected to different stimuli or plated with low or high numbers of cells. Besides the cholesterol level, the molecular species compositions of several lipid classes were determined. Figure 3 shows the principal components represented in a two-dimensional scaling model. The principal component P1 extracted those molecular species which can be clearly attributed to the lipid changes due to the distinct initial cell densities, while the lipid molecular species belonging in component P2 were mainly altered by other stimuli. Our parallel investigations on the gene expression patterns by the chip method and the induction of HSPs by RT-PCR revealed that the lipid status plays a regulatory role in the stress response (Balogh G. and Vigh L, unpublished). The combination of gene expression and lipidome data by appropriate statistical methods will trigger the postulation of new theories with higher explanatory relevance.

Computational Methods for the Design of Subtle Interactions between Lipids and Proteins of Membranes

The interest of computational biologists has recently focused on the active roles of phospholipids in affecting the behavior of membrane proteins,[84] the assembly of protein-lipid arrays and the modulation of protein-protein interactions.[85] Lipid and protein-lipid- phase equilibria are also believed to be relevant for membrane fusion and raft formation. Computational methods can provide useful information on the inner organization of membranes, which are normally inaccessible by direct experimental investigations. During the past twenty years, atomistic simulation and, in particular, full atom Molecular Dynamics (MD), have been developed to a degree

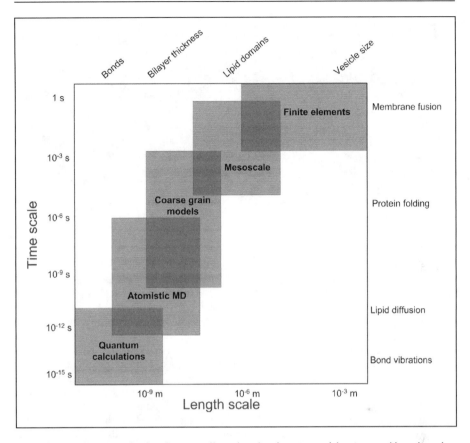

Figure 4. Simulation methods schematically ordered as functions of the time and length scales. Some relevant membrane events are indicated.

that allowss prediction of the structural characteristics with acceptable precision. Besides MD, computer modeling also utilizes coarse grain (CG) dynamics, quantitative structure analysis relationship methods (QSAR) and quantum mechanics/molecular mechanics (QM/MM), which have a large potential for the investigation of biological structures (Fig. 4).

A number of atomistic simulations have been performed to reproduce and predict many fundamental properties of lipid membranes.[86,87] However, with today's computers and algorithms, the size of the systems and the time scales for which phospholipid bilayers can be studied preclude the scrupulous examination of many phenomena such as the cooperative motion in phase transitions, or the interaction of proteins with the membrane or the flip-flop. For these reasons, numerous strategies have been conceived to extend the temporal and spatial limits of MD, which are capable of covering a large spatial and time scale, but with a reduced level of detail in the representation of the membrane. CG models are an improvement toward the study of collective phenomena in membranes[88] which are not accessible by the MD approach.

CG simulation are extremely effective in simulating both the formation and the behavior of a mixed bilayer. In Figure 5 the spontaneous formation of a bilayer of dimirystoyl-phospahtidylcholine/dimirystoyl-phosphatidylethanolamine (DMPC/DMPE) was calculated through the use of CG dynamics. The simplification introduced with CG can strongly affect the possibilities of investigations of protein folding and protein-membrane interactions.

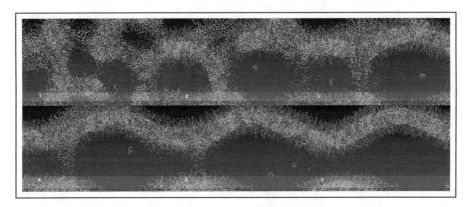

Figure 5. Spontaneous organization of DMPC/DMPE and water into a bilayer structure (CG simulation).

The continuous development of new parameterization techniques will hopefully permit the simulation of ever more complicated mixtures of lipids, and consequently a better mimicking of membranes. Evidently, simulations and fast computers are not sufficient for the generation of new understanding: new models are necessary as well. A model is a simple way of describing and/or predicting scientific results, which is known to yield an incorrect or incomplete description. Models may be simple mathematical descriptions or completely nonmathematical. In recent decades, continuous efforts have been made to reformulate biological concepts into a more formal, mathematical language, allowing the prediction and understanding of phenomena without the performance of difficult experiments. The most difficult problem in the simulation of lipid membrane behavior is the lack of correct models. Implicit in most membrane studies is the idea of an "average membrane" or, in other words, the idea that a membrane with a given lipid composition exists for each cell (which is not true, of course). As described earlier, a cell membrane is organized in discrete domains, and the knowledge of the lipid composition is therefore necessary for the simulation of membrane suprastructures.[89] A tremendous improvement of the computer simulations of lipid rafts will be possible in the coming years with the accumulation of experimental data on lipid composition and with the availability of faster computers. The design of new membrane-interacting drugs (such as the stress protein coinducer bimoclomol and its analogs) will be assisted by the molecular simulation of membranes and proteins. It will be possible to highlight slight conformational changes induced in a membrane protein induced in the target raft by a chemical, physical (e.g., temperature) or biological alteration of the bilayer. Nevertheless, the generation of new models for membrane behavior is a task which cannot be performed via molecular (or CG) dynamics, since these are not intrinsically able to observe properties which are not encoded in the simulation. There are computational methods, which to some extent, are able to generate models that can ultimately be validated via MD. Examples come from **g**enetic **a**lgorithms (GAs) and neural networks.

GAs permit computers to generate new, random data. In combination with a fitness function, only the best data are randomly exchanged and mutated until a new generation of data is produced. In all cells, the genetic material is DNA, which is described in terms of base pairs. A string of three base pairs makes up a codon, which corresponds to a specific amino acid in the coded protein. As in real genes, GAs operate on a sort of genetic material, but as a substitute of DNA; the "genetic material" is a linear string of symbols. The computer equivalent of the base pairs and codon is replaced by a symbol or numbers that represent any structural unit.

We recently (Piotto S, Bianchino E., unpublished) extracted information on the antimicrobial activities of short peptides (Fig. 6) through the successful application of GA techniques.

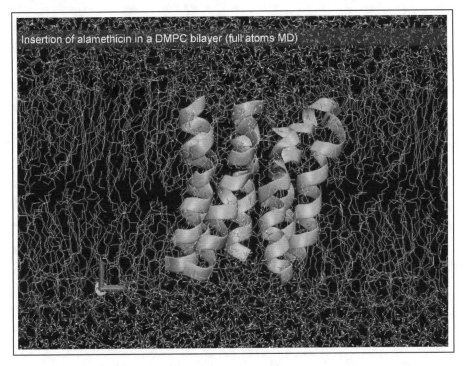

Figure 6. Insertion of alamethicin in a DMPC bilayer (full atoms MD).

Natural antimicrobial peptides generally consist of 10-20 amino acids and afford the first line of defense against infections in all metazoans and plants. Several mechanisms have been proposed to explain their activity, but only computational analysis of the similarity, carried out with a GA, permits extrapolation of the chemical-physical parameters responsible for the corresponding perturbation of a cell membrane.

Today, more than ever, experiments are becoming electronic; this approach allows not only the management and analysis of vast data sets, but also their acquisition in the first place. There is a widespread tendency to perform in silico experiments, not only to investigate the interactions of small molecules with a protein, but also to analyze changes in the membrane physical state. The development of computer modeling requires both new formalisms and new methods to implement these formalisms in physical devices.

Conclusions

In the past decade, a membrane sensor model has evolved which predicts the existence of a membrane-associated stress sensing and signaling mechanism from prokaryotes to mammalian cells. Changes in the physical state or composition of lipid molecular species with the concomitant reorganization of such membrane microdomains (rafts) may serve as the molecular switch for the operation of these "cellular thermometers". Single-cell fluorescence imaging technologies, real-time detection and the monitoring of rafts and combination of these technologies with lipidomics and computational methods may allow the identification and characterization of these hypothetical stress-sensory domains. HSPs could be induced or inhibited by drugs that specifically change the operation of such sensory lipid-protein membrane hyperstructures.

Acknowledgements

This work was supported by grants from the Hungarian National Scientific Research Foundation (OTKA: TS 044836, T 038334) and Agency for Research Fund Management and Research Exploitation (RET OMFB00067/2005 and Bio- 00120/2003 KPI).

References

1. Kultz D. Molecular and evolutionary basis of the cellular stress response. Annu Rev Physiol 2005; 67:225-257.
2. Ananthan J, Goldberg AL, Voellmy R. Abnormal proteins serve as eukaryotic stress signals and trigger the activation of heat shock genes. Science 1986; 232:522-524.
3. Lepock JR. Measurement of protein stability and protein denaturation in cells using differential scanning calorimetry. Methods 2005; 35:117-125.
4. Pockley GA. Heat shock proteins in health and disease: Therapeutic targets or therapeutic agents? Expert Rev Mol Med 2001; 2001:1-21.
5. Soti C, Nagy E, Giricz Z et al. Heat shock proteins as emerging therapeutic targets. Br J Pharmacol 2005; 146:769-780.
6. Vigh L, Literati NP, Horvath I et al. Bimoclomol: A novel nontoxic, hyroxylamine derivative with stress protein inducing activity and wide cytoprotective effects. Nature Med 1997; 3:1150-1154.
7. Török Z, Tsvetkova NM, Balogh G et al. Heat shock protein coinducers with no effect on protein denaturation specifically modulate the membrane lipid phase. Proc Natl Acad Sci USA 2003; 100:3131-3136.
8. Vigh L, Escriba P, Sonnleitner A et al. The significance of lipid composition for membrane activity: New concepts and ways of assessing function. Progr Lipid Res 2005; 44:303-344.
9. Escriba PV. Membrane-lipid therapy: A new approach in molecular medicine. Trends Mol Med 2006; 12:34-43.
10. Yatvin MB, Cramp WA. Role of cellular membranes in hyperthermia: Some observations and theories reviewed. Int J Hyperthermia 1993; 9:165-185.
11. Hazel JR, Williams EE, Livermore R et al. Thermal adaptation in biological membranes: Functional significance of changes in phospholipids molecular species composition. Lipids 1991; 26:277-282.
12. Vigh L, Horváth I, van Hasselt PR et al. Effect of frost hardening on lipid and fatty acid composition of chloroplast thylakoid membranes in two wheat varieties of contrasting hardiness. Plant Physiol 1985; 79:756-759.
13. Glatz A, Vass I, Los DA et al. The Synechocystis model of stress: From molecular chaperones to membranes. Plant Physiol Biochem 1999; 37:1-12.
14. Vigh L, Joó F. Modulation of membrane fluidity by catalytic hydrogenation affects the chilling susceptibility of the blue-green alga, Anacystis nidulans. FEBS L 1983; 162:423-426.
15. Vigh L, Gombos Z, Joó F. Selective modification of cytoplasmic membrane fluidity by catalytic hydrogenation provides evidence on its primary role in chilling susceptibility of the blue-green alga, Anacystis nidulans. FEBS L 1985; 191:200-204.
16. Benkö S, Hilkmann H, Vigh L et al. Catalytic hydrogenation of fatty acyl chains in plasma membranes; effect of membrane fluidity and expression of cell surface antigens. Biochim Biophys Acta 1987; 896:129-135.
17. Quinn PJ, Joó F, Vigh L. The role of unsaturated lipids in membrane structure and stability. Progr Biophys Mol Biol 1989; 53:71-103.
18. Joó F, Balogh N, Horváth I et al. Complex hydrogenation/oxidation reactions of the water-soluble hydrogenation catalyst palladium di(sodium alizarinmonosulfonate) and details of homogeneous hydrogenation of lipids in isolated membranes and living cells. Anal Biochem 1991; 194:34-40.
19. Vigh L, Horváth I, Thomspon GA. Recovery of Dunaliella salina cells following hydrogenation of lipids in specific membranes by a homogeneous palladium catalyst. Biochim Biophys Acta 1988; 937:42-50.
20. Pak Y, Joó F, Vigh L. Action of a homogeneous hydrogenation catalyst on living Tetrahymena mimbres cells. Biochim Biophys Acta 1990; 1023:230-238.
21. Vigh L, Los DA, Horváth I et al. The primary signal in the biological perception of temperature: Pd-catalyzed hydrogenation of membrane lipids stimulated the expression of the desA gene in Synechocystis PCC 6803. Proc Natl Acad Sci USA 1993; 90:9090-9094.
22. Los D, Horváth I, Vigh L et al. The temperature-dependent expression of the desaturase gene desA in Synechocystis PCC6803. FEBS L 1993; 318:57-60.
23. Suzuki I, Los DA, Kanesaki Y et al. The pathway for perception and transduction of low-temperature signals in Synechocystis. EMBO J 2000; 19:1327-1334.

24. Murata N, Los DA. Histidine kinase Hik33 is an important participant in cold-signal transduction in cyanobacteria. Physiol Plant 2006; 126:17-27.
25. Maresca B, Cossins AR. Fatty feedback and fluidity. Nature 1993; 365:606-607.
26. Carratù L, Franceschelli S, Pardini CL et al. Membrane lipid perturbation sets the temperature of heat shock response in yeast. Proc Natl Acad Sci USA 1996; 93:3870-3875.
27. Chatterjee MT, Khalawan SA, Curran BPG. Alterations in cellular lipids may be responsible for the transient nature of the yeast heat shock response. Microbiology 1997; 143:3063-3068.
28. Chatterjee MT, Khalawan SA, Curran BPG. Cellular lipid composition influences stress activation of the yeast general stress response element (STRE). Microbiology 2000; 146:877-884.
29. Nguyen AN, Shiozaki K. Heat shock-induced activation of MAP kinase is regulated by threonine- and tyrosine-specific phosphatases. Genes Dev 1999; 13:1353-1363.
30. Madrid M, Soto T, Khong HK et al. Stress-induced response, localization, and regulation of the Pmk1 cell integrity pathway in Schizosaccharomyces pombe. J Biol Chem 2006; 281:233-243.
31. Shiozaki K, Russell P. Cell-cycle control linked to extracellular environment by MAP kinase pathway in fission yeast. Nature 1995; 378:739-743.
32. Bimbó A, Jia Y, Poh SL et al. Systematic deletion analysis of fission yeast protein kinases. Eukaryotic Cell 2005; 4:799-813.
33. Toda T, Dhut S, Superti-Furga G et al. The fission yeast pmk1+ gene encodes a novel mitogen-activated protein kinase homolog which regulates cell integrity and functions coordinately with the protein kinase C pathway. Mol Cell Biol 1996; 16:6752-6764.
34. Gallo GJ, Prentice H, Kingston RE. Heat shock factor is required for growth at normal temperatures in the fission yeast. Mol Cell Biol 1993; 13:749-761.
35. Saltsman HA, Prentice HL, Kingston RE. Mutations in the Schizosaccharomyces pombe heat shock factor that differentially affect responses to heat and cadmium stress. Mol Gen Genet 1999; 261:161-169.
36. Horváth I, Glatz A, Varvasovszki V et al. Membrane physical state controls the signaling mechanism of the heat shock response in Synechocystis PCC 6803: Identification of hsp17 as a "fluidity gene". Proc Natl Acad Sci USA 1998; 95:3513-3518.
37. Inoue K, Matsuzaki H, Matsumoto K et al. Unbalanced membrane phospholipids, compositions affect transcriptional expression of certain regulatory genes in Escherichia coli. J Bacteriol 1997; 179:2872-2878.
38. Wilkinson WO, Bell RM. Sn-glycerol-3-phosphate acyl-transferase tubule formation is dependent upon heat shock proteins (htpR). J Biol Chem 1988; 263:14505-14510.
39. Shigapova N, Török Z, Balogh G et al. Membrane fluidization triggers membrane remodeling which affects the thermotolerance in Escherichia coli. Biochem Biophys Res Comm 2005; 328:1216-1223.
40. De Marco A, Vigh L, Diamant S et al. Native folding of aggregation-prone recombinant proteins in Escherichia coli by osmolytes, plasmid- or benzyl alcohol-overexpressed molecular chaperones. Cell Stress Chaperones 2005; 10:329-339.
41. Saidi Y, Finka A, Chakhporanian M et al. Controlled expression of recombinant proteins in Physcomitrella patents by a conditional heat-shock promoter: A tool for plant research and biotechnology. Plant Mol Biol 2005; 59:697-711.
42. Mikami K, Murata N. Membrane fluidity and the perception of environmental signals in cyanobacteria and plants. Progr Lipid Res 2003; 42:527-543.
43. Mileykovskaya E, Dowhan WJ. The Cpx two-component signal transduction pathway is activated in Escherichia coli mutant strains lacking phosphatidylethanolamine. J Bacteriol 1997; 179:1029-34.
44. Lehel C, Los D, Wada H et al. A second GroEL-like gene, organized in a groESL operon is present in the genome of Synechocystis sp. PCC 6803. J Biol Chem 1993; 268:1799-1804.
45. Balogh G, Horváth I, Nagy E et al. The hyperfluidization of mammalian cell membranes acts as a signal to initiate the heat shock preotein response. FEBS J 2005; 272:6077-6086.
46. Yan D, Saito K, Ohmi Y et al. Paeoniflorin, a novel heat shock protein-inducing compound. Cell Stress Chaperones 2004; 9:378-389.
47. Hargitai J, Lewis H, Boros I et al. Bimoclomol, a heat shock protein coinducer, acts by the prolonged activation of heat shock factor-1. Biochem Biophys Res Comm 2003; 307:689-695.
48. Kieran D, Kalmar B, Dick J et al. Treatment with arimoclomol, a coinducer of heat shock proteins, delays disease progression in ALS mice. Nature Med 2004; 10:402-405.
49. Park HG, Han SI, Oh SY et al. Cellular responses to mild heat stress. Cell Mol Life Sci 2005; 62:10-23.
50. Bornfeldt KE. Stressing Rac, Ras, and downstream heat shock protein 70. Circ Res 2000; 86:1101-1103.

51. Xu Q, Schett G, Li C et al. Mechanical stress-induced heat shock protein 70 expression in vascular smooth muscle cells is regulated by Rac and Ras small G proteins but not mitogen-activated protein kinases. Circ Res 2000; 86:1124-1130.
52. Del Pozo MA, Alderson NB, Kiosses WB et al. Integrins regulate Rac targeting by internalization of membrane domains. Science 2004; 303:839-842.
53. Han SI, Oh SY, Woo SH et al. Implication of a small GTPase Rac1 in the activation of c-Jun N-terminal kinase and heat shock factor in response to heat shock. J Biol Chem 2001; 276:1889-1895.
54. Kaspler P, Horowitz M. Heat acclimation and heat stress have different effects on cholinergic muscarinic receptors. Ann Ny Acad Sci 1997; 813:620-627.
55. Vigh L, Maresca B, Harwood J. Does the membrane physical state control the expression of heat shock and other genes? Trends Biochem Sci 1998; 23:369-373.
56. Vigh L, Maresca B. Dual role of membranes in heat stress: As thermosensors they modulate the expression of stress genes and, by interacting with stress proteins, reorganize their own lipid order and functionality. In: Storey KB, Storey JM, eds. Cell and Molecular Responses to Stress. Amsterdam: Elsevier, 2002:173-188.
57. Westerheide SD, Morimoto RI. Heat shock response modulators as therapeutic tools for diseases of protein conformation. J Biol Chem 2005; 280:33097-33100.
58. Bijur GN, Jope RS. Opposing actions of phosphatidylinositol 3-kinase and glycogen synthase kinase-3beta in the regulation of HSF-1 activity. J Neurochem 2000; 75:2401-2408.
59. Hooper PL, Hooper JJ. Loss of defense against stress: Diabetes and heat shock proteins. Diabetes Technol Ther 2005; 7:204-208.
60. Kürthy M, Mogyorósi T, Nagy K et al. Effect of BRX-220 against peripheral neuropathy and insulin resistance in diabetic rat models. Ann NY Acad Sci 2002; 967:482-489.
61. Meijer L, Flajolet M, Greengard P. Pharmacological inhibitors of glycogen synthase kinase 3. Trends Pharmacol Sci 2004; 25:471-480.
62. Khaleque MA, Bharti A, Sawyer D et al. Induction of heat shock proteins by heregulin beta1 leads to protection from apoptosis and anchorage-independent growth. Oncogene 2005; 24:6554-6573.
63. Pike LJ, Han X, Gross RW. Epidermal growth factor receptors are localized to lipid rafts that contain a balance of inner and outer leaflet lipids: A shotgun lipidomics study. J Biol Chem 2005; 280:26796-26804.
64. Gur G, Yarden Y. Enlightened receptor dynamics. Nature Biotechnology 2004; 22:169-170.
65. Koijman EE, Chupin V, Fuller NL et al. Spontaneous curvature of phosphatidic acid and lysophosphatidic acid. Biochemistry 2005; 44:2097-2102.
66. Vereb G, Szöllösi J, Matko J et al. Dynamic, yet structured: The cell membrane three decades after the Singer-Nicolson model. Proc Natl Acad Sci USA 2003; 100:8053-8058.
67. Hinderliter A, Biltonen RL, Almeida PF. Lipid modulation of protein-induced membrane domains as a mechanism for controlling signal transduction. Biochemistry 2004; 43:7102-7110.
68. Török Z, Horváth I, Goloubinoff P et al. Evidence for a lipochaperonin: Association of active protein-folding GroESL oligomers with lipids can stabilize membranes under heat shock conditions. Proc Natl Acad Sci USA 1997; 94:2192-2197.
69. Török Z, Goloubinoff P, Horváth I et al. Synechocystis HSP17 is an amphitropic protein that stabilizes heat-stressed membranes and binds denatured proteins for subsequent chaperone-mediated refolding. Proc Natl Acad Sci USA 2001; 98:3098-3103.
70. Tsvetkova NM, Horváth I, Török Z et al. Small heat shock proteins regulate membrane lipid polymorphism. Proc Natl Acad Sci USA 2002; 99:13504-13509.
71. Harder T. Formation of functional cell membrane domains: The interplay of lipid- and protein-mediated interactions. Phil Trans R Soc Lond B 2003; 358:863-868.
72. Douglass AD, Vale RD. Single-molecule microscopy reveals plasma membrane microdomains created by protein-protein networks that exclude or trap signaling molecules in T cells. Cell 2005; 121:937-950.
73. Sanchez SA, Gratton E. Lipid-protein interactions revealed by two-photon microscopy and fluorescence correlation spectroscopy. Acc Chem Res 2005; 6:469-477.
74. Levsky JM, Shenoy SM, Pezo RC et al. Single-cell gene expression profiling. Science 2002; 297:836-840.
75. Sako Y, Yanagida T. Single-molecule visualization in cell biology. Nature Reviews Mol Cell Biol 2003; 4:SS1-SS5.
76. Balogi Z, Török Z, Balogh G et al. "Heat shock lipid" in cyanobacteria during heat/light-acclimation. Arch Biochem Biophys 2005; 436:346-354.
77. Kunimoto S, Murofushi W, Kai H et al. Steryl glucoside is a lipid mediator in stress-responsive signal transduction. Cell Struct Funct 2002; 27:157-162.

78. Kunimoto S, Murofushi W, Yamatsu I et al. Cholesteryl glucoside-induced protection against gastric ulcer. Cell Struct Funct 2003; 28:179-186.
79. van Meer G. Cellular lipidomics. EMBO J 2005; 24:3159-3165.
80. Han X, Gross RW. Shotgun lipidomics: Multidimensional MS analysis of cellular lipidomes. Expert Rev Proteomics 2005; 2:253-264.
81. Pulfer M, Murphy RC. Electrospray mass spectrometry of phospholipids. Mass Spectrom Rev 2003; 22:332-364.
82. van der Greef J, Stroobant P, van der Heijden R. The role of analytical sciences in medical systems biology. Curr Opin Chem Biol 2004; 8:559-565.
83. Forrester JS, Milne SB, Ivanova PT et al. Computational lipidomics: A multiplexed analysis of dynamic changes in membrane lipid composition during signal transduction. Mol Pharmacol 2004; 65:813-821.
84. de Kruijff B. Biomembranes. Lipids beyond the bilayer. Nature 1997; 386:129-130.
85. Mouritsen OG, Sperotto MM et al. Computational approach to lipid-protein interactions in membranes. Advances in Computational Biology 1996; 2:15-64.
86. Tieleman DP, Marrink SJ, Berendsen HJ. A computer perspective of membranes: Molecular dynamics studies of lipid bilayer systems. Biochim Biophys Acta 1997; 1331:235-270
87. Feller SE. Molecular dynamics simulations of lipid bilayers. Curr Op Coll Int Sc 2000; 5:217-223.
88. Goetz R, Lipowsky R. Computer simulations of bilayer membranes: Self-assembly and interfacial tension. J Chem Phys 1998; 108:7397-7409.
89. Pralle A, Keller P, Florin EL. Sphingolipid-cholesterol rafts diffuse as small entities in the plasma membrane of mammalian cells. J Cell Biol 2000; 148:997-1007.

CHAPTER 12

Beyond the Lipid Hypothesis:
Mechanisms Underlying Phenotypic Plasticity in Inducible Cold Tolerance

Scott A.L. Hayward, Patricia A. Murray, Andrew Y. Gracey and Andrew R. Cossins*

Abstract

The physiological adjustment of organisms in response to temperature variation is a crucial part of coping with environmental stress. An important component of the cold response is the increase in membrane lipid unsaturation, and this has been linked to an enhanced resistance to the debilitating or lethal effects of cold. Underpinning the lipid response is the upregulation of fatty acid desaturases (des), particularly those introducing double bonds at the 9-10 position of saturated fatty acids. For plants and microbes there is good genetic evidence that regulation of des genes, and the consequent changes in lipid saturation, are causally linked to generation of a cold-tolerant phenotype. In animals, however, supporting evidence is almost entirely limited to correlations of saturation with cold conditions. We describe our recent attempts to provide a direct test of this relationship by genetic manipulation of the nematode *Caenorhabditis elegans*. We show that this species displays a strong cold tolerant phenotype induced by prior conditioning to cold, and that this is directly linked to upregulated des activity. However, whilst genetic disruption of des activity and lipid unsaturation significantly reduced cold tolerance, animals retained a substantial component of their stress tolerant phenotype produced by cold conditioning. This indicates that mechanisms other than lipid unsaturation play an important role in cold adaptation.

Introduction

Cold presents a major problem for all living organisms by reducing the rates of molecular processes that underpin life, and by causing damaging and lethal effects at low extremes.[1,2] Poikilotherms in particular, whose tissues, cells and molecules are exposed to the full effects of any environmental temperature variation, often display powerful adaptive responses to cold, which can be classified into two broad groups: (i) capacity adaptations, which occur over the normal range of temperatures and promote the constancy of life functions; and (ii) resistance adaptations, which enhance resistance to the debilitating and even lethal effects of extreme cold.[1] Resistance adaptations thus underlie the adaptive transition to an enhanced stress tolerant phenotype and can occur within minutes/hours—*rapid cold hardening*, or take days/weeks of conditioning to sub-lethal temperatures—*acclimation*. Understanding the magnitude and limitations of these responses, their impact upon fitness and identifying the underlying

---

*Corresponding Author: Andrew Cossins—School of Biological Sciences, Liverpool University, The Biosciences Building, Crown St., Liverpool, L69 7ZB, U.K. Email: cossins@liv.ac.uk

Molecular Aspects of the Stress Response: Chaperones, Membranes and Networks, edited by Peter Csermely and László Vígh. ©2007 Landes Bioscience and Springer Science+Business Media.

mechanisms are central to ecophysiology. In addition, the study of stress adaptation provides the basis for addressing environmental health problems, toxicological risk assessment, utilizing bioindication processes to monitor climate change, and the improvement of clinical techniques.[3]

A historically important and central concept in cold adaptation research is the idea that bodily lipids alter their physical properties in response to cold exposure, and that this mediates cold adaptation in its various forms. This originated over a century ago, when Henriques and Hansen[4] noted the higher melting point of subcutaneous fats in pigs raised wearing clothing, which resulted in increased tissue temperature. From these somewhat bizarre beginnings, the 'lipid hypothesis of temperature adaptation' has developed to encompass the idea of an optimal physical condition (variously termed fluidity or hydrocarbon physical ordering) of cellular and sub-cellular membranes for supporting critical membrane-associated functions, and a homeostatic mechanism for preserving this property in the face of environmental disturbance. Thus, acute cold causes some disturbance to membrane properties which leads to disrupted function or injury, but prior conditioning alters physical properties to offer protection.[5] While the primary lesion inflicted through cold exposure has not been identified with any great precision, it is thought to involve either a disruption in the performance of membrane-bound proteins, or a disturbance to normal membrane permeability. The process of conserving membrane physical state under fluctuating environmental conditions has been termed 'homeoviscous adaptation'[6] and has found application in a wide variety of organisms from microbes to mammals. The lipid hypothesis has also been invoked in a variety of other phenomena including environmentally adaptive dormancies, such as insect diapause,[7] and mechanisms of anesthesia.[8]

Here we review evidence in favour of the lipid hypothesis and explore the extent to which this mechanism accounts for inducible increases in cold tolerance, and for the inducible cold resistant phenotype of the whole organism. We highlight recent gene manipulation work in plants and microorganisms that strongly support the hypothesis, and comment on the paucity of equivalent data in animals.

Cold Adaptation and the Lipid Hypothesis

The recognition that lipid compositional adjustments underpin the acquired protection to chilling (i.e., nonfreeze), through preconditioning or acclimation, originates from the early work of Heilbrunn,[9] and membrane adaptation is now widely regarded as a central contributor to low temperature survival of all organisms, including bacteria,[10] plants[11] and poikilothermic animals.[12] Prior exposure to cool conditions typically leads to an increase in the proportion of unsaturated fatty acids in membrane phospholipids, at the expense of saturated fatty acids, the increased hydrocarbon disorder compensating for the rigidifying effects of cooling.[12] There is now a huge literature since the 1950s supporting this basic observation using progressively more sophisticated analytical techniques. As a result, the syndrome now also includes changes in phospholipid headgroup class,[13] molecular species composition,[14] cholesterol content,[15] as well as a distinction between lipid raft and nonraft regions.[16] Most attention however has focused on the mechanisms of changing fatty acid unsaturation, and evidence has accumulated in all domains of life for a key role of fatty acid desaturases, since they alone insert double bonds at specific positions of the fatty acyl chain. Evidence that acyl desaturase activity contributes directly to chilling resistance is strong, at least for prokaryotes and plants, where genetic or gene expression manipulation has allowed this hypothesis and its implications to be evaluated. In contrast, while considerable circumstantial evidence exists supporting the role of membrane lipid unsaturation in acquired cold tolerance in animals, clear indications of a causal relationship are at present limited.

Evidence in Prokaryotes

The unicellular cyanobacterium *Synechocystis* sp. PCC 6803 (hereafter *Synechocystis*) is perhaps the most intensively studied prokaryote with respect to membrane adaptation in response to cold. Systematic mutagenesis of potential cold sensors and signal transducers, in combination with DNA microarrays, have advanced considerably our understanding of the cold stress response[17] and have also highlighted a critical homeoviscous function for fatty acid desaturases.

Inactivation of the Δ9 desaturase gene in *Synechocystis* is lethal.[18] In contrast, Δ12-desaturase mutants are viable, but have a dramatically different fatty acid composition relative to wild type cells.[19-21] The resultant strains were characterised by a considerable increase in the level of oleic acid, at the expense of polyunsaturated fatty acids, and by a strong sensitivity to cold exposure.[18] Combined inactivation of Δ12 and Δ6 genes resulted in the complete inability to synthesise polyunsaturated fatty acids, and cells were unable to recover from cold-induced photosystem damage.[22] Further evidence supporting the importance of polyunsaturated fatty acids to cold tolerance was provided by experiments with *Synechococcus* sp PCC 7942, which has a single Δ9 desaturase and can only synthesize monounsaturated fatty acids. When manipulated to express the Δ12 gene of *Synechocystis*, considerable amounts of di-unsaturated fatty acids were produced and the cold tolerance of cells increased.[23] In contrast, targeted inactivation of the ω-3 desaturase in *Synechocystis*, which introduces a double bond into fatty acids that already have a double bond at the Δ12 position, had a limited influence on cold tolerance,[20] unless combined with conditions of nutrient limitation.[21] Thus cold protection was dependent upon position-specific desaturase reactions. Experimentally increasing membrane rigidity, through a catalyst-mediated hydrogenation reaction, provided direct support that desaturase transcript induction was linked to membrane physical condition through diminishing quantities of polyunsaturated fatty acids.[24] Vigh and colleagues[24] thus suggested that membrane fluidity is the first signal in the perception of cold, and that desaturase induction occurs when this change exceeds a critical threshold.

Evidence in Plants

The use of compounds which inhibit the synthesis of polyunsaturated fatty acids provided perhaps the first direct evidence of their importance cold tolerance in plants.[25] However, clear genetic proof of the relationship between desaturase enzyme activity and cold tolerance in plants did not emerge until the isolation of fatty acid desaturase (*fad*) mutants in *Arabidopsis thaliana*.[26] Several mutants, defective in the desaturation of membrane lipids, have been identified,[27-29] each having reduced amounts of polyunsaturated fatty acids and greater chilling susceptibility relative to wild type plants.[30,31] *fad* genes from *A. thaliana*, e.g., *fad7* (the gene for chloroplast ω-3 fatty acid desaturase), have also been successfully introduced into tobacco plants with a resultant increase in trienoic fatty acids and enhanced cold tolerance.[32] In addition, transgenic tobacco plants expressing Δ9-desaturase genes from *A. thaliana*[33] or cyanobacterium[34] have greatly reduced saturated fatty acid content and significantly increased chilling resistance. Other genes involved in fatty acid desaturation pathways have generated similar results, for example glycerol-3-phosphate acyltransferase, a key enzyme determining the extent of *cis*-unsaturation in the phosphatidylglycerol. Plants overexpressing this gene had increased levels of saturated and trans-monounsaturated molecular species of phosphatidylglycerol in chloroplast membranes[35,36] and demonstrated reduced recovery from chilling in both the photosynthetic machinery[37] and reproductive organs.[36] Interestingly, silencing of chloroplast ω-3 fatty acid desaturase in tobacco plants enhanced their ability to photosynthesize at high temperatures, while mutants lacking ω-3 fatty acid desaturase in the endoplasmic reticulum showed wild type photosynthetic activity.[38] Thus, adaptation to temperature change in the chloroplast fatty acids may have a greater influence on survival of temperature stress than in nonchloroplast lipids in plants.[38]

Evidence in Animals

The principal evidence in favour of the lipid hypothesis in animals comes from correlated changes in lipid unsaturation and cold tolerance. Examples of this relationship continue to emerge from the literature and, for invertebrates at least, a reasonable diversity of organisms and tissues have been studied to date.[39-44] In contrast, studies investigating homeoviscous adaptation in poikilothermic vertebrates appear to be exclusively restricted to fish.[45]

Typically these works have mainly been directed at following the effects of cold acclimation, but Logue and colleagues[46] demonstrated a clearcut increase in unsaturation of membrane phosphoglycerides from fish living at high latitudes, including in polar waters, compared to warm water fish species and homeotherms. This effect was linked particularly to the *sn-2* position of the ethanolamine phosphoglycerides with the replacement of a saturate with a Δ9-monunsaturate, a change that has the greatest effect upon membrane physical properties. Wodtke and colleagues[47] first convincingly showed that a hepatic Δ9-desaturase in fish was substantially induced in the 5-7 days following transfer to the cold. Wodtke and Cossins[48] later found that the time course of induction was broadly similar to that of changes in membrane physical structure, the implication being that homeoviscous adaptation was a direct result of the changing expression of this gene. Tiku et al[49] cloned a Δ9-desaturase from carp and demonstrated that cold induction of enzyme activity was caused by both post-translational and transcriptional mechanisms, which together accounted for the graded, but transient, response to increasing cold. Polley et al[50] subsequently identified a second hepatic isoform and showed that only one of the two hepatic isoforms was induced transcriptionally by cold.

Using cDNA microarrays to screen for changing gene expression in multiple tissues, Gracey et al[51] detected desaturase transcripts in a variety of tissues other than liver in the common carp. Thus, the key gene accounting for increased unsaturation in the cold has a widespread distribution in animals and earlier ideas on its restriction to hepatic tissues are giving way to a much wider tissue distribution. Most recently, the Cossins group has discovered new 9-desaturase isoforms in fugu and the zebrafish, which arose from an ancient duplication event that preceded that of the carp hepatic isoforms (Evans H, Cossins AR, Berenbrink M, unpublished studies). One of the new isoforms was brain-specific but at present its cold-inducibility is not known.

In animals the most direct evidence in favour of a role for desaturases enzymes in establishing the cold tolerant phenotype comes from genetic manipulation of *Drosophila melanogaster*. Greenberg et al[52] used site-directed gene replacement to transfer a geographically differentiated desaturase allele between 'African' and 'Cosmopolitan' races of *Drosophila* within an otherwise identical genetic background. A significant decrease in cold tolerance was noted when the 'African' allele was artificially introduced into 'Cosmopolitan' flies, and it was suggested the loss of this allele facilitated the spread of this species out of Africa to more temperate latitudes. Recent studies with *Drosophila* have also suggested an important role for homeoviscous adaptation during rapid cold hardening.[53]

Caenorhabditis elegans Cold Tolerance and the Contribution of Desaturases

Thus, whilst persuasive, the evidence to date in favour of the lipid hypothesis in animals is generally restricted to compositional studies and new information on the cold induction of desaturase enzymes. Together these support the hypothesis but do not constitute strong tests, since their relationship may be correlative rather than causative. We sought to address this problem by seeking to manipulate the expression of the desaturase gene(s), and thus enzymatic activity, to control the saturation of membrane lipids. We then determined whether this treatment influenced cold tolerance.

The nematode *C. elegans* offers many well-appreciated advantages as a subject for experimental study,[54] particularly the ability to control gene expression using interfering double stranded RNA.[55] In unpublished work we have established that this species possesses a strong cold tolerance phenotype, with an increase in the time to 50% mortality (LT_{50}) at 0°C from ~20h, when reared at 25°C, to almost 80h after 12 days acclimation to 10°C (Fig. 1). Of the 7 acyl CoA desaturases in *C. elegans* only *fat-7* transcripts were up-regulated (20-fold) upon transfer of worms to 10°C.[56] This was blocked by amanitin, indicating that increased transcript levels resulted from de novo transcription rather than just the increased stabilization of the mRNA. Knockdown of *fat-7* by RNAi was >90% effective, and this partially ablated the

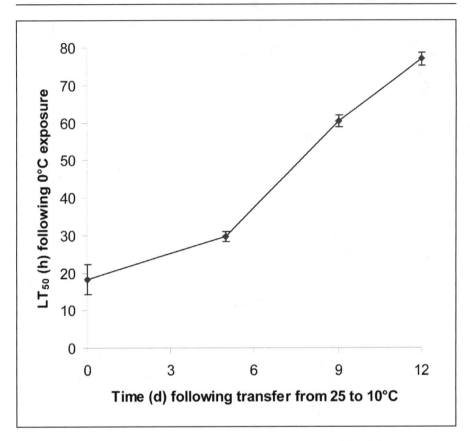

Figure 1. Effect of acclimation period at 10°C on *C. elegans* mortality (LT_{50}) following exposure to 0°C. *N* ranges from 3-13. Means ± SD are presented.

chill-induced change in lipid composition and the degree of cold tolerance. However, a compensatory increase in *fat-5* transcript abundance was noted, indicating a functional linkage in the expression of these two genes. By combining RNAi for *fat-7* with a genetic knockout (mutant strain) of the associated *fat-5*, lipid saturation was brought under experimental control and acquired cold tolerance was substantially reduced. At both 10 and 25°C the manipulation of desaturase genes caused substantial changes in lipid saturation, and this was directly linked to changes in cold tolerance. However, the change in cold tolerance caused by transfer to 10°C was substantially greater than could be explained by this lipid effect alone, i.e., 25 and 10°C-cultured worms had very different tolerance phenotypes even when manipulated to express the same lipid saturation. Indeed, we calculated that only 15% of the cold tolerance response could be quantitatively attributed to changes in lipid saturation.

Thus, our work suggests that other stress adaptive mechanisms must operate independently of homeoviscous adaptation at least, as indicated by lipid saturation measurements. Equally, cold tolerance mechanisms initiated via the membrane response cannot be contributing to the cold tolerant phenotype in manipulated worms, as all mechanisms downstream of homeoviscous adaptation are made redundant. This has caused us to focus on other response mechanisms and to explore potential interactions between alterations in lipid composition and these other stress responses.

Nonlipid Mechanisms of Cold Tolerance

While heat stress induces a highly conserved heat shock protein (Hsp) response in all organisms, a similar set of cold inducible proteins have not been identified. Indeed, the application of microarrays to study genome-wide responses to cold in different organisms has highlighted considerable diversity,[51,57-60] and a conserved cold response, certainly in eukaryotes, is not clearly evident. Despite this, several common mechanisms appear to underpin low temperature adaptation at the cellular level, and represent the most likely candidates contributing to a stress tolerant phenotype in conjunction with homeoviscous adaptation.

One well studied component of cold adaptation in prokaryotes is the preservation of ribosome function, with strong evidence suggesting the cold shock response is initiated by an inhibition of translation.[61] In the bacterium *Escherichia coli* a family of structurally related proteins, CspA, CspB and CspG show the highest induction response to cold, and act as RNA chaperones.[61] Homologues in other prokaryotes and eukaryotes have been identified and, while these are not always cold inducible,[62,63] clearly the preservation of ribosomal function is critical to organism survival post cold exposure.

Another common component of the cold stress response is the accumulation low-molecular-mass polyhydric alcohols and sugars, such as glycerol and trehalose.[64] While these have primarily been studied with respect to protection against freezing, i.e., as cryoprotectants,[64] they can also contribute to chill tolerance.[65] In addition, microarrays have consistently identified genes regulating the synthesis or transport of cryoprotectants as cold shock stimulons.[58,60]

The role of HSPs during cold acclimation has received limited attention, but HSP-induced cross-tolerance between different environmental stressors[66] suggests that some underlying mechanisms of cold adaptation may be similar to those of heat stress. To date, however, the results are somewhat equivocal with evidence for both the cold downregulation[58] and up-regulation[59] of Hsp transcripts.

Thus, multiple mechanisms underpin the phenotypic transition within acquired cold tolerance. Furthermore, these mechanisms probably do not operate in isolation, but interact closely in generating the stress tolerant phenotype. Understanding their relative importance and the extent to which they interact, and so offer a redundant capacity, remains an important goal.

Interaction and Compensatory Mechanisms

As outlined earlier, interactions between physiological mechanisms of stress adaptation can result in compensatory responses when one mechanism is perturbed, e.g., between *fat-5* and *fat-7* in *C. elegans*. Such interactions and compensatory responses have received limited attention, especially in multi-cellular organisms, but we now know that they are important, and identifying their complexity is fundamental to understanding ecophysiological adaptability. Studies targeting these interactions could also identify cross-tolerance between stress mechanisms, and highlight the best gene targets for the more complete disruption of stress responses.

There is growing evidence to indicate that membrane lipid order, as affected by lipid saturation, is a critical parameter regulating Hsp gene expression. For example, manipulation of membrane composition and/or fluidity altered the threshold temperature of Hsp expression in both yeast[67] and bacteria.[68] The yeast study also noted that the Δ9-desaturase activity directly influenced Hsp transcription. Recent, as yet unpublished, work from our lab has generated similar results with *C. elegans*. As described earlier, combining RNAi for *fat-7* with a genetic knockout (mutant strain) of the associated *fat-5* allowed lipid saturation to be controlled, and the resultant worms had distinctly lower Hsp70 transcript abundance under heat stress, when compared to wild type animals (Fig. 2). Circumstantial evidence supporting this relationship is provided by data from a variety of organisms indicating that temperature acclimation, which influences membrane lipid composition, also changes Hsp expression thresholds.[69] A membrane-Hsp association is further supported by data suggesting that

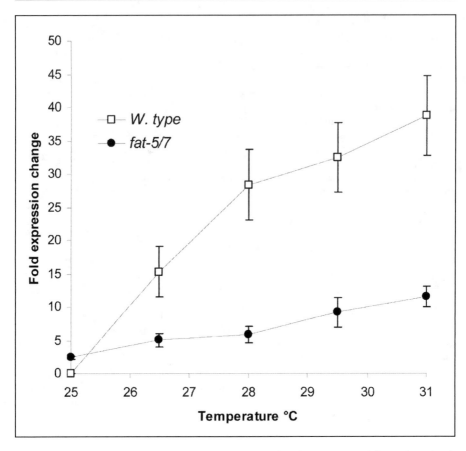

Figure 2. *Hsp70* (Wormbase gene C12C8.1) expression after 1h exposure to different heat shock temperatures: a comparison between N2 (W. type) *C. elegans,* and *fat-5* mutants fed *fat-7* RNAi (*fat-5/7*). Fold expression change calculated from Real-time quantitative RT-PCR. *N*=9 for each data point, means ± SEM are presented.

lipid interactions differentiate the constitutive (Hsc70) and stress-induced (Hsp70) heat shock protein response.[70] Recent studies have shown that Hsps, in turn, may be involved in a feedback process by their stabilizing effects on membrane proteins and lipids.[71]

Interestingly, trehalose and glycerol are also likely to influence both membrane adaptation and Hsp transcription. Trehalose interacts with, and directly protects, lipid membranes and proteins during cold stress,[72] and glycerol can also influence Hsp expression thresholds.[73] Thus any alteration in levels of these cryoprotectants during acclimation may interfere with homeoviscous adaptation and/or the Hsp response.

Conclusions

The question of cold tolerance in animals remains enigmatic. Much of the work over the past 50 years has focused on changing lipid composition in the cold, yet this has never been unequivocally linked to changes in cold tolerance. The evidence is much more persuasive in microbes and plants where genetic manipulation has established that lipid biosynthetic genes, notably the desaturases, mediate this effect. In animals, recent work has identified a network of desaturase genes expressed in many different tissues, and that they are cold-inducible. Our work with the genetic manipulation of Δ9-desaturase in *C. elegans* offers the first direct test in an animal species of the relationship between lipid saturation and the cold tolerance phenotype; and our results indicate that, at best, the lipid response accounts for only a small fraction of acquired cold tolerance. *C. elegans* in many ways is an ideal organism to study the mechanisms underpinning environmental adaptation, with access to RNAi libraries and mutant strains etc., providing the opportunity to quantify the impact of any gene upon environmentally-relevant stress tolerance phenotypes. We can also manipulate genes in groups to overcome the problem of redundant or overlapping mechanisms to study how interactions impact on the phenotype. The key and as yet unanswered question is whether the outcome of this work is directly translatable to other animal species, and particularly to vertebrates.

Acknowledgements

The authors would like to thank Gregor Govan for technical assistance in conducting this research. Work in Prof. Cossins' laboratory was supported by grants from the Natural Environment Research Council (NERC) and the Biotechnology and Biological Science Research Council (BBSRC).

References

1. Cossins AR, Bowler K. The Temperature Biology of Animals. London: Chapman and Hall, 1987.
2. Hochachka PW, Somero GN. Biochemical adaptation. Princeton University Press, 2002.
3. Kültz D. Molecular and evolutionary basis of the cellular stress response. Annu Rev Physiol 2005; 67:225-257.
4. Henriques V, Hansen C. Vergleichende Untersuchungen über die chemische Zusammenstzung des thierischen Fettes. Skand Arch Physiol 1901; 11:151-165.
5. Cossins AR. The adaptation of membrane structure and function to changes in temperature. In: Cossins AR, Sheterline P, eds. Cellular acclimatisation to environmental change. Cambridge University Press, 1983:3-32.
6. Sinensky M. Homeoviscous adaptation - A homeostatic process that regulates the viscosity of membrane lipids in E. coli. Proc Nat Acad Sci USA 1974; 71:522-525.
7. Hodkova M, Simek P, Zahradnickova H et al. Seasonal changes in the phospholipid composition in thoracic muscles of a heteropteran, Pyrrhocoris apterus. Insect Biochem Mol Biol 1999; 29:367-376.
8. In: Aloia RC, Curtain CC, Gordon LM eds. Drug and anesthetic effects on membrane structure and function. New York: Wiley, 1991.
9. Heilbrunn LV. The colloid chemistry of protoplasm. IV. The heat coagulation of protoplasm. Am J Physiol 1924; 69:190-199.
10. Suzuki I, Los DA, Kanesaki Y et al. The pathway for perception and transduction of low-temperature signals in Synechocystis. EMBO J 2000; 19:1327-1334.
11. Orr GR, Raison JK. The effect of changing the composition of phosphatidylglycerol from thylakoid polar lipids of oleander and cucumber on the temperature of the transition related to chilling injury. Planta 1990; 181:137-143.
12. Cossins AR. Homeoviscous adaptation of biological membranes and its functional significance. In: Cossins AR, ed. Temperature adaptation of biological membranes. London: Portland Press, 1994:63-75.
13. Hazel JR, Landrey SR. Time course of thermal adaptation in plasma membranes of trout kidney. I. Headgroup composition. Am J Physiol Regul Int Comp Physiol 1988; 255:R622-627.
14. Hazel JR, Zerba E. Adaptation of biological membranes to temperature: Molecular species compositions of phosphatidylcholine and phosphatidylethanolamine in mitochondrial and microsomal membranes of liver from thermally-acclimated rainbow trout. J Comp Physiol B Biochem Syst Environ Physiol 1986; 156:665-674.
15. Robertson JC, Hazel JR. Cholesterol content of trout plasma membranes varies with acclimation temperature. Am J Physiol Regul Int Comp Physiol 1995; 269:R1113-1119.
16. Zehmer JK, Hazel JR. Thermally induced changes in lipid composition of raft and nonraft regions of hepatocyte plasma membranes of rainbow trout. J Exp Biol 2005; 208:4283-4290.
17. Murata N, Suzuki I. Exploitation of genomic sequences in a systematic analysis to access how cyanobacteria sense environmental stress. J Exp Bot 2006; 57:235-247.

18. Los DA, Murata N. Structure and expression of fatty acid desaturases. Biochim Biophys Acta 1998; 1394:3-15.
19. Murata N, Wada H. Acyl-lipid desaturases and their importance in the tolerance and acclimatization to cold of cyanobacteria. Biochem J 1995; 308:1-8.
20. Tasaka Y, Gombos Z, Nishiyama Y et al. Targeted mutagenesis of acyl-lipid desaturases in Synechocystis: Evidence for the important roles of polyunsaturated membrane lipids in growth, respiration and photosynthesis. Embo J 1996; 15:6416-6425.
21. Sakamoto T, Shen GZ, Higashi S et al. Alteration of low-temperature susceptibility of the cyanobacterium Synechocystis sp. PCC 7002 by genetic manipulation of membrane lipid unsaturation. Arch Microbiol 1997; 169:20-28.
22. Kanervo E, Tasaka Y, Murata N et al. Membrane lipid unsaturation modulates processing of the photosystem II reaction-center protein D1 at low temperatures. Plant Physiol 1997; 114:841-849.
23. Wada H, Gombos Z, Murata N. Enhancement of chilling tolerance of a cyanobacterium by genetic manipulation of fatty acid desaturation. Nature 1990; 347:200-203.
24. Vigh L, Los DA, Horvath I et al. The primary signal in the biological perception of temperature: Pd-catalysed hydrogenation of membrane lipids stimulates the expression of the desA gene in Synechocystis PCC 6803. Proc Nat Acad Sci USA 1993; 90:9090-9094.
25. St. John JB, Christiansen MN, Ashworth EN et al. Effect of BASF 13-338, a substituted pyridazinone, on linolenic acid levels and winterhardiness of cereals. Crop Sci 1979; 19:65-69.
26. Somerville C, Browse J. Plant lipids: Metabolism, mutants and membranes. Science 1991; 252:80-87.
27. Browse J, Kunst L, Anderson S et al. A mutant of Arabidopsis deficient in the chloroplast 16:1/18:1 desaturase. Plant Physiol 1989; 90:522-529.
28. Kunst L, Browse J, Somerville C. A mutant of Arabidopsis deficient in desaturation of palmitic acid in leaf lipids. Plant Physiol 1989; 90:943-947.
29. Lemieux B, Miquel M, Somerville C et al. Mutants of Arabidopsis with alterations in seed lipid fatty acid composition. Theor Appl Genet 1990; 80:234-240.
30. Hugly S, Somerville C. A role for membrane lipid polyunsaturation in chloroplast biogenesis at low temperature. Plant Physiol 1992; 99:197-202.
31. Miquel M, James D, Dooner H et al. Arabidopsis requires polyunsaturated lipids for low-temperature survival. Proc Nat Acad Sci USA 1993; 90:6208-6212.
32. Kodama H, Hamada T, Horiguchi G et al. Genetic enhancement of cold tolerance by expression of a gene for chloroplast ω-3 fatty acid desaturase in transgenic tobacco. Plant Physiol 1994; 105:601-605.
33. Ishizaki-Nishizawa O, Fujii T, Azuma M et al. Low-temperature resistance of higher plants is significantly enhanced by nonspecific cyanobacterial desaturase. Nat Biotechnol 1996; 14:1003-1006.
34. Orlova IV, Serebriiskaya TS, Popov V et al. Transformation of tobacco with a gene for the thermophilic acyl-lipid desaturase enhances the chilling tolerance of plants. Plant cell Physiol 2003; 44:447-450.
35. Murata N, Ishizaki-Nishizawa O, Higashi S et al. Genetically engineered alteration in the chilling sensitivity of plants. Nature 1992; 356:710-713.
36. Sakamoto A, Sulpice R, Hou CX et al. Genetic modification of the fatty acid unsaturation of phosphatidylglycerol in chloroplasts alters the sensitivity of tobacco plants to cold stress. Plant Cell Environ 2003; 27:99-105.
37. Moon BY, Higashi S, Gombos Z et al. Unsaturation of the membrane lipids of chloroplasts stabilizes the photosynthetic machinery against low-temperature photoinhibition in transgenic tobacco plants. Proc Nat Acad Sci USA 1995; 92:6219-6223.
38. Murakami Y, Tsuyama M, Kobayashi Y et al. Trienoic fatty acids and plant tolerance of high temperature. Science 2000; 287:476-479.
39. Eguchi E, Ogawa Y, Okamoto K et al. Fatty acid compositions of arthropod and cephalopod photoreceptors - Interspecific, seasonal and development studies. J Comp Physiol B 1994; 164:94-102.
40. Kostal V, Simek P. Changes in fatty acid composition of phospholipids and triacylglycerols after cold-acclimation of an aestivating insect prepupa. J Comp Physiol B 1998; 168;453-460.
41. Cuculescu M, Pearson T, Hyde D et al. Heterothermal acclimation: An experimental paradigm for studying the control of thermal acclimation in crabs. Proc Nat Acad Sci USA 1999; 96:6501-6505.

42. Lahdes E, Balogh G, Fodor E et al. Adaptation of composition and biophysical properties of phospholipids to temperature by the crustacean, Gammarus spp. Lipds 2000; 35:1093-1098.
43. Bayley M, Peterson SO, Knigge T et al. Drought acclimation confers cold tolerance in the soil collembolan Folsomia candida. J Insect Physiol 2001; 47:1197-1204.
44. Hall JM, Parrish CC, Thompson RJ. Eicosapentaenoic acid regulates scallop (Placopecten magellanicus) membrane fluidity in response to cold. Biol bull 2002; 202:201-203.
45. Cossins AR, Crawford DL. Fish as models for environmental genomics. Nat Rev Gen 2005; 6:324-333.
46. Logue JA, de Vries AL, Fodor E et al. Lipid compositional correlates of temperature-adaptive interspecific differences in membrane physical structure. J Exp Biol 2000; 203:2105-2115.
47. Wodtke E, Teichert T, Konig, A. Control of membrane fluidity in carp upon cold stress: Studies on fatty acid desaturases. In: Heller HC, ed. Living in the Cold: Physiological and Biochemical Adaptations. Elsevier, 1986:35-42.
48. Wodtke E, Cossins AR. Rapid cold-induced changes of membrane order and Δ^9-desaturase activity in endoplasmic reticulum of carp liver: A time-course study of thermal acclimation. Biochim Biophys Acta 1991; 1064:343-350.
49. Tiku PE, Gracey AY, Macartney AI et al. Cold-induced expression of Δ^9-desaturase by transcriptional and post-translational mechanisms. Science 1996; 271:815-818.
50. Polley SD, Tiku PE, Trueman RT et al. Differential expression of cold- and diet-specific genes encoding two carp liver Δ^9-acyl-CoA desaturase isoforms. Am J Physiol 2003; 284:R41-R50.
51. Gracey AY, Fraser EJ, Li W et al. Coping with cold: An integrative, multi-tissue analysis of the transcriptome of a poikilothermic vertebrate. Proc Nat Acad Sci USA 2004; 101:6970-16975.
52. Greenberg AJ, Moran JR, Coyne JA et al. Ecological adaptation during incipient speciation revealed by precise gene replacement. Science 2003; 302:1754-1757.
53. Overgaard J, Sørensen JG, Petersen SO et al. Changes in membrane lipid composition following rapid cold hardening in Drosophila melanogaster. J Insect Physiol 2005; 51:1173-1182.
54. C. elegans II. Riddle DL, Blumenthal T, Meyer BJ, Priess JR, eds. Cold Spring Harbor Laboratory Press, 1997.
55. Fire A, Xu SQ, Montgomery MK et al. Potent and specific genetic interference by double-stranded RNA in Caenorhabditis elegans. Nature 1998; 391:806-811.
56. Murray PA, Hayward SAL, Gracey AY et al. Proc Natl Acad Sci USA, (In submission).
57. Phadtare S, Inouye M. Genome-wide transcriptional analysis of the cold shock response in wild-type and cold-sensitive, quadrupole-csp-deletion strains of Escherichia coli. J Bacteriol 2004; 186:7007-7014.
58. Han Y, Zhou D, Pang X et al. DNA microarray analysis of the heat- and cold-shock stimulons in Yersinia pestis. Microb Infect 2005; 7:335-348.
59. Qin W, Neal SJ, Robertson RM et al. Cold hardening and transcriptional change in Drosophila melanogaster. Insect Mol Biol 2005; 14:607-613.
60. Hannah MA, Heyer AG, Hincha DK. A global survey of gene regulation during cold acclimation in Arabidopsis thaliana. PLoS Gen 2005; 1:179-196.
61. Thieringer HA, Jones PG, Inouye M. Cold shock and adaptation. BioEssays 1998; 20:49-57.
62. Schroder K, Zuber P, Willimsky G et al. Mapping of the Bacillus subtilis cspB gene and cloning of its homologs in thermophilic, mesophilic and psychrotropic bacilli. Gene 1993; 136:277-280.
63. Wolffe AP. Structural and functional properties of the evolutionary ancient Y-box family of nucleic acid binding protein. BioEssays 1994; 16:245-250.
64. Denlinger DL, Lee RE. Physiology of cold sensitivity. In: Hallman GJ, Denlinger DL, eds. Temperature sensitivity in insects and application in integrated pest management. Boulder: Westview Press, 1998:55-95.
65. Slachta M, Vambera J, Zahradnickova H et al. Entering diapause is a prerequisite for successful cold-acclimation in adult Graphosoma lineatum (Heteroptera: Pentatomidae). J Insect Physiol 2002; 48:1031-1039.
66. Sabehat A, Weiss D, Lurie S. Heat-shock proteins and cross tolerance in plants. Physiol Plant 1998; 103:437-441.
67. Carratù L, Franceschelli S, Pardini CL et al. Membrane lipid perturbation modifies the set point of the temperature of heat shock response in yeast. Proc Nat Acad Sci USA 1996; 93:3870-3875.
68. Horváth I, Glatz A, Varvasovszki V et al. Membrane physical state controls the signaling mechanism of the heat shock response in Synechocystis PCC 6803: Identification of hsp17 as a "fluidity gene". Proc Nat Acad Sci USA 1998; 95:3513-3518.

69. Barua D, Heckathorn SA. Acclimation of the temperature set-points of the heat-shock response. J Therm Biol 2004; 29:185-193.
70. Arispe N, Doh M, De Maio A. Lipid interaction differentiates the constitutive and stress-induced heat shock proteins Hsc70 and Hsp70. Cell Stress Chap 2002; 7:330-338.
71. Vigh L, Maresca B, Harwood JL. Does the membrane's physical state control the expression of heat shock and other genes? Trends Biochem Sci 1998; 23:369-374.
72. Behm CA. The role of trehalose in the physiology of Nematodes. Int J Parasitol 1997; 27:215-229.
73. Brown CR, Hong-Brown LQ, Doxsey SJ et al. Molecular chaperones and the centrosome. J Biol Chem 1996; 271:833-840.

CHAPTER 13

Trehalose As a "Chemical Chaperone":
Fact and Fantasy

John H. Crowe*

Abstract

Trehalose is a disaccharide of glucose that is found at high concentrations in a wide variety of organisms that naturally survive drying in nature. Many years ago we reported that this molecule has the remarkable ability to stabilize membranes and proteins in the dry state. A mechanism for the stabilization rapidly emerged, and it was sufficiently attractive that a myth grew up about trehalose as a universal protectant and chemical chaperone. Many of the claims in this regard can be explained by what is now known about the physical properties of this interesting sugar. It is emerging that these properties may make it unusually useful in stabilizing intact cells in the dry state.

Sugars and Stabilization of Biological Materials

We reported two decades ago that biomolecules and molecular assemblages such as membranes and proteins can be stabilized in the dry state in the presence of a sugar found at high concentrations in many anhydrobiotic organisms, trehalose.[1] We also showed that trehalose was clearly superior to other sugars in this regard.[2] This effect seemed so clear it quickly led to wide-spread, and often uncritical, use of the sugar for preservation and other purposes. In fact, an array of applications for trehalose have been reported, ranging from stabilization of vaccines and liposomes to hypothermic storage of human organs.[3] Other workers showed that it might even be efficacious in treatment of dry eye syndrome or dry skin in humans.[4,5] Trehalose is prominently listed as an ingredient in cosmetics, apparently because it is reputed to inhibit oxidation of certain fatty acids in vitro that might be related to body odor.[6] Trehalose has been shown by several groups to suppress free radical damage, protect against anoxia, inhibit dental caries, enhance ethanol production during fermentation, stabilize the flavor in foods, and to protect plants against physical stress.[7-13] According to one group, trehalose inhibits bone resorption in ovariectomized mice, apparently by suppressing osteoclast differentiation; the suggestion followed that trehalose might be used to treat osteoporosis in humans.[14,15] More recently, Tanaka et al[16] reported that trehalose could be used to inhibit the protein aggregation associated with Huntington's disease in vivo in a rat model for this disease. That report that has already led to an unorthodox clinical trial in humans.[17]

A myth has grown up about trehalose and its properties, as a result of which it is being applied, sometimes rather uncritically, to a myriad of biological and clinical problems. Thus, we are making special efforts in the literature to clarify the properties of trehalose that make it useful for stabilization of biomaterials and to dispel the most misleading aspects of this myth.

*Corresponding Author: Dr. John H. Crowe—Section of Molecular and Cellular Biology, University of California, Davis, California 95616, U.S.A. Email: jhcrowe@ucdavis.edu

Molecular Aspects of the Stress Response: Chaperones, Membranes and Networks, edited by Peter Csermely and László Vígh. ©2007 Landes Bioscience and Springer Science+Business Media.

Origins of the Trehalose Myth

We recently reviewed the history of this field (see ref. 3) and provide only a brief summary here. The key observations were: (1) The first model membrane investigated was sarcoplasmic reticulum, isolated from lobster muscle (reviewed in ref. 18). We found that trehalose was without question superior to all other sugars tested at preserving these membranes during drying. However, we later obtained evidence that these SR membranes have a mechanism for translocating trehalose across the bilayer. We suggest that other sugars such as sucrose might preserve the membranes at concentrations similar to those seen with trehalose if they had access to the aqueous interior. (2) Initial studies with liposomes, from the mid-1980s (reviewed in ref. 19), were done with a phospholipid with low T_m. When the liposomes were freeze dried with trehalose and rehydrated, the vesicles were seen to be intact, and nearly 100% of the trapped solute was retained. It quickly emerged that stabilization of these liposomes, and other vesicles prepared from low melting point lipids, had two requirements, as illustrated in Figure 1: (a) inhibition of fusion between the dry vesicles; and (b) depression of T_m in the dry state. In the hydrated state, T_m for egg PC is about -1°C and rises to about + 70°C when it is dried without trehalose. In the presence of trehalose, T_m is depressed in the dry state to - 20°C. Thus, the lipid remains in liquid crystalline phase in the dry state, and phase transitions are not seen during

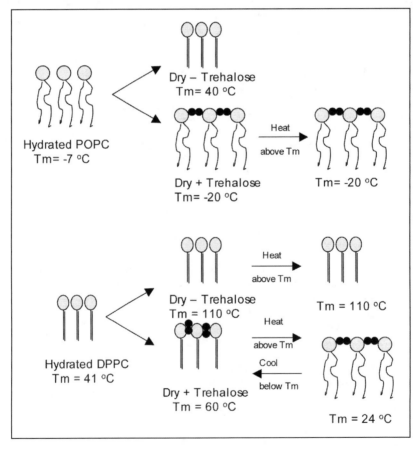

Figure 1. Mechanism for stabilization of phospholipid bilayers in the dry state. Adapted from reference 18.

rehydration. The significance of this phase transition during rehydration is that, when phospholipids pass through such transitions, the bilayer becomes transiently leaky. The physical basis for this leakiness has recently been investigated in some detail.[20] These effects were reported first for trehalose (reviewed in ref. 19). When we compared the effects of other sugars and polymers on the preservation, we found that, with vesicles made from lipids with low T_m, trehalose appeared to be significantly superior to the best of the additives tested. Oligosaccharides larger than trisaccharides did not work at all.[21] Other sugars, particularly disaccharides, did provide good stabilization of egg PC vesicles in the dry state, but much higher concentrations than trehalose were required, at least according to initial reports. However, as freeze-drying technology improved, the differences between disaccharides tended to disappear, and the myth eventually got modified to encompass disaccharides in general. Nevertheless, the observation that trehalose was significantly more effective at low concentrations under suboptimal conditions for freeze drying requires explanation, which we provide later. (3) At first it appeared that the ability to preserve liposomes in the dry state is restricted to disaccharides. Subsequently, we found this is not the case. For example, DPPC is a lipid with saturated acyl chains and thus an elevated T_m (41°C). When it is dried without trehalose T_m rises to about 110°C; with trehalose present T_m rises to about 65°C (reviewed in ref. 22). Thus, DPPC is in gel phase at all stages of the freeze-drying and rehydration process, and one would expect that inhibition of fusion might be sufficient for the stabilization. In other words, any inert solute that would separate the vesicles in the dry state and thus prevent aggregation and fusion should stabilize the dry vesicles. That appears to be the case; a high molecular weight (450,000) HES has no effect on T_m in dry DPPC, but preserves the vesicles, nevertheless.

The Mechanism of Depression of Tm

The Mechanism of Depression of Tm has received a great deal of attention since the discovery of this effect.[23] Three main hypotheses have emerged: The water replacement hypothesis suggests that sugars can replace water molecules by forming hydrogen binds with polar residues, thereby stabilizing the structure in the absence of water.[3,22,24-26] The water entrapment hypothesis suggests that sugars concentrate water near surfaces, thereby preserving its salvation.[27-29] The vitrification hypothesis suggests that the sugars form amorphous glasses, thus reducing structural fluctuations.[30-31]

A consensus has emerged that these three mechanisms are not mutually exclusive (reviewed in ref. 3). Vitrification may occur simultaneously with direct interactions between the sugar and polar residues. Direct interaction, on the other hand, has been demonstrated by a wide variety of physical techniques, including infrared spectroscopy, NMR, and X-ray.[23,32-36]

Theoretical analyses have contributed greatly to this field in recent years. Chandrasekhar and Gaber[37] and Rudolph et al[38] in the earliest studies, showed that trehalose can form energetically stable conformations with phospholipids, binding three adjacent phospholipids in the dry state. Similarly, trehalose-protein interactions have been studied by simulations, with similar conclusions.[28,29] More recently, Sum et al[39] showed by molecular simulations that the sugars adapt molecular conformations that permit them to fit onto the surface topology of the bilayer through hydrogen bonds. The sugars interact with up to three adjacent phospholipids. Pereira et al[40] produced complementary results from molecular dynamics simulations, with comparable conclusions.

Trehalose Stabilizes Microdomains in Membranes

Phase separation is segregation of membrane components in the plane of the bilayer. Although there are lingering doubts about whether or not phase separated domains in native membranes are real (see refs. 41-43) or artifacts (see refs. 44-46), there is abundant evidence that these domains, known as "rafts", are involved in such processes (among others) as signaling, endocytosi, and viral assembly.[47-51] Although several forces are involved, one of the main driving forces for phase separation is the hydrophobic mismatch, which arises from a difference in membrane thickness between two species within a bilayer, such as a protein and a lipid or a lipid and a lipid.[52,53] The differences in thickness lead to exposure of hydrophobic residues to water and, consequently, to a decrease in entropy of the system resulting from ordering of the water. Thus, the assembly of components of similar thickness into relatively homogeneous domains is entropically driven. The net increase in entropy driving the process is contributed by water.

Phase separated domains in lipid bilayers are becoming increasingly well understood (see ref. 54 for a recent review). Thus, we have investigated whether the domains can be maintained in freeze-dried liposomes. DLPC (Tm = 0°C) and DSPC (Tm = 50°C) are well known to undergo complete phase separation in the fully hydrated state.[55,56] When these liposomes were dried, the two lipids underwent extensive mixing. In samples dried with trehalose, by contrast, the DLPC transition is depressed to about -20°C, and the DSPC transition increases by about 10°C and becomes more cooperative, suggesting that it is more like pure DSPC. Thus, the phase separation—and the domain structure—are maintained by the trehalose in the dry state. Other pairs of lipids that phase separate when fully hydrated give similar results.

We propose that trehalose maintains phase separation in this mixture of lipids in the dry state by the following mechanism.[36] The DLPC fraction, with its low Tm in the hydrated state, might be expected to behave like unsaturated lipids, in that Tm in the dry state is reduced to a minimal and stable value immediately after drying with trehalose, regardless of the thermal history. That appears to be the case. The DSPC fraction, by contrast, would be expected to behave like DPPC, as described earlier. DSPC is in gel phase in the hydrated state at room temperature, and it remains in gel phase when it is dried with trehalose. In other words, we are proposing that by maintaining one of the lipids in liquid crystalline phase during drying, while the other remains in gel phase, trehalose maintains the phase separation (Fig. 2). We suggest that this is the fundamental mechanism by which trehalose maintains phase separated domains in membranes drying.

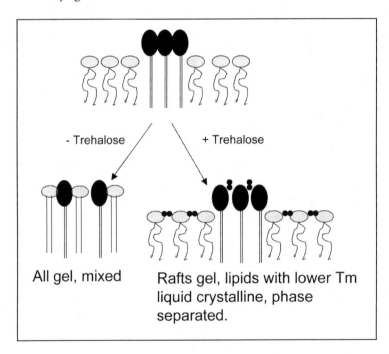

Figure 2. Mechanism of preservation of phase separated microdomains in dry membranes. Adapted from reference 54.

There Is More Than One Way to the Same End

Although the occurrence of trehalose at high concentrations is common in anhydrobiotic animals, some such animals have only small amounts of trehalose or none at all.[57-60] It is tempting to construe these findings as evidence against a role for sugars in anhydrobiosis.[61] We suggest that it is not the sugars per se that are of interest in this regard, but rather the physical principles of the requirements for stabilization, as described above. There are multiple ways to achieve such stabilization: (1) Hydroxyethyl starch (HES) alone will not stabilize dry membrane vesicles composed of lipids with low Tm, but a combination of a low molecular weight sugar such as glucose and HES can be effective.[62] Here is the apparent mechanism: glucose depresses T_m in the dry lipid, but has little effect on inhibiting fusion, except at extremely high concentrations. On the other hand, the polymer has no effect on the phase transition, but inhibits fusion. Thus, the combination of the two meets both requirements, while neither alone does so.[62] It seems likely that such combinations of molecules might be found in anhydrobiotes in nature. (2) In fact, a glycan isolated from the desiccation tolerant alga *Nostoc* apparently works in conjunction with oligosaccharides.[63] Similarly, certain proteins have been shown to affect the phase state of the sugars and either enhance or are required for stabilization (reviewed in ref. 64). (3) Hincha et al[65] have shown that fructans from desiccation tolerant higher plants will by themselves both inhibit fusion and reduce T_m in dry phospholipids such as egg PC. The mechanism behind this effect is still unclear. The interaction is similar to that shown by sugars, but it is also specific to fructans and is not shown by other polymers.[66] In a related study, Hincha et al[67] reported that a series of raffinose family oligosaccharides are all capable of stabilizing dry liposomes. (4) Hincha and Hagemann[68] recently studied effects of other compatible solutes on stabilization of liposomes by sugars. This approach is in its earliest stages, but those authors found that some compatible solutes improve the stabilization in the presence of sugars, suggesting that the solutes might decrease the amount of sugar required in vivo. (5) Hoekstra and Golovina[69] have reported that amphiphiles that are free in the cytoplasm in fully hydrated cells of anhydrobiotes apparently insert into membranes during dehydration. The role of this phenomenon in stabilization is uncertain, but presumably the amphiphiles alter the order of the acyl chains. (6) Goodrich et al[70] reported that disaccharides tethered to the bilayer surface by a flexible linker esterified to cholesterol has an effect on membrane stability similar to that seen in the free sugar. Such molecules could provide stability in anhydrobiotes, although they have not yet been reported. However, Popova and Hincha[71] recently found that digalactosyl diacylglycerol depressed Tm in dry phospholipids, perhaps in keeping with this suggestion.

The point is there are many ways to achieve stability. Once an understanding of the physical requirements for preservation was achieved, it became apparent that many routes can lead to the same end. Similar observations on the stability of dry proteins have been made by Carpenter and his group, with similar conclusions.[24,25]

Trehalose Has Useful Properties, Nevertheless

We implied above that trehalose works well for freeze-drying liposomes under less than optimal conditions. The same applies for storage under conditions that would normally degrade the biomaterial. Bacteria freeze-dried in the presence of trehalose showed remarkably high survival immediately after freeze drying. Furthermore, t the bacteria freeze-dried with trehalose retained high viability even after long exposure to moist air.[72] By contrast, when the bacteria were freeze-dried with sucrose they showed lower initial survival, and when they were exposed to moist air viability deceased rapidly. Further, when immunoconjugates were freeze dried with trehalose or other disaccharides all the sugars provided reasonable levels of preservation. However, when the dry samples were stored at high relative humilities and temperatures, those dried with trehalose were stable for much longer than those dried with other sugars.[73] This finding is of some considerable significance since there is a need for shipping immunoconjugates, vaccines, antisera and the like to locales where they would be exposed to high temperatures and humidities as soon as they are exposed to air.

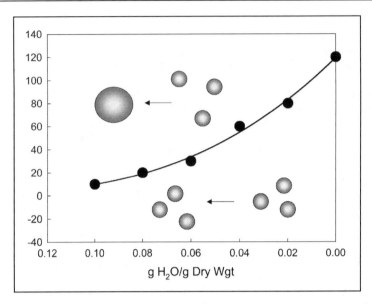

Figure 3. The relationship between the glass transition temperature (Tg) and water content of trehalose, known as a state diagram. Below Tg membrane vesicles are rigidly held in the glassy matrix, but above Tg mobility of the system increases such that the vesicles may come in contact and undergo fusion. Data for Tg from reference 74. Fusion studies from reference 31.

Glass Transitions and Stability

Using liposomes as a model, we attempted to find a mechanism for long term stability in the presence of trehalose. As with the bacteria and immunoconjugates, the dry liposomes exposed to increased relative humidity rapidly leaked their contents when they were dried with sucrose, but not when they were dried with trehalose.[31,74] The liposomes underwent extensive fusion in the moist air when dried with sucrose, but not with trehalose.

Trehalose, along with many other sugars, forms a glass when it is dried. This glass undergoes a transition from a highly viscous fluid to a highly mobile system when it is heated above a characteristic temperature, Tg, which increases sharply as dehydration progresses, resulting in what is known as a state diagram (Fig. 3). The importance of the state diagram is as follows. It has become widely accepted that stability of dry materials in which close approach of surfaces must be prevented requires that the material remain below the curve for the state diagram, i.e., it must be maintained in the glassy state. Above the curve the mobility of the system increases, while below it the materials are held in a relatively rigid matrix (Fig. 3). For instance, heating a sample containing liposomes above Tg results in increased mobility to the point where fusion occurs in the concentrated solution. Brief excursions above the curve are not necessarily damaging, since the surface to surface interaction has a kinetic component. Because of this kinetic component, there is a lot of confusion in the literature concerning whether the glassy state is even required for stabilization.

T_g for trehalose is much higher than that for sucrose (Fig. 4), a finding first reported by Green and Angell.[75] As a result, one would expect that addition of small amounts of water to sucrose by adsorption in moist air would decrease T_g to below the storage temperature, while at the same water content T_g for trehalose would be above the storage temperature. Indeed, at water contents around 5%, Tg for trehalose is about 40°C, while that for sucrose is about 15°C. Tg for glucose at a similar water content is about -10°C (Fig. 4). One would predict that at such water contents trehalose would be the only one of these three sugars that would stabilize

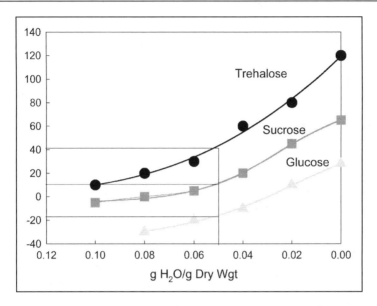

Figure 4. Comparison of state diagrams for trehalose, sucrose, and glucose. Data from reference 74.

the sample, and this appears to be the case. This would seem to provide an explanation for the superior stability of, for example, the immunoconjugates stored in sucrose or trehalose described above. We stress, however, that the elevated T_g seen in trehalose is not anomalous. Indeed, trehalose lies at the end of a continuum of sugars that show increasing T_g (40), although the basis for this effect is not understood.

Nonenzymatic Browning and Stability of the Glycosidic Bond

The Maillard (browning) reaction between reducing sugars and proteins in the dry state has often been invoked as a major source of damage,[76] and the fact that both sucrose and trehalose are nonreducing sugars may explain at least partly why they are the natural products accumulated by anhydrobiotic organisms. However, the glycosidic bonds linking the monomers in sucrose and trehalose have very different susceptibilities to hydrolysis.[77,78] For instance, the activation energy for acid hydrolysis in aqueous solution is nearly twice that that for other disaccharides.[77] When O'Brien[77] and subsequently Schebor et al[78] incubated a freeze dried model system (albumin, with the addition of lysine) with sucrose, trehalose, and glucose at relative humidities in excess of 20%, the rate of browning seen with sucrose approached that of glucose—as much as 2000 times faster than that with trehalose, although they observed a distinct lag in the onset of browning (Fig. 5). Schebor et al found that a peak in the appearance of monosaccharides occurs prior to the onset of browning, after which free monosaccharides decline, coincidentally with the onset of browning (Fig. 5). These observations strongly suggest that the browning seen with sucrose—but not with trehalose—is due to hydrolysis of the glycosidic bond during storage.

The glassy state is undoubtedly related to these effects; if the samples are stored at very low humidities only minimal amounts of hydrolysis and subsequent browning were seen in the sucrose preparations.[77,78] Nevertheless, since sucrose is the major sugar associated with desiccation tolerance in higher plants, consideration of the mechanisms by which devitrification at moderate water contents and hydrolysis of the glycosidic bond in sucrose glasses are obviated in anhydrobiotic plants is instructive.

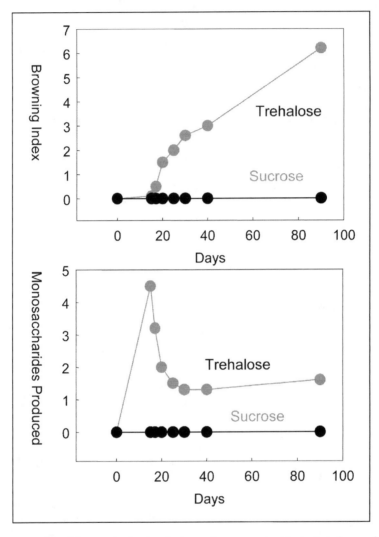

Figure 5. Browning of dry proteins (top) and release of monosaccharides by trehalose and sucrose during storage. Adapted from reference 78.

Sugar Glasses in Plant Anhydrobiotes

Buitink has published an elegant series of studies of the properties of glasses in vivo in anhydrobiotic plants (see refs. 79,80), along with a superb review of the work (see ref. 81). Briefly, Koster[82] found that mixtures of sugars similar to those found in desiccation tolerant (corn) embryonic axes (85% (w/w) sucrose, 15% (w/w) raffinose) formed glasses at temperatures above 0°C, while sugar mixtures similar to those found in desiccation-sensitive axes (75% (w/w) glucose, 25% (w/w) sucrose) formed glasses only at subzero temperatures. These and similar data suggested that sugar mixtures form glasses in plant anhydrobiotes, but subsequent studies indicated that the intracellular glasses are not composed of sugars alone. The state diagram for intact pollen of cat tail at first glance seems to agree reasonably well with that for sucrose (the major sugar in these pollen grains). However, subtle differences can be seen that have turned out to be significant: at low water contents Tg in the intact pollen is higher than predicted based on the state diagram for the sugar, while at low water contents it is lower than predicted.[79] Furthermore, the temperature at which the glass collapses (Tc), which occurs several degrees above Tg, is elevated by as much as 40°C in intact pollen and other anhydrobiotes.[80,81] The outcome of these studies is a clear indication that glasses in intact anhydrobiotic plants are not composed simply of mixtures of sugars.

Wolkers et al[83] developed a powerful approach based on infrared spectroscopy that permitted characterization of cytoplasmic glasses. The measurement—vibrational frequency of the -OH stretch in sugars—permitted an estimate of the length and strength of hydrogen bonds within the glass. Using this technique, Wolkers et al[84,85] found that the molecular density of the cytoplasm resembled that of protein glasses more than that of sucrose, a finding that initially suggested that sucrose may be a relatively minor player in formation of the cytoplasmic glass. However, studies on molecular motion in protein glasses have shown that rotational mobility of the proteins is almost twice that seen in the cytoplasmic glass.[80] The conclusion is that the cytoplasmic glass is likely to consist of a mixture of sucrose and proteins. The most likely candidates for the protein component are the late embryogenesis abundant (LEA) proteins that are accumulated in seeds and pollen late in development, and there is some evidence suggesting that this is the case; when extracts are made from wheat embryos a sucrose in large amounts was coisolated with the LEA proteins.[86] Exhaustive dialysis removed only a fraction of the sucrose, indicating that it is tightly bound to the protein.

The conclusion from these studies is that at least in plants cytoplasmic glasses consist of sugar-protein mixtures. The apparent elevation of Tg and the collapse temperature by addition of the protein to the glass is likely to lead to increased stability of the kind seen in trehalose alone in vitro, owing to its elevated Tg. Thus, devitrification at moderate water contents is obviated. The problem of stability of the glycosidic bond in sucrose during storage in the dry state is somewhat more problematic, but it seems likely that the association with the protein fraction, leading to the elevated Tg, could limit accessibility of water to the bond, thus limiting hydrolysis.

Lessons from Nature Can Be Used to Preserve Intact Cells in the Dry State

Clearly, trehalose must be introduced into the cytoplasm of a cell if it is to be effective at stabilizing intracellular proteins and membranes during dehydration. Previous efforts centered around this fundamental problem involving molecular engineering have not been particularly successful.[87,88] More recently, Wolkers et al[89] made the surprising discovery that when human blood platelets are placed in the presence of modest amounts of trehalose, they take it up by fluid phase endocytosis, and the intact sugar ends up in the cytoplasm. Wolkers et al were able to show only indirectly that trehalose is in the cytoplasm, but subsequent studies have shown that this is so. Oliver et al[90] have followed the fate of the sugar once it enters the endocytotic pathway in a stem cell line, using fluorescence microscopy. The fluorescence initially appears in vesicles, but with time it becomes diffuse, suggesting that the sugar is released into the cytoplasm. The mechanism of release is not entirely clear, but we have proposed that the sugars follow the pathway described in Figure 6. The endocytotic vesicles progress through the normal pathway to lysosomes. It is well established that low molecular weight compounds such as glucose readily cross the lysosomal membrane into the cytoplasm, but there is very little evidence concerning the fate of disaccharides in lysosomes.[91] Incubation of cells in sucrose led to persistent vesiculation, suggesting that, apparently unlike trehalose, sucrose is retained in the lysosomes. This seems surprising because even though lysosomes lack invertase and thus cannot break the glycosidic bond in sucrose enzymatically, the glycosidic bond in sucrose should be hydrolyzed at the pH known to occur in lysosomes, while that of trehalose should not.[91] Thus, if anything, one would expect that the sucrose, broken down into component glucose and fructose, should cross the lysosomal membrane into the cytosol by means of the glucose carrier, while the still intact trehalose should be retained. This matter is unresolved, but we

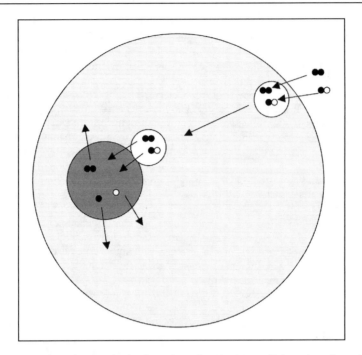

Figure 6. Proposed mechanism for loading disaccharides into cell, based on data of (ref. 90) and unpublished data of Auh et al. Trehalose (solid double dots, representing two glucose monomers) and sucrose (open and solid mixed dots, representing glucose and fructose monomers) enter the cell by fluid phase endocytosis (A) and are passed down the lysosomal pathway (B). We suggest that the stability of the glycosidic bond in trehalose will permit trehalose to survive at lysosomal pH, but that sucrose might be hydrolyzed. We further propose that trehalose, sucrose, glucose, and fructose will all leak into the cytoplasm due to the known effects of pH on permeability of phospholipid bilayers.[20]

suspect that the lysosomal pH itself might lead to leakage; when liposomes loaded with a polar fluorescent marker about the same size as trehalose were subjected to lysosomal pH, the marker leaked across the bilayer into the external medium.[20] We suggest that the pH gradient across the lysosomal membrane might lead to leakage of trehalose and other low molecular weight molecules into the cytosol. At any rate, we have found that trehalose can be introduced into the cytoplasm of every cell we have tested, so long as the cell has a functional fluid phase endocytotic pathway.

Successful Freeze-Drying of Trehalose-Loaded Cells

We have recently reported successful freeze-drying of platelets, with a detailed discussion of the procedure, which results in survival exceeding 90%.[89] We started this project at the invitation of the Department of Defense, where there is an obvious need for platelets for use in severe trauma cases. At present, platelets are stored in blood banks for a maximum of 3-5 days, by Federal regulation, after which they are discarded. Furthermore, the platelets are stored at room temperature; they cannot even be refrigerated without rendering them useless therapeutically, a phenomenon for which we have provided an explanation.[92-97] There is a chronic shortage of platelets in hospitals, and field hospitals operated by the military rarely have access to platelets at all. Thus, prolonging the shelf life of platelets would be a valuable contribution. The freeze-dried platelets have the following properties: (a) The dry platelets are stable for at least two years when stored at room temperature, under vacuum. During that time we have seen no loss of platelets. (b) The freeze-dried, rehydrated cells respond to normal platelet agonists including thrombin, ADP, collagen, and ristocetin. (c) Studies on the morphology of the trehalose-loaded, freeze-dried, and rehydrated platelets show that they are affected by the drying, but are morphologically similar to fresh platelets. When they were dried without trehalose, on the other hand, most of the platelets disintegrated during the rehydration event, but of the small number that were left, most had fused with adjacent cells, forming an insoluble clump. (d) We have extended the freeze-drying to mouse and pig platelets as animal models for in vivo testing.

The rehydrated platelets are far from perfect, but they nevertheless show surprisingly good regulation of key elements of cellular physiology such as intracellular calcium (Auh et al, 2004).[98] For instance, when fresh platelets are challenged with thrombin they show an increase in $[Ca_i]$ that is dose dependent. The rehydrated platelets show a similar response, although it is strongly attenuated. Nevertheless, the increase in $[Ca_i]$ appears to be sufficient to trigger morphological and physiological changes necessary for coagulation.[98]

Can Nucleated Cells Be Stabilized in the Dry State?

Platelets are admittedly specialized cellular fragments, so it seemed likely at the outset that the single perturbation of adding trehalose might not be sufficient to stabilize more complex living cells. Indeed, this appears to be the case. When 293 cells were dried without prior loading with trehalose by the fluid phase endocytosis described previously they all died at fairly high water contents.[99] When they were loaded with trehalose survival was extended considerably, but the cells nevertheless died when water content was reduced below about 0.3 g H_2O/g dry wgt. Thus, we have begun studies on effects of stress proteins on improving survival of nucleated cells at lower water contents and settled on p26 a small α-crystallin stress protein from desiccation tolerant Artemia cysts, discovered by Clegg and his colleagues.[100-103] In *Artemia*, p26 protects against many different stresses.[104-107] In addition, p26 has been shown to protect synergistically with trehalose in vitro conditions or when loaded artificially into mammalian cells.[108,109] Along the same lines, Singer and Lindquist[110,111] previously showed that trehalose acts synergistically with heat shock proteins in protein folding.

Sun et al[107] isolated and cloned the gene for this protein and transfected 293 cells with it. They generously supplied us with the transfected cells. To our surprise, this protein significantly improved the survival to low water contents, even though the levels of expression have been very low—much less than that reported by Clegg for Artemia cysts.[109] The protein alone does nothing to improve survival; trehalose is required as well, and the two appear to act synergistically. The effects of expression of this gene become even more pronounced in the time following rehydration; the cells expressing p26 showed a ten fold increase in colony growth over those without the protein. Metabolism, expressed in terms of Alamar blue reduction, improved at least five fold compared with cells dried with trehalose alone.

What Is the Role of p26 in Stabilizing Dry Nucleated Cells?

One hypothesis is that p26 participates in modulating the structure of the sugar glass, as suggested from the findings of Wolkers[83] and Buitink and Leprince.[81] However, the expression levels are so low in this case that we doubt that the protein would have much effect on the glass. We favor instead the catalytic-like functions ascribed to stress proteins. A possible role for heat shock proteins in the protection of mammalian cells during dehydration stress has not been addressed, but there are indications from other organisms implicating HSPs in this regard. For instance, a drought-resistant form of maize expresses a 45 Kd HSP that is not found in drought-sensitive lines.[112] Further, crossing the drought-resistant and sensitive lines led to F2 plants in which tolerance to soil drying was associated with expression of the 45Kd HSP.[112] In addition, the flesh fly *Sarcophaga crassipalpis* expresses two inducible HSPs (HSP23 and HSP70) during dehydration of nondiapausing pupae.[113] In diapausing pupae, these proteins are already highly expressed, and desiccation does not cause a further increase in expression.

The most important mechanism by which heat shock proteins protect cells from various stresses has traditionally been considered the protein chaperone function, assisting nascent and misfolded proteins to gain their proper folded configuration.[114,115] However, an association of stress proteins with membranes has more recently been described.[116-120] In fact, the "membrane trigger" hypothesis suggests that the membrane may serve as an indicator, sensing the initial stress and leading to the expression of heat shock proteins within the cell.[121,122] Finally, in addition to the protein and membrane effects, heat shock proteins have also been implicated in the inhibition of apoptosis and oxidative damage.[123-128] Finally, even more recent results have shown that arbutin, a glycosylated hydroquinone found at high concentrations in certain resurrection plants, also enhances survival of mammalian cells, in concert with trehalose.[129] This small molecule is known to have antioxidant properties and to interact with membranes in the dry state, both of which observations seem consistent with the suggestions above.

Summary and Conclusions

Under ideal conditions for drying and storage, trehalose is probably no more effective than other oligosaccharides at preserving biomaterials. However, under suboptimal conditions it can be very effective and is thus still a preferred excipient. There is growing evidence that additional modifications to the cellular milieu will probably be required if we are to achieve a stable, freeze-dried mammalian cell, including expression of stress proteins, as reported here, and administration of antioxidants and inhibitors of enzyme activity, as described elsewhere.[130] Thus, we have come full circle over the past decades on the requirements for stabilization of cells in the dry state; 34 years ago, we suggested that survival of living cells in the dry state is a complex phenomenon that is likely to involve multiple adaptations.[131] With the discovery that membranes and proteins, and human platelets can be stabilized by the single perturbation of adding trehalose, we suggested that this single lesson from nature might be sufficient, at least under ideal storage conditions. However, the studies on nucleated cells summarized here indicate that the original viewpoint on this matter[131] is most likely the correct one.

Acknowledgments

This work was supported by grants HL57810 and HL98171 from NIH, 98171 from ONR, and N66001-00-C-8048 from DARPA.

References

1. Crowe JH, Crowe LM, Jackson SA. Preservation and functional activity in lyophilized sarcoplasmic reticulum. Arch Biochem Biophys 1983; 220:477-484.
2. Crowe M, Mouradian R, Crowe JH et al. Effects of carbohydrates on membrane stability at low water activities. Biochim Biophys Acta 1984; 769:141-150.
3. Crowe JH, Crowe LM, Oliver AE et al. The trehalose myth revisited: Introduction to a symposium on stabilization of cells in the dry state. Cryobiology 2001; 43:89-105.
4. Matsuo T. Trehalose protects corneal epithelial cells from death by drying. British J Ophthalmol 2001; 85:610-612.
5. Norcia M. Compositions and methods for wound management. Off Gaz US Patent and Trademark Office 2000; 1232:424-448.
6. Higashiyama T. Novel functions and applications of trehalose. Pure Appl Chem 2002; 74:1263-1269.
7. Benaroudj N, Lee DL, Goldberg AL. Trehalose Accumulation during cellular stress protects cells and cellular proteins from damage by oxygen radicals. J Biol Chem 2001; 276:24261-24267.
8. Chen Q, Haddad GG. Role of trehalose phosphate synthase and trehalose during hypoxia: From flies to mammals. J Exp Biol 2004; 207:3125-3129.
9. Neta T, Takada K, Hirasawa M. Low-cariogenicity of trehalose as a substrate. Dent 2000; 28:571-6.
10. Gimeno-Alcañiz JV, Pèrez-Ortìn JE, Matallana E. Differential pattern of trehalose accumulation in wine yeast strains during the microvinification process. Biotechnology Lett 1999; 21:271-274.
11. Pataro C, Guerra JB, Gomes FCO et al. Trehalose accumulation, invertase activity and physiological characteristics of yeasts isolated from 24 h fermentative cycles during the production of artisanal Brazilian cachaça. Brazilian J Microbiol 2002; 33:202-208.
12. Komes D, Lovri T, Kovaevi Gani T et al. Study of trehalose addition on aroma retention in dehydrated strawberry puree. Food Technol Biotechnol 2003; 41:111-119.

13. Garg AK, Kim JK, Owens TG et al. Trehalose accumulation in rice plants confers high tolerance levels to different abiotic stresses. Proc Nat Acad Sci 2002; 99:15898-15903.
14. Nishizaki Y, Yoshizane C, Toshimori Y et al. Disaccharide-trehalose inhibits bone resorption in ovariectomized mice. Nutr Res 2000; 20:653-664.
15. Yoshizane C, Arai N, Arai C et al. Trehalose suppresses osteoclast differentiation in ovariectomized mice: Correlation with decreased in vitro interleukin-6 production by bone marrow cells. Nutr Res 2000; 20:1485-1491.
16. Tanaka M, Machida Y, Niu S et al. Trehalose alleviates polyglutamine-mediated pathology in a mouse model of Huntington disease. Nat Med 2004; 10:148-54.
17. Couzin J. Huntington's disease. Unorthodox clinical trials meld science and care. Science 2004; 304:816-817.
18. Crowe JH, Crowe LM, Carpenter JF et al. Stabilization of dry phospholipid bilayers and proteins by sugars. Biochem J 1987; 242:1-10.
19. Crowe JH, Crowe LM. Preservation of liposomes by freeze drying. In: Gregoriadis G, ed. Liposome Technology, 2nd ed. CRC Press Inc., 1992.
20. Hays LM, Crowe JH, Wolkers W et al. Factors affecting leakage of trapped solutes from phospholipid vesicles during thermotropic phase transitions. Cryobiology 2001; 42:88-102.
21. Crowe JH, Crowe LM. Factors affecting the stability of dry liposomes. Biochim Biophys Acta 1988; 939:327-334.
22. Crowe JH, Carpenter JF, Crowe LM. The role of vitrification in anhydrobiosis. Annu Rev Physiol 1998; 6:73-103.
23. Crowe JH, Crowe LM, Chapman D. Preservation of membranes in anhydrobiotic organisms: The role of trehalose. Science 1984; 223:701-703.
24. Allison SD, Manning MC, Randolph TW et al. Optimization of storage stability of lyophilized actin using combinations of disaccharides and dextran. J Pharm Sci 2000; 89:199-214.
25. Anchordoquy TJ, Izutsu KI, Randolph TW et al. Maintenance of quaternary structure in the frozen state stabilizes lactate dehydrogenase during freeze-drying. Arch Biochem Biophys 2001; 390:35-41.
26. Cleland JL, Lam X, Kendrick B et al. A specific molar ratio of stabilizer to protein is required for storage stability of a lyophilized monoclonal antibody. J Pharm Sci 2000; 90:310-321.
27. Belton PS, Gil AH. IR and Raman spectroscopic studies of the interaction of trehalose with hen egg lysozyme. Biopolymers 1994; 34:957-961.
28. Cottone G, Cicotti G, Cordone L. Protein-trehalose-water structures in trehalose coated carboxy-myoglobn. J Cell Phys 2002; 117:9862-9866.
29. Lins RD, Pereira CS, Hunenberger PH. Trehalose-protein interactions in aqueous solutions. Proteins 2004; 55:177-186.
30. Sun WQ, Leopold AC. Cytoplasmic vitrification and survival of anhydrobiotic organisms. Comp. Biochem Physiol 1997; 117A:327-333.
31. Sun WQ, Leopold AC, Crowe LM et al. Stability of dry liposomes in sugar glasses. Biophys J 1996; 70:1769-1776.
32. Lee CWB, Waugh JS, Griffin RG. Solid-state NMR study of trehalose/1,2-dipalmitoyl-sn-phosphatidylcholine interactions. Biochemistry 1986; 25:3737-3742.
33. Nakagaki M, Nagase H, Ueda H. Stabilization of the lamellar structure of phosphatidylcholine by complex-formation with trehalose. J Mem Sci 1992; 73:173-180.
34. Tsvetkova NM, Phillips BL, Crowe LM et al. Effect of sugars on headgroiup mobility in freeze-dried dipalmitoylphosphatidylcholine bilayers: Solid-state P-31 NMR and FTIR studies. Biophys J 1998; 75:2947-2955.
35. Luzardo MD, Amalfa F, Nunez AM et al. Effect of trehalose and sucrose on the hydration and dipole potential of lipid bilayers. Biophys J 2000; 78:2452-2458.
36. Ricker JV, Tsvetkova NM, Wolkers WF et al. Trehalose maintains phase separation in an air-dried binary lipid mixture. Biophys J 2003; 84:3045-3051.
37. Chandrasekhar I, Gaber BP. Stabilization of the biomembrane by small molecules: Interaction of trehalose with the phospholipid bilayer. J Biomol Struct Dyn 1988; 5:1163-1171.
38. Rudolph BR, Chandrasekhar I, Gaber BP et al. Molecular modeling of saccharide-lipid interactions. Chem Phys Lipids 1990; 53:243-261.
39. Sum AK, Faller F, de Pablo JJ. Molecular simulation of phospholipid bilayers and insights of the interactions with disaccharides. Biophys J 2003; 85:2830-2844.
40. Pereira CS, Lins RD, Chandrasekhar I et al. Interaction of the disaccharide trehalose with a phospholipid bilayer: A molecular dynamics study. Biophys J 2004; 86:2272-2285.
41. Brown DA. Seeing is believing: Visualization of rafts in model membranes. Proc Nat Acad Sci 2001; 98:10517.
42. Brown DA, London E. Structure and origin of ordered lipid domains in biological membranes. J Mem Biol 1998; 164:103-114.

43. London E. Insights into lipid raft structure and formation from experiments in model membranes. Curr Opinion Struct Biol 2002; 12:480.
44. Horejs V. Membrane rafts in immunoreceptor signaling: New doubts, new proofs? Trends Immunol 2002; 23:562-564.
45. Heerklotz H. Triton promotes domain formation in lipid raft mixtures. Biophys J 2002; 83:2693-2697.
46. Shogomori H, Brown DA. Use of detergents to study membrane rafts: The good, the bad, and the ugly. Biol Chem 2003; 384:1259-1263.
47. Draber P, Draberova L. Lipid rafts in mast cell signaling. Mol Immunol 2002; 38:1247-1253.
48. Horejsi V. The roles of membrane microdomains (rafts) in T cell activation. Immunol Rev 2003; 191:148-154.
49. FukI IV, Meyer ME, Williams KJ. Transmembrane and cytoplasmic domains of syndecan mediate a multi-step endocytic pathway involving detergent-insoluble membrane rafts. Biochem J 2001; 351:607-613.
50. Ono A, Freed EO. Plasma membrane rafts play a critical role in HIV-1 assembly and release. Proceedings of the National Academy of Sciences 2001; 98:13925-13929.
51. Vincent S, Gerlier D, Manié SN. Measles virus assembly within membrane rafts. J Virology 2000; 74:9911-9916.
52. Gil T, Ipsen JH, Mouritsen OG et al. Theoretical analysis of protein organization in lipid membranes. Biochim Biophys Acta 1998; 1376:245-262.
53. Killian JA. Hydrophobic mismatch between proteins and lipids in membranes. Biochim Biophys Acta 1998; 137:401-416.
54. Leidy C, Gousset K, Ricker JV et al. Lipid phase behavior and stabilization of domains in membranes of platelets. Cell Biochem Biophys 2004; 40:123-135.
55. Mabrey S, Sturtevant JM. Investigation of phase transitions of lipids and lipid mixtures by sensitivity differential scanning calorimetry. Proc Nat Acad Sci 1976; 73:3862-3879.
56. Mabrey S, Mateo PL, Sturtevant JM. High-sensitivity scanning calorimetric study of mixtures of cholesterol with dimyristoyl- and dipalmitoylphosphatidylcholines. Biochemistry 1978; 17:2464-2468.
57. Womersley C. Dehydration survival and anhydrobiotic potential of entomopathogenic nematodes. In: Gaugler R, Kaya HK, eds. Entomopathogenic Nematodes in Biological Control. Boca Raton, FL: CRC Press, 1990:117-130.
58. Westh P, Ramløv H. Trehalose accumulation in the tardigrade Adorybiotus coronifer during anhydrobiosis. J Exp Zool 1991; 258:303-311.
59. Lapinski J, Tunnacliffe A. Anhydrobiosis without trehalose in bdelloid rotifers. FEBS Lett 2003; 553:387-390.
60. Caprioli M, Katholm AK, Melone G et al. Trehalose in desiccated rotifers: A comparison between a bdelloid and a monogonont species. Comp Biochem Physiol A 2004; 139:527-532.
61. Tunnacliffe A, Lapinski J. Resurrecting Van Leeuwenhoek's rotifers: A reappraisal of the role of disaccharides in anhydrobiosis. Philos Trans R Soc Lond B 2003; 358:1755-1771.
62. Crowe JH, Oliver AE, Hoekstra FA et al. Stabilization of dry membranes by mixtures of hydroxyethyl starch and glucose: The role of vitrification. Cryobiology 1997; 3:20-30.
63. Hill DR, Keenan TW, Helm RF et al. Extracellular polysaccharide of Nostoc commune (Cyanobacteria) inhibits fusion of membrane vesicles during desiccation. J Applied Phycol 1997; 9:237-248.
64. Buitink J, Walters-Vertucci C, Hoekstra FA et al. Calorimetric properties of dehydrating pollen: Analysis of a desiccation-tolerant and an -intolerant species. Plant Physiol 1996; 111:235-242.
65. Hincha DK, Hellwege EM, Meyer AG et al. Plant fructans stabilize phosphatidylcholine liposomes during freeze-drying. Eur J Biochem 2000; 267:535-540.
66. Vereyken IJ, Chupin V, Hoekstra FA et al. The effect of fructan on membrane lipid organization and dynamics in the dry state. Biophys J 2003; 84:3759-3766.
67. Hincha DK, Zuther E, Heyer AG. The preservation of liposomes by raffinose family oligosaccharides during drying is mediated by effects on fusion and lipid phase transitions. Biochim Biophys Acta 2003; 1612:172-177.
68. Hincha DK, Hagemann M. Stabilization of model membranes during drying by compatible solutes involved in the stress tolerance of plants and microorganisms. Biochem J 2004; 383:277-83.
69. Hoekstra FA, Golovina EA. The role of amphiphiles. Comp Biochem Physiol 2002; 131A:527-533.
70. Goodrich RP, Crowe JH, Crowe LM et al. Alteration in membrane surfaces induced by attachment of carbohydrates. Biochemistry 1991; 30:2313-2318.
71. Popova AV, Hincha DK. Effects of the sugar headgroup of a glyoglycerolipid on the phase behavior of phospholipid model membranes in the dry state. Glycobiology 2005; 15:1150-1155.
72. Leslie SB, Israeli E, Lighthart B et al. Trehalose and sucrose protect both membranes and proteins in intact bacteria during drying. Econ Env Microbiol 1995; 61:3592-3597.
73. Esteves MI, Quintilio W, Sato RA et al. Stabilisation of immunoconjugates by trehalose. Biotechnol Lett 2001; 22:417-420.

74. Crowe LM, Reid DS, Crowe JH. Is trehalose special for preserving dry biomaterials? Biophys J 1996; 71:2087-2093.
75. Green JL, Angell CA. Phase relations and vitrification in saccharide-water solutions and the trehalose anomaly. J Phys Chem 1989; 93:2880-2882.
76. Li S, Patapoff TW, Overcashier et al. Effects of reducing sugars on the chemical stability of human relaxin in the lyophilized state. J Pharm Sci 1996; 85:873-877.
77. O'Brien J. Stability of trehalose, sucrose and glucose to nonenzymatic browning in model systems. J Food Sci 1996; 61:679-682.
78. Schebor C, Burin L, del Pilar Bueras M et al. Stability to hydrolysis and browning of trehalose, sucrose and raffinose in low-moisture systems in relation to their use as protectants of dry biomaterials. Lebensm-Wiss u-Technol 1999; 32:481-485.
79. Buitink J, van den Dries IJ, Hoekstra FA et al. High critical temperature above Tg may contribute to the stability of biological systems. Biophys J 2000; 79:1119-1128.
80. Buitink J, Heminga MA, Hoekstra FA. Is there a role for oligosaccharides in seed longevity? An assessment of intracellular glass stability. Plant Physiology 2000; 122:1217-1224.
81. Buitink J, Leprince O. Glass formation in plant anhydrobiotes: Survival in the dry state. Cryobiology 2004; 48:215-228.
82. Koste KL. Glass formation and desiccation tolerance in seeds. Plant Physiol 1991; 96:302-304.
83. Wolkers WF, Tetteroo FAA, Alberda M et al. Changed properties of the cytoplasmic matrix associated with desiccation tolerance of dried carrot somatic embryos. An in situ Fourier transform infrared spectroscopic study. Plant Physiol 1999; 120:153-163.
84. Wolkers WF, Alberda M, Koornneef M et al. Properties of proteins and the glassy matrix in maturation-defective mutant seeds of Arabidopsis thaliana. Plant J 1998; 16:133-143.
85. Wolkers WF, Oldenhof H, Alberda M et al. A Fourier transform infrared study of sugar glasses: Application to anhydrobiotic higher plant cells. Biochim Biophys Acta 1998; 1379:83-96.
86. Walters C, Reid JL, Walker-Simmons MK. Heat soluble proteins extracted from wheat embryos have tightly bound sugars and unusual hydration properties. Seed Sci Res 1997; 7:125-134.
87. Eroglu A, Russo MJ, Bieganski R et al. Intracellular trehalose improves the survival of cryopreserved mammalian cells. Nat Biotechnol 2000; 18:163-167.
88. Guo NI, Puhlev DR, Brown J et al. Trehalose expression confers desiccation tolerance on human cells. Nat Biotechnol 2000; 18:168-171.
89. Wolkers WF, Walker NJ, Tablin F et al. Human platelets loaded with trehalose survive freeze-drying. Cryobiology 2001; 42:79-87.
90. Oliver AE, Jamil K, Crowe JH et al. Loading human mesenchymal stem cells with trehalose by fluid-phase endocytosis. Cell Preservation Tech 2004; 2:35-49.
91. Lloyd JB. Lysosome membrane permeability: Implications for drug delivery. Advanced Drug Delivery Reviews 2000; 41:189-200.
92. Tablin F, Oliver AE, Walker NJ et al. Membrane phase transitions of intact human platelets: Correlation with cold-induced activation. J Cell Physiol 1996; 168:305-313.
93. Tablin F, Wolkers WF, Walker NJ et al. Membrane reorganization during chilling: Implications for long term storage. Cryobiology 2001; 43:114-123.
94. Oliver AE, Tablin F, Walker NJ et al. The internal calcium concentration of human platelets increases during chilling. Biochim Biophys Acta 1999; 1416:349-60.
95. Crowe JH, Tablin F, Tsvetkova NM et al. Are lipid phase transitions responsible for chilling damage in human platelets? Cryobiology 1999; 38:180-191.
96. Crowe JH, Tablin F, Wolkers WF et al. Stabilization of membranes in human platelets freeze-dried with trehalose. Chem Phys Lipids 2003; 122:41-52.
97. Tsvetkova NM, Walker NJ, Crowe JH et al. Lipid phase separation correlates with activation in platelets during chilling. Mol Mem Biol 2001; 17:209-218.
98. Auh JH, Wolkers WF, Looper SA et al. Calcium mobilization in freeze-dried human platelets. Cell Preservation Tech 2004; 2:180-187.
99. Ma X, Jamil K, MacRae TH et al. A small stress protein acts synergistically with trehalose to confer desiccation tolerance on mammalian cells. Cryobiology 2005; 51:15-28.
100. Clegg JS, Jackson SA, Warner AH. Extensive intracellular translocations of a major protein accompany anoxia in embryos of Artemia franciscana. Exp Cell Res 1994; 212:77-83.
101. Liang P, Amons R, Clegg HS et al. Purification, structure and in vitro molecular-chaperone activity of Artemia p26, a small heat- shock/alpha-crystallin protein. Eur J Biochem1997; 243:225-32.
102. Liang P, Amons R, Clegg HS et al. Molecular characterization of a small heat shock/alpha-crystallin protein in encysted Artemia embryos. J Biol Chem 1997; 272:19051-19058.
103. Liang P, MacRae TH. The synthesis of a small heat shock/α-crystallin protein in Artemia and its relationship to stress tolerance during development. Devel Biol 1999; 207:445-456.

104. MacRae TH. Molecular chaperones, stress resistance and development in Artemia franciscana. Semin Cell Dev Biol 2003; 14:251-258.
105. Willsie JK, Clegg JS. Nuclear p26, a small heat shock/α-crystallin protein, and its relationship to stress resistance in Artemia franciscana embryos. J Exp Biol 2001; 204:2339-2350.
106. Day RM, Gupta JS, MacRae TH. A small heat shock/alpha-crystallin protein from encysted Artemia embryos suppresses tubulin denaturation. Cell Stress and Chaperones 2003; 8:183-193.
107. Sun Y, Mansour M, Crack JA et al. Oligomerization, chaperone activity, and nuclear localization of p26, a small heat shock protein from Artemia franciscana. J Biol Chem 2004; 279:39999-40006.
108. Viner RI, Clegg JS. Influence of trehalose on the molecular chaperone activity of p26, a small heat shock/α-crystallin protein. Cell Stress and Chaperones 2001; 6:126-135.
109. Collins CH, Clegg JS. A small heat-shock protein, p26, from the crustacean Artemia protects mammalian cells (Cos-1) against oxidative damage. Cell Biol Intl 2004; 28:449-455.
110. Singer MA, Lindquist S. Multiple effects of trehalose on protein folding in vitro and in vivo. Mol Cell 1998; 1:639-648.
111. Singer MA, Lindquist S. Thermotolerance in Saccharomyces cerevisiae: The Yin and Yang of trehalose. Trends Biotech 1998; 1:460-468.
112. Ristic Z, Williams G, Yang G et al. Dehydration, damage to cellular membranes, and heat-shock proteins in maize hybrids from different climates. J Plant Physiol 1996; 149:424-432.
113. Hayward SAL, Rinehart JP, Denlinger DL. Desiccation and rehydration elicit distinct heat shock protein transcript responses in the flesh fly pupae. J Exp Biol 2004; 207:963-971.
114. Hartl FU, Hayer-Hartl M. Molecular chaperones in the cytosol: From nascent chain to folded protein. Science 2002; 295:1852-1858.
115. Barral JM, Broadley SA, Schaffar G et al. Roles of molecular chaperones in protein misfolding diseases. Sem Cell Devel Biol 2004; 15:17-29.
116. Trent JD, Kagawa HK, Paavola CD et al. Intracellular localization of a group II chaperonin indicates a membrane-related function. Proc Natl Acad Sci 2003; 100:15589-15594.
117. Torok Z, Horvath I, Goloubinoff P et al. Evidence for a liposhaperonin: Association of active protein-folding GroESLj oligomers with lipids can stabilize membranes under heat shock conditions. Proc Natl Acad Sci 1997; 94:2192-2197.
118. Torok Z, Tsvetkova NM, Balogh G et al. Heat shock protein coinducers with no effect on protein denaturation specifically modulate the membrane lipid phase. Proc Nat Acad Sci 2003; 100:3131-3136.
119. Torok Z, Goloubinoff P, Horvath I et al. Synechocystis HSP17 is an amphitropic protein that stabilizes heat-stressed membranes and binds denatured proteins for subsequent chaperone-mediated refolding. Proc Natl Adad Sci 2001; 98:3098-3103.
120. Tsvetkova NM, Horvath I, Torok Z et al. Small heat-shock proteins regulate membrane lipid polymorphism. Proc Natl Acad Sci 2002; 99:13504-13509.
121. Vigh L, Maresca B, Harwood JL. Does the membrane's physical state control the expression of heat shock and other genes? Trends Bio Sci1998; 23:369-374.
122. Horvath I, Glatz A, Varvasovszki V et al. Membrane physical state controls the signaling mechanism of the heat shock response in Synechocystis PCC 6803: Identification of hsp17 as a "fluidity gene". Proc Nat Acad Sci 1998; 95:3513-3518.
123. Beere HM, Green DR. Stress management - Heat shock protein-70 and the regulation of apoptosis. Trends Cell Biol 2001; 11:6-10.
124. Concannon CG, Gorman AM, Samali A. On the role of Hsp27 in regulating apoptosis. Apoptosis 2003; 8:61-70.
125. Samali A, Orrenius S. Heat shock proteins: Regulators of stress response and apoptosis. Cell Stress and Chaperones 1998; 3:228-236.
126. Downs CA, Jones LR, Heckathorn SA. Evidence for a novel set of small heat-shock proteins that associates with the mitochondria of murine PC12 cells and protects NADH:ubiquinone oxidoreductase from heat and oxidative stress. Arch Biochem Biophys 1999; 365:344-350.
127. Gill RR, Gbur CJ, Fisher BJ et al. Heat shock provides delayed protection against oxidative injury in cultured human umbilical nein endothelial cells. J Molec Cell Cardiol 1998; 30:2739-2749.
128. Park YM, Han MY, Blackburn RV et al. Overexpression of HSP25 reduces the level of TNFalpha-induced oxidative DNA damage biomarker, 8-hydroxy-2'-deoxyguanosine, in L929 cells. J Cell Physiol 1998; 174:27-34.
129. Jamil K, Crowe JH, Tablin F et al. Arbutin enhances recovery and osteogenic differentiation in dried and rehydrated human mesenchymal stem cells. Cell Preservation Tech 2005; 3:244-255.
130. Oliver AE, Hincha DK, Crowe JH. Looking beyond sugars: The role of amphiphilic solutes in preventing adventitious reactions in anhydrobiotes at low water contents. Comp Biochem Physiol 2002; 131A:515-525.
131. Crowe JH. Anhydrobiosis: An unsolved problem. Am Nat 1971; 105:563-573.

CHAPTER 14

Chaperones As Part of Immune Networks

Zoltán Prohászka*

Abstract

Network theory is increasingly accepted as a basic regulatory mechanism in diverse immunological functions. Heat shock proteins (Hsps) are involved in multiple networks in the immune system. Hsps themselves (foreign or endogenous) activate innate immunity and play important roles to deliver self or nonself materials to antigen presenting cells. However, Hsps are immunodominant antigens during infectious diseases making self Hsps endangered targets of autoimmunity by cross-reactive clones. Therefore, it is not surprising that the mechanism of protection of self Hsps is not clonal deletion in natural self tolerance; rather, self Hsps are protected by active regulating natural autoimmunity. The active regulatory/protective immunity is accomplished by natural autoantibodies and regulatory T cells, both recognizing Hsps. The multiple involvements of Hsps in immune networks make them ideal targets of therapy in autoimmune diseases. Indeed, immunotherapy with Hsps was recently reported to be effective treatment modality against cancer, arthritis or diabetes mellitus.

Introduction

Owing to their highly conserved and inducible nature stress proteins are ideal messengers of cellular stress. Nearly all pathogen microorganisms studied till now possess heat inducible stress protein genes and respond to the thermal (and other) stresses of infection with increased heat shock protein (Hsp) expression. Higher organisms have innate sense, by means of innate immunity, to respond stress signals, on one hand. On the other hand, basically the same molecules may messenger stress of the host organism, making altered self 'dangerous', in case, for example, of cell necrosis.[1] However, the innate role of Hsps may become problematic in the new world of the adaptive immunity. With the appearance of specific receptors (i.e., antibodies and T cell receptors) the overexpressed and conserved stress proteins turn into first class targets of autoimmunity by means of infection induced molecular mimicry. The highly conserved Hsps, which are present in all mammalian cells, need therefore special protection in the 'adaptive world'. In this Chapter recent data on the relationship of Hsps with innate immunity, their protection by regulating autoimmunity will be summarized and a new concept on their embedment in multiple networks of the organism will be presented.

Activation of Innate Immunity by Heat Shock Proteins

Mammalian heat shock proteins were described to activate innate immunity by nonclonal receptors belonging to the family of pattern recognition receptors such us Toll-like receptors. Members belonging to the 70- and 60-kDa Hsp family were mostly investigated and TLR2 and TLR4 were reported as important players in the transduction of signals. However, the

*Zoltán Prohászka—IIIrd Department of Internal Medicine, Semmelweis University, H-1125 Budapest, Kútvölgyi st. 4, Hungary. Email: prohoz@kut.sote.hu

Molecular Aspects of the Stress Response: Chaperones, Membranes and Networks, edited by Peter Csermely and László Vígh. ©2007 Landes Bioscience and Springer Science+Business Media.

binding receptors for Hsps are not yet completely known or debated (see ref. 2) and some of the effects Hsps on the target cells seemed to be indistinguishable from effects of endotoxins (LPS). This similarity between the effects of Hsps and LPS lend credence to the hypothesis that the so-called cytokine effects of Hsps are due to contaminants copurified with the recombinant proteins used in such experiments (reviewed in ref. 3). In some recent studies this hypothesis was investigated in details and Gao and Tsan pinpointed out the importance for caution while interpreting the results of studies with recombinant Hsps.[4] We learned from the works of Gao and Tsan that usually applied methods to control endotoxin contamination (polymyxin B treatment and boiling of proteins) might be inappropriate and much more attention has to be paid to control endotoxin effects in studies on innate immunity. However, several recent independent lines of evidence obtained in well conducted and controlled experiments suggest that Hsps, nevertheless, do manifest innate immune activities that are not due to bacterial contaminants. These include: activation of DCs and macrophages by mammalian Hsp70 and gp96 purified by a pyrogen-free method (see ref. 5); activation of macrophages via TLR2- and TLR4-signaling by mammalian cells displaying Hsp70 on their surface.[6] Furthermore, Habich et al reported results of elegant binding studies using fluorochrome labeled Hsp60 preparations. In these studies specific epitopes of Hsp60 were identified for binding to primary macrophages.[7] Importantly, these epitopes overlap with B-cell epitopes of the molecule (see below). In addition, the same group determined the binding site of Hsp60 for lipopolysaccharide as well (see ref. 8), indicating that endotoxin is not just a simple contaminant in recombinant Hsp60 preparation, rather, it might reflect an inherent property of this protein as a sensor and messenger of danger. Interestingly, Hsp70 and Hsp90 have also been reported to act as receptors for LPS.[9,10] In conclusion, all of the above experiments indicate that Hsps play important roles in the activation of innate immunity, but critical works rightly call our attention to the fact that Hsps are likely found in complexes with self or nonself material.

Immunological Protection of Heat Shock Proteins

As mentioned above, Hsps are structurally highly conserved molecules from eubacteria till humans what is true for their molecular structure, biochemical properties but partly even for their immunological (epitope) structure. It is not surprising, therefore, that several autoantigens characterized in different autoimmune conditions share some of their epitopes with Hsps.[11] These findings indeed support the long lasting observation that infections may induce autoimmune processes. More surprising is the fact, however, that autoimmune diseases are rare diseases and are characterized by well defined autoimmune reaction to only a few number of autoantigens in a conserved nature. Furthermore, according to the idea that the source of self tolerance is the clonal deletion, one may speculate that the closer a molecule is to self, the less immunogenic it should be. It is surprising, therefore, to find that among the major antigens recognized during a wide variety of infections many belong to conserved protein families sharing extensive sequence identity with host's molecules, for example Hsps. One interesting hypothesis, dealing with the above listed controversies, was presented by Irun Cohen and coworkers, called 'the immunological homunculus'.[12] The idea is that some, perhaps all, major autoantigens are indeed dominant because each of them is encoded in the organizational structure of the immune system. Each dominant self antigen is served by an interacting set of T and B cells that includes cells with receptors for the antigen (antigen-specific) and cells with receptors for the antigen-specific receptors (anti-idiotypic). Some of these lymphocytes suppress and others stimulate. Owing to the mutual connections between the various interacting lymphocytes in the network, some lymphocytes become activated even without being driven by contact with specific antigen in an immunogenic form. The state of autonomous activity defined by a pattern of interconnected lymphocytes constitutes a functional representation of the particular self antigen around which the network is organized. In other words, the picture of the self antigen is encoded within a cohort of lymphocytes.

At the time of the presentation of the immunological homunculus, the information that cells of innate immunity, and even B cells, possess more abundant and effective receptors (as compared to the specific ones), the pattern recognition receptors (PRRs, i.e., Toll-like receptors), for the conserved antigens, for example for Hsps, was not available. Most probable the recognition of the conserved molecules by PRRs is also a fundamental mechanism why these molecules became immunodominant during infection. Although the idea on 'immunological homunculus' was debated in the meantime and the conception is far from a general acceptance in the immunologists' community, some recent results support the existence of regulating natural autoimmunity. Heat shock proteins seem to be an important contributor to, and target of, these regulatory mechanisms.

Role of Natural Autoantibody Networks in Regulation of Autoimmunity

Natural antibodies refer to antibodies that are present in the serum of healthy individuals in the absence of deliberate immunization with the target antigen.[13] Autoantibodies are immunoglobulins that react with at least one self antigen, whether they originate form healthy individuals or patients with autoimmune disease. The importance of natural antibodies reactive with self antigens (natural autoantibodies, NAA) has earlier been neglected, as tolerance to self was thought to be primarily dependent on the deletion of autoreactive clones during ontogeny. It is now established, that NAAs are present in almost all vertebrates investigated until now.[14] NAAs may belong to IgM, IgG or IgA isotypes, whereas in humans IgG dominates. Natural autoreactive B cells are endowed with some switching ability in the absence of cognate interactions with T cells. NAAs are characterized by polyreactivity resulting in specific reactivity patterns. NAAs are characterized by a broad range of affinities having mainly dissociation constants in the micromolar range. The notion that an antibody has to be of high affinity in order to be biologically relevant originates primarily from the analysis of the requirements for an efficient immune response against pathogens. This concept does not necessarily apply to natural antibodies. Specific polyreactive recognition patterns of low affinity antibodies may result in novel biological properties that emerge from the organization of networks. The biological activities of such networks may not be predicted from activities of individual members of the network.

Evidence for the intrinsic network regulation of autoreactive antibody repertoires include (reviewed in ref. 15): (i) the V-region-dependent connectivity between NAA; (ii) the restricted pattern of reactivity to a set of self antigens that is conserved among healthy individuals and throughout life; (iii) the spontaneous fluctuations of concentrations of NAAs over time; and iv, the suppression of pathogenic autoreactive clones by infusions of normal polyvalent IgG (IVIG) in patients with autoimmune diseases.

Polyreactivity and V-region-dependent connectivity are basic features of NAAs comprising interactions with other IgG- and immunological active molecules. Reactivities of NAAs towards cytokines and cytokine receptors, cell surface receptors (such as CD4 or MHC class I molecules), adhesion molecules, coagulation factors and heat shock proteins have been reported.[15]

Anti-Hsp90 reacting antibodies were recently shown to be part of the natural autoantibody repertoire and were characterized as broadly cross-reacting antibodies mainly belonging to the IgG2 subclass.[16] Furthermore, in the study of Pashov and coworkers,[16] although the anti-Hsp90 antibodies bound to the same set of antigens in the patients, there were quantitative differences among the antibodies tested. Thus, the natural autoantibody repertoire qualitatively seems to be invariable, directed against a highly conserved set of immunodominant antigens including Hsp90. Work in our laboratory in the past years was focused on the analysis of anti-Hsp60 and anti-Hsp70 IgG autoantibodies. In a recent review paper supporting evidence was summarized on the view that anti-Hsp60 IgG autoantibodies belong to NAAs.[17] These antibodies are present in all healthy individuals investigated and in IVIG preparations as well.[18] In our recent studies

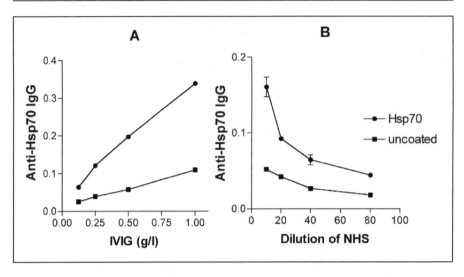

Figure 1. IgG anti-Hsp70 autoantibodies are present in the normal immunoglobulin preparation (IVIG) and in normal human serum (NHS). ELISA plates were coated with human heat shock protein 70 (2 μg/ml) and incubated with serial dilutions of IVIG (A) or NHS (B). Uncoated wells were used as control. Anti-Hsp70 IgG antibodies were determined as described earlier.[23] Means of two parallel measurements and SD are plotted, the experiment was repeated four times with identical results.

anti-Hsp60 autoantibody level-associated genetic factors were determined. Allelic variants of promoter polymorphism -174 of interleukine-6 (IL-6) were shown to associate with anti-Hsp60 concentrations in two independent populations (ref. 19 and unpublished studies from the author's lab). Furthermore, allelic determinants of immunoglobulin constant regions (KM determinants) have also been shown to associate with increased autoantibody levels.[20] Most importantly, IL-6 and Ig KM determinants were in epistatic interaction in the regulation of anti-Hsp60 levels.[21] IgM and IgD natural autoantibodies to Hsp70 have earlier been reported in normal mice.[22] This observation is strengthened by recent results on the presence of anti-Hsp70 IgG reactivity in all healthy human serum samples tested (see refs. 23,24) and in IVIG preparation (Fig. 1).

Taken together natural autoantibodies to different families of Hsps are present in sera of healthy individuals. These anti-Hsp antibodies—although directly not yet investigated—most probably operate in networks like other NAAs. Since Hsp70 and Hsp60 has been reported to be present in the circulation of healthy individuals (see refs. 24-26) and heat shock proteins might be expressed on the surface of stressed or transformed cells, this setting preferentially give rise to the formation of self regulating networks. Importantly, B cell- and receptor binding epitopes of Hsp60 are partly overlapping (see refs. 7,18) indicating functional relevance of NAAs in the possible regulation of biological activities of Hsp60.

Heat Shock Proteins as Negotiators between Promotion of Inflammation or Control of Autoimmunity

Heat shock proteins present in the circulation of healthy individuals and patients might originate from cells of the organism or from colonizing or pathogenic microorganisms. The source of Hsps, their complexes with self- or foreign materials and the nature of the host's regulating immune network is decisive on the shape of subsequent immunological events.

Hsps originating from self tissues (endogenous Hsps) are present in healthy and diseased subjects.[24-26] The source of these Hsps is likely cellular necrosis (passive release), but active release (i.e., secretion of small membrane vesicles) can not be excluded. Indeed, exosomes have repeatedly been shown to contain cytoplasmic members of Hsp families.[27] These molecules might carry endogenous peptides and deliver them to antigen presenting cells thus signalling the state of self to other cells. Srivastava and colleagues recently reviewed the potential of Hsps eliciting anti-tumor immune responses.[28] Furthermore, endogenous Hsps may stimulate regulatory T-cells that control pathogenic T cells for self antigens other than Hsps.[29] However, Hsps originating from infecting agents might be complexed with endotoxins and other bacterial materials making these complexes highly active in initiating inflammation (see above). Thus, depending on the source of Hsp material and the surrounding of recognition processes immunological responses of the host may be very diverse. In other words Hsps may both promote or regulate inflammatory processes having fundamental relevance to pathological conditions.

The best studied examples in this field are adjuvant arthritis (see ref. 30) and autoimmune diabetes mellitus type I.[31] In adjuvant arthritis a pathogenic role of T cell reactivity to a nonconserved epitope of Hsp65 (AA180-188) has been shown.[32] Later studies confirmed that the self antigen recognized by this clone is not Hsp60 but the cartilage proteoglycan. Furthermore, Hsp65 can be used effectively to prevent different experimental arthritis types, and this is based on a general anti-inflammatory pathway induced by Hsp65.[33-34] The protective effects of Hsp65 depend on cell mediated processes that induce regulatory cells turning down inflammation. In case of diabetes mellitus a beta-cell target antigen in nonobese diabetic (NOD) mice is a molecule cross-reactive with the Hsp65. The importance of hsp65 in the pathogenesis of the disease was confirmed by the ability of clones of anti-hsp65 T cells to cause insulitis and hyperglycemia in young NOD/Lt mice. Moreover, hsp65 antigen could be used either to induce diabetes or to vaccinate against diabetes, depending on the form of its administration to prediabetic NOD/Lt mice.[35] Subsequent research recognized the minimum structure (AA437-460 called p277) of Hsp60 arresting the destruction of the beta-cells in NOD mice.[36] The mechanism of protection against diabetes in the NOD mice is characterized by the shift of Th1/Th2 balance towards Th2, by the downregulation of T-cell and IgG responses against self antigens but no modification of spontaneous Th1 cytokine secretion to Hsp65 peptide. The successful formulation of the p277 peptide (DiaPep277) led to the initiation of phase I and II human studies with prosperous results.[37]

Taken together the above summarized experimental data strongly support the view that the conserved heat shock protein (self or foreign) molecules are indeed negotiators between promotion of inflammation or control of autoimmunity.[38] Translation of this knowledge into practical usage in human health will most probably help to control some of the autoimmune diseases.

Heat Shock Proteins as Elements of Multiple Networks

Jerne suggested some 25 years ago that immunoglobulin idiotypes are organized as network of complementary shapes.[39] Network regulation in immunology was neglected for longer time and tolerance was regarded as result of the elimination of all self-reactive clones. However, accumulating data suggested that basic functions of regulatory T cells and natural autoantibodies are related to network operation, thus more and more light was shed onto this area of immunology.[40-41] Conventional immune responses and much of their regulation are satisfactorily explained by clonal selection principles, such as somatic mutation and 'induced fit' and clonal selection also contributes the rational basis for anti-infectious protection. However, core operation of immune networks has little to do with immune responses, but is fundamental to the understanding of questions that were not solved by the clonal selection theory. These unresolved areas include internal lymphocyte activities, natural antibody

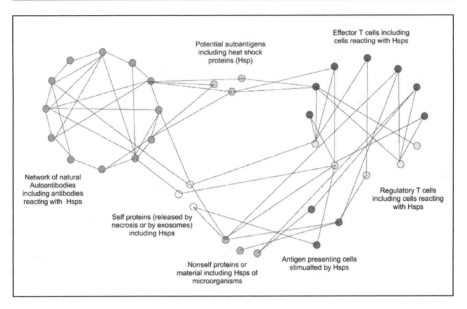

Figure 2. Heat shock proteins in multiple immunological networks and presentation of their putative complex roles in promotion or control of inflammation and autoimmunity. Hsps present in mammalian host organisms may originate from endogenous or exogenous sources. Endogenous Hsps may be released by host cells during necrosis passively or by active secretion by means of small membrane vesicles (exosomes). Endogenously released Hsps carry self peptides and may 'signal' the composition of self to other tissues in the body. Pathogen microbes or organisms of the body's normal flora may constitute sources of exogenous Hsps. Hsps from both sources can form complexes with hydrophobic materials such as endotoxins resulting in highly proinflammatory motifs. Hsps are potent inducers of regulatory T cells, however, they may also contribute to the activation of effector T cells via interactions with antigen presenting cells. The type of reaction (proinflammatory or regulatory) depends on the composition of Hsp material present. 'Dangerous' complexes induce maturation and activation of APCs via pattern recognition receptors whereas 'nondangerous' Hsps promote regulatory T cell activity downregulating inflammation. The network of natural antibodies contains polyreactive, low-affinity antibodies reacting with Hsps as well. The interaction of these antibodies with Hsps is the fine-tuning mechanism in the regulation of inflammation during infection, stress, trauma or development of tissues. (Please note that the various networks on the figure are only illustrative).

production in unimmunized animals, preimmune repertoire selection, tolerance and self-nonself discrimination, memory and the evolution of immune systems.[42] Basic features of Hsps make them core elements (or using the network term: hubs) in immunological networks. These features include their presence in almost all living organisms in conserved manner, their immunodominance and their pivotal immunological activities (recognition by several classes of receptors). Figure 2 illustrates the complex role of Hsps in immunological processes. Research in the past years could focus only on parts of this complex picture. Future research, however, has to deal with the challenges illustrated on Figure 2, i.e., upcoming studies need to use all of the modern technological, genomical and bioinformatical approaches to task on understanding the complete processes. Taking Hsps as parts of multiple networks will help to delineate their multiple and decisive roles in promotion or regulation of inflammatory and autoimmune processes.

Acknowledgements

Work in the authors' laboratory was supported by research grant from the National Research Fund of Hungary (OTKA) T46837. The skillful technical assistance of Szigeti Antalné is acknowledged with many thanks.

References

1. Matzinger P. The danger model: A renewed sense of self. Science 2002; 296:301-305.
2. Binder RJ, Vatner R, Srivastava PK. The heat-shock protein receptors: Some answers and more questions. Tissue Antigens 2004; 64:442-451.
3. Tsan MF, Gao B. Endogenous ligands of Toll-like receptors. J Leukoc Biol 2004; 76:514-519.
4. Gao B, Tsan MF. Induction of cytokines by heat shock proteins and endotoxin in murine macrophages. Biochem Biophys Res Commun 2004; 317:1149-1154.
5. Panjwani NN, Popova L, Srivastava PK. Heat shock proteins gp96 and hsp70 activate the release of nitric oxide by APCs. J Immunol 2002; 168:2997-3003.
6. Korbelik M, Sun J, Cecic I. Photodynamic therapy induced cell surface expression and release of heat shock proteins: Relevance for tumor response. Cancer Res 2002; 65:1018-1026.
7. Habich C, Kempe K, Gomez FJ et al. Heat shock protein 60: Identification of specific epitopes for binding to primary macrophages. FEBS Lett 2006; 580:115-120.
8. Habich C, Kempe K, van der Zee R et al. Heat shock protein 60: Specific binding of lipopolysaccharide. J Immunol 2005; 174:1298-1305.
9. Byrd CA, Bornmann H, Erdjument-Bromage P et al. Heat shock protein 90 mediates macrophage activation by taxol and bacterial lipopolysaccharide. Proc Natl Acad Sci USA 1999; 96:5645-5651.
10. Triantafilou K, Triantafilou M, Dedrick RL. A CD14-independent LPS receptor complex. Nat Immunol 2001; 2:338-342.
11. Jones DB, Coulson AF, Duff GW. Sequence homologies between Hsp60 and autoantigens. Immunol Today 1993; 14:115-8.
12. Cohen IR, Young DB. Autoimmunity, microbial immunity and the immunological homunculus. Immunol Today 1992; 12:105-110.
13. Coutinho A, Kazatchkine MD, Avrameas S. Natural autoantibodies. Curr Opin Immunol 1995; 7:812-821.
14. Lacroix-Desmazes S, Kaveri SV, Mouthon L et al. Self-reactive antibodies (natural autoantibodies) in healthy individuals. J Immunol Meth 1998; 216:117-137.
15. Lacroix-Desmazes S, Mouthon L, Spalter SH et al. Immunoglobulins and the regulation of autoimmunity through the immune network. Clin Exp Rheumatol 1996; 14:S9-S15.
16. Pashov A, Kenderov A, Kyurkchiev S et al. Autoantibodies to heat shock protein 90 in the human natural antibody repertoire. Int Immunol 2002; 14:453-461.
17. Prohászka Z, Füst G. Immunological aspects of heat shock proteins - The optimum stress of life. Mol Immunol 2004; 41:29-44.
18. Uray K, Hudecz F, Füst G et al. Comparative analysis of linear antibody epitopes on human and mycobacterial 60 kDa heat shock proteins using samples of healthy blood donors. Int Immunol 2003; 15:1229-1236.
19. Veres A, Prohászka Z, Kilpinen S et al. The promoter polymorphism of the IL-6 gene is associated with levels of antibodies to 60 kDa heat shock proteins. Immunogenetics 2002; 53:851-856.
20. Pandey JP, Prohászka Z, Veres A et al. Epistatic effects of genes encoding immunoglobulin GM allotypes and interleukin-6 on the production of autoantibodies to 60-kDa and 65 kDa heat shock proteins. Genes and Immunity 2004; 5:68-71.
21. Prohászka Z. Gene-gene interactions in immunology as exemplified with studies on autoantibodies against 60 kDa heat shock protein. In: Falus A, ed. Immunogenomics. London: Wiley and Sons, 2005.
22. Menoret A, Chandawarkar RY, Srivastava PK. Natural autoantibodies against heat-shock proteins hsp70 and gp96: Implications for immunotherapy using heat-shock proteins. Immunology 2000; 101:364-370.
23. Kocsis J, Veres A, Vatay Á et al. Antibodies against the human heat shock protein hsp70 in patients with severe coronary artery disease. Immunol Invest 2002; 31:219-231.
24. Pockley AG, Shepherd J, Corton JM. Detection of heat shock protein 70 (Hsp70) and anti-Hsp70 antibodies in the serum of normal individuals. Immunol Invest 1998; 27:367-377.
25. Pockley AG, Bulmer J, Hanks BM et al. Identification of human heat shock protein 60 (Hsp60) and anti-Hsp60 antibodies in the peripheral circulation of normal individuals. Cell Stress Chap 1999; 4:29-35.

26. Dybdahl B, Wahba A, Lien E et al. Inflammatory response after open heart surgery: Release of heat-shock protein 70 and signaling through toll-like receptor-4. Circulation 2002; 105:685-90.
27. Clayton A, Turkes A, Navabi H et al. Induction of heat shock protein sin B-cell exosomes. J Cell Sci 2005; 118:3631-3638.
28. Srivastava PK. Immunotherapy of human cancer: Lessons from mice. Nat Immunol 2000; 1:363-6.
29. Quintana FJ, Cohen IR. Heat shock proteins as endogenous adjuvants in sterile and septic inflammation. J Immunol 2005; 175:2777-2782.
30. Prakken BJ, Roord S, Ronaghy A et al. Heat shock protein 60 and adjuvant arthritis: A model for T cell regulation in human arthritis. Springer Semin Immunopathol 2003; 25:47-63.
31. Cohen IR. Peptide therapy for Type I diabetes: The immunological homunculus and the rationale for vaccination. Diabetologia 2002; 45:1468-1474.
32. van Eden W, Thole JE, van der Zee R et al. Cloning of the mycobacterial epitope recognized by T lymphocytes in adjuvant arthritis. Nature 1988; 331:171-173.
33. Anderton SM, van der Zee R, Noordzij A et al. Differential mycobacterial 65-kDa heat shock protein T cell epitope recognition after adjuvant arthritis-inducing or protective immunization protocols. J Immunol 1994; 152:3656-3664.
34. Paul AG, van Kooten PJ, van Eden W et al. Highly autoproliferative T cells specific for 60-kDa heat shock protein produce IL-4/IL-10 and IFN-gamma and are protective in adjuvant arthritis. J Immunol 2000; 165:7270-7277.
35. Elias D, Markovits D, Reshef T et al. Induction and therapy of autoimmune diabetes in the nonobese diabetic (NOD/LT) mouse by a 65-kDa heat shock protein. Proc Natl Acad Sci USA 1990; 87:576-1580.
36. Elias D, Reshef T, Birk OS et al. Vaccination against autoimmune mouse diabetes with a T-cell epitope of the human 65 kD heat shock protein. Proc Natl Acad Sci USA 1991; 88:3088-3091.
37. Raz I, Elias D, Avron A et al. Beta-cell function in new-onset type 1 diabetes and immunomodulation with heat-shock protein peptide (DiaPep277): A randomized, double-blind, phase II trial. Lancet 2001; 358:1749-1753.
38. van Eden W, Koets A, van Kooten P et al. Immunopotentiating heat shock proteins: Negotiatiors between innate danger and control of autoimmunity. Vaccine 2003; 21:897-901.
39. Jerne N. Towards a network theory of the immune system. Ann Inst Pasteur Immunol 1974; 125C:435-441.
40. Varela FJ, Coutinho A. Second generation immune networks. Immunol Today 1991; 12:159-166.
41. Avrameas S. Natural autoantibodies: From 'horror autotoxicus' to 'gnothi seauton'. Immunol Today 1991; 12:154-158.
42. Coutinho A. Will the idiotypic network help to solve natural tolerance? Trends Immunol 2003; 24:53-54.

CHAPTER 15

The Stress of Misfolded Proteins:
C. elegans Models for Neurodegenerative Disease and Aging

Heather R. Brignull, James F. Morley and Richard I. Morimoto*

Abstract

A growing number of human neurodegenerative diseases are associated with the expression of misfolded proteins that oligomerize and form aggregate structures. Over time, accumulation of misfolded proteins leads to the disruption of cellular protein folding homeostasis and eventually to cellular dysfunction and death. To investigate the relationship between misfolded proteins, neuropathology and aging, we have developed models utilizing the nematode *C. elegans*. In addition to being genetically tractable, *C. elegans* have rapid growth rates and short life-cycles, providing unique advantages for modeling neurodegenerative diseases of aging caused by the stress of misfolded proteins. The *C. elegans* models described here express polyglutamine expansion-containing proteins, as occur in Huntington's disease. Through the use of tissue-specific expression of different lengths of fluorescently tagged polyglutamine repeats, we have examined the dynamics of aggregate formation both within individual cells and over time throughout the lifetime of individual animals, identifying aging and other genetic modifiers as an important physiologic determinant of aggregation and toxicity.

Introduction

Misfolded proteins, aggregates, and inclusion bodies are hallmarks of a range of neurodegenerative disorders including Alzheimer's disease (AD), Parkinson's disease (PD), prion disorders, amyotrophic lateral sclerosis (ALS), and polyglutamine (polyQ) diseases that include Huntington's disease (HD) and related ataxias.[1-3] Each of these disorders exhibits aging-dependent onset and selective neuronal vulnerability despite widespread expression of the related proteins, and a progressive, usually fatal clinical course. The deposition of intra- or extracellular protein aggregates is a well-conserved pathological feature and has been the focus of extensive investigation.[3] Despite differences in the underlying genes involved, inheritance and clinical presentation, the similarities observed have led to the proposal of shared pathogenic mechanisms and the hope that insights into one process may be generalized to others.

In support of this premise is growing evidence that the cellular protein quality control system appears to be an underlying common denominator of these diseases.[4] For example, genes involved in protein folding and degradation, including molecular chaperones and components of the proteasome, have been shown to modulate onset, development and progression in models of multiple neurodegenerative diseases.[5-7] Further, it has been suggested that despite

*Corresponding Author: Richard I. Morimoto—Department of Biochemistry, Molecular Biology, and Cell Biology. Rice Institute for Biomedical Research, Northwestern University, 2153 North Campus Drive, Evanston, Illinois 60208, U.S.A. Email: r-morimoto@northwestern.edu

Molecular Aspects of the Stress Response: Chaperones, Membranes and Networks, edited by Peter Csermely and László Vígh. ©2007 Landes Bioscience and Springer Science+Business Media.

the absence of sequence homology, different disease-related proteins share a common ability to adopt similar proteotoxic conformations and that these might be used as therapeutic targets.[8,9]

Models of Neurodegenerative Disease

Some of these disorders, including the polyQ diseases, exhibit familial inheritance, allowing the use of genetic studies and positional cloning to identify single gene alterations underlying the disorders.[10-13] Other diseases are sporadic, but rare familial forms have allowed the identification of candidate genes that could reveal insights into pathology. These include mutations of amyloid precursor protein in AD, parkin and α-synuclein in PD and superoxide dismutase in ALS.[14-19] Identification of these genes has led to the development of many model systems to investigate the underlying pathology and to identify factors and pathways that modify the disease process.

In addition to transgenic mouse and cell culture models has been the development of invertebrate models using *Drosophila* and *C. elegans* for the study of neurodegenerative disease.[20-24] As described here for *C. elegans*, these systems provide a genetic approach for identification of modifiers of both cellular and behavioral phenotypes which together with a relative ease of technical manipulation facilitates rapid, high-throughput testing of hypotheses.

In a number of cases, disease gene orthologs have been identified and their loss of function phenotypes observed. For example, inactivation of the fly *parkin* gene results in a degenerative phenotype.[25,26] However, pathogenic mechanisms in neurodegeneration often involve a gain of function toxicity allowing these disorders to be modeled using transgenic overexpression of human disease-related proteins regardless of whether a clear ortholog can be identified. Expression of polyQ containing proteins is neurotoxic in both *Drosophila* retinal neurons and *C. elegans* chemosensory or mechanosensory neurons despite the absence of clear disease gene orthologs.[27-30] Similar strategies have been used to examine the toxicity of APP, SOD, α-synuclein in flies and worms.[21,31-36] Despite the idiosyncrasies of different models, each provides unique insights that clearly validate the general approach.

C. elegans Model of polyQ Disease

Our studies have focused initially on polyQ expansions as occur in Huntington's Disease and related disorders including several spinocerebellar ataxias and Kennedy's Disease.[37-40] At least nine human neurodegenerative diseases are associated with polyQ expansions within otherwise unrelated genes. In addition to a shared pathogenic motif, gene products associated with this class of neurodegenerative diseases are ubiquitously expressed but affect only neurons. Even within neuronal tissues, diseases show subset-specific aggregation, toxicity, and death.[40,41]

Expression of expanded polyQ, with or without flanking sequences from the endogenous proteins—or when inserted into an unrelated protein—is sufficient to recapitulate pathological features of the diseases in multiple model systems including the appearance of protein aggregates, loss of cell function, and cell death.[40,42-45] This suggests a central role for the polyQ expansion in these disorders and supports an approach whereby expression of isolated polyQ expansions without flanking sequences could lend insight into shared features of these disorders.

In vivo, pathogenesis is length-dependent and results from a toxic gain of function.[37,40,46,47] Genetic studies have established that Huntingtin alleles from normal chromosomes contain fewer than 30-34 CAG repeats, whereas those from affected chromosomes contain greater than 35-40 repeats.[48] Analysis of patient databases has established a strong inverse correlation between repeat length and age of onset.[48,49] Similar length dependencies are seen for the other polyQ repeat diseases suggesting a 35-40 residue threshold at which the disease gene products are converted to a proteotoxic state.

C. elegans, as a model has certain advantages for the study polyQ expansions. *C. elegans* is a roundworm that in its free-living form can be found in a variety of soil habitats. In the laboratory, the animals can be readily cultured in large numbers on agar plates seeded with a lawn of *E. coli*, as wild-type adult animals reach a maximum length of approximately 1mm.

The adult stage is preceded by progression through embryonic development, four stereo-typed larval stages and an additional alternative quiescent form—termed the dauer larva—that can be accessed under conditions of low food, high population density or conditions otherwise unfavorable for growth. This life cycle is completed in approximately 2 days under typical growth conditions (20-25°C).

At all stages of development, *C. elegans* are transparent, permitting easy detection of fluorescent proteins in live animals. The hermaphrodite body plan is relatively simple, comprised of 959 somatic cells of which 302 cells are neurons. Despite the small number of cells *C. elegans* have multiple complex tissues types including intestine, muscle, hypodermis and a fully differentiated nervous system. Furthermore, *C. elegans* display conservation of basic molecular mechanisms enabling comparison with vertebrate models.[50-52] Thus, despite its simplicity and ease of handling, studies in *C. elegans* can offer insight into processes unique to complex multicellular organisms.

The *C. elegans* polyQ Series in Neurons

To determine the effect of polyQ proteins in the neurons of *C. elegans*, we took a broad approach and expressed polyQ proteins throughout the nervous system. Phenotypes such as aggregation and neurotoxicity would therefore be the result of an integrated, system wide stress response rather than the stress response of a few isolated cells. Transgenic lines were generated using a range of polyQ expansion proteins (Q0, Q19, Q35, Q40, Q67, Q86) tagged with a fluorescent protein, and then expressed in neurons.

In young adult animals, expression of Q19 resulted in a pattern of diffuse neuronal distribution that persisted throughout the life of the animal, similar to control animals expressing the fluorescent marker (Q0). In Q19 animals cell bodies of commissural neurons (Fig. 1B, arrow) and the dorsal nerve cord (DNC) (Fig. 1B, open arrow), can be observed. The DNC is formed almost exclusively of neuronal processes and therefore displays a smooth, diffuse fluorescent intensity in Q19 animals. By comparison, the ventral nerve cord (VNC) contains visible cell bodies scattered along its length (Fig. 1B, triangle).

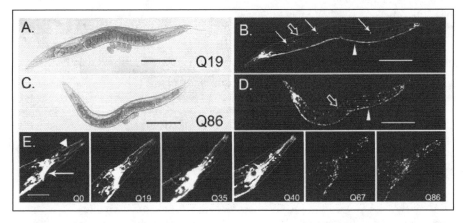

Figure 1. Poly-Q length dependent changes in protein distribution patterns. Pan-neuronal Q19::CFP has a soluble distribution pattern (A,B) while Q86::CFP is distributed into discreet foci (C,D). Arrows indicate commissural neurons. Open arrow indicates DNC, triangle indicates VNC, scale bar = 100 μm. E) Flattened z-stacks of *C. elegans* head, scale bar = 50 μm. Expressing a range of polyQ lengths reveals that proteins with tracts of ≤Q40 maintain a soluble distribution pattern while those ≥Q67 form foci. All animals depicted are young adults, four days post-hatch.

Distribution of Q86 protein was strikingly different from the diffuse distribution of Q19 protein. Throughout the nervous system, Q86 protein was limited to discrete foci, indicating polyQ length-dependent aggregation in *C. elegans* neurons (Fig. 1C-D). Q86 foci were detected in the earliest larval stage and persisted throughout the lifespan of the animal suggesting that length-dependent changes in distribution of proteins with long polyQ expansions occurred independent of neuronal subtype in *C. elegans*.

To determine the threshold length required to change the distribution pattern from diffuse (Q19) to foci (Q86), we expressed Q35, Q40, and Q67 fused to a fluorescent reporter. Animals expressing proteins with ≤Q40 display clearly delineated neuronal cell bodies and processes, suggesting the presence of soluble proteins. The soluble distribution pattern of Q0, Q19, Q35 and Q40 can be clearly observed in the head of *C. elegans* where the distinctive neuroanatomy of chemosensory processes (Fig. 1E, arrow) and the circumpharyngeal nerve ring can be identified (Fig. 1E, triangle). Q67 formed foci exclusively, similar to those observed in neurons of Q86 *C. elegans* (Fig. 1E). In contrast, the distribution pattern of proteins ≤Q40 never resembled that of Q67 or Q86. The polyQ length-dependent changes in protein distribution shown here, together with a threshold for foci formation of >Q40, recapitulate two of the major features of most polyQ-repeat diseases.

Biophysical Properties of polyQ Proteins in Neurons of Live Animals

The soluble distribution pattern of protein with ≤Q40 was visually distinct from Q67 and Q86. To assess whether the visual changes in distribution correspond to changes in protein solubility, we employed fluorescence recovery after photobleaching (FRAP) analysis. Determining the rate of recovery after photobleaching in an individual neuron provides a direct measure of protein solubility and mobility and therefore enabled us to discriminate between changes in distribution due to protein aggregation, in which interacting proteins are stably associated and immobile, or changes due to restricted subcellular localization.[53,54]

FRAP experiments on live animals that expressed a soluble control, Q0 or Q19, showed rapid recovery from photobleaching (Fig. 2A,B) consistent with soluble protein (Fig. 2D). However, foci bleached in Q86 neurons did not recover, indicating that the protein was immobile, from which we concluded that protein aggregates had formed (Fig. 2C,D). FRAP results were confirmed by Western blot analysis showing that Q86 aggregates were resistant to 5% SDS treatment, characteristic of polyQ aggregates.[55,56] Therefore, aggregates in our analysis of *C. elegans* neurons were visually distinct foci formed by stably associated, immobile proteins. Q86 aggregates failed to recover from photobleaching in all neurons examined, a result which suggests that polyQ aggregates form independently of cell-type. This observation is consistent with the extensive distribution of Q86 proteins into foci as described (Fig. 1C-E).

It has been proposed that polyQ expansions result in the formation of β-sheet structures that self-associate and lead to aggregation.[57,58] This hypothesis predicts that in addition to being immobile and resistant to SDS, polyQ proteins in aggregates would be closely associated in an ordered structure. Fluorescence Resonance Energy Transfer (FRET) is a technique developed for in vitro biochemical studies and applied to cell culture to determine in vivo, the proximity of two proteins. The resolution of this technique is in the nanometer range; FRET is maximal at 50Å and will not occur if proteins are more than 100Å apart. This technique has been widely employed to show protein interactions in vitro and in cell culture.[59,60] We therefore performed FRET experiments to determine whether polyQ proteins in *C. elegans* neurons are sufficiently ordered, and in molecular proximity for energy transfer between CFP and YFP.

FRET efficiencies were determined for neurons of a live animals using the acceptor photobleaching technique in which donor (CFP) intensity is compared before and after acceptor (YFP) photobleaching.[61] In *C. elegans* expressing both Q19::CFP and Q19::YFP, acceptor photobleaching had no effect on donor intensity, similar to negative control animals (Fig. 3C-F). These results show that, indirectly, FRET was not occurring between Q19 proteins. In contrast, FRET was observed in neuronal aggregates of expanded polyQ proteins. *C. elegans*

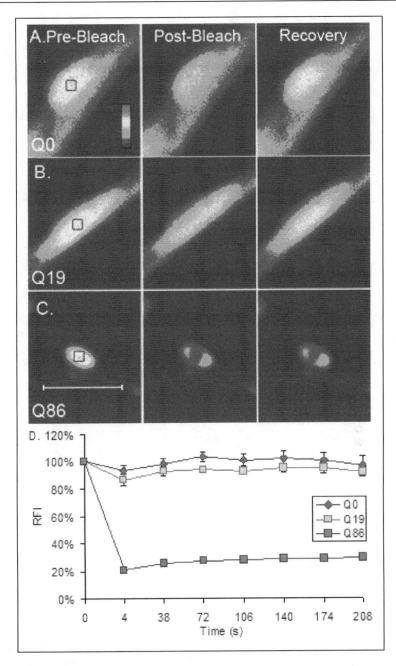

Figure 2. Solubility of PolyQ foci is consistent with aggregation. FRAP in *C. elegans* neurons expressing Q19::YFP, a photobleached area (box) recovers rapidly (B,D), similar to Q0 (A,D) indicating soluble proteins while bleached Q86::YFP foci do not recover (C,D). Therefore, Q86::YFP foci are insoluble, consistent with aggregation. Quantification (D) is ≥5 experiments with SEM. Signal intensity is measured in the color scale (A) where blue is least intense and red is most intense. Bar = 5 μm. A color version of this figure is available online at www.Eurekah.com.

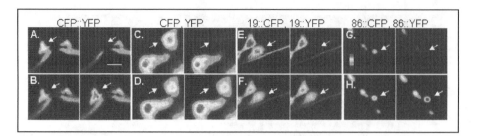

Figure 3. Q86 protein in neuronal aggregates exhibits FRET, indicating close and roughly ordered interactions at the molecular level. YFP photobleaching is seen on the YFP channel, (panels A,C,E,G) and its effect on CFP donor intensity in the same cell (panels B,D,F,H). Control animals expressing CFP::YFP FRET (A,B), $E_i = 0.248(\pm 0.089)$ while CFP and YFP coexpression (C,D) do not FRET, $E_i = 0.001$ (± 0.025). Neurons coexpressing Q19::CFP and Q19::YFP (E,F) do not FRET, $E_i = -0.090(\pm -0.057)$ while coexpression of Q86::CFP and Q86::YFP (G,H) does produce FRET, $E_i = 0.224(\pm 0.076)$. Cells shown are representative of FRET experiments. Intensity is by a color scale (G) where blue is least intense and red is most intense. Scale bar = 2 μm. A color version of this figure is available online at www.Eurekah.com.

coexpressing Q86::CFP and Q86::YFP showed an increase in donor intensity following acceptor photobleaching, similar to positive of animals expressing a chimeric protein of CFP linked to YFP (Fig. 3A-B,G-H).[59] FRET positive aggregates were detected in a wide range of neurons with no clear correlation between FRET intensity and cell-type. The visible redistribution of Q67 and Q86 into foci, combined with SDS resistance, FRAP and FRET data, shows that large polyQ expansions form insoluble, ordered aggregates in neurons throughout the nervous system of *C. elegans* with no evidence of cell-specific solubility.

PolyQ Length-Dependent Aggregation Correlates with Neuronal Dysfunction

In human disease, and in mouse, *Drosophila*, and cell culture studies of polyQ pathogenesis, aggregation is often accompanied by cellular dysfunction, although it remains unknown whether the toxic species are polyQ aggregates visible by light microscopy or some intermediate oligomeric species along the pathway to aggregate formation.[62] To test whether polyQ aggregation in *C. elegans* neurons was accompanied by neurotoxicity, we examined behavioral phenotypes regulated by the nervous system. The most striking phenotype was a polyQ length-dependent loss of coordinated movement leading to nearly complete paralysis. More than 60 neurons in *C. elegans* enervate muscle cells. Dysfunction or loss of these neurons results in lack of coordination or paralysis.[63-65] Q0 or Q19 animals with no visible polyQ aggregates showed rapid movement similar to wild type (N2) animals and Q19 showed a slight decrease. Animals with visible aggregates, Q67 and Q86, had limited capacity for coordinated movement exhibiting a decrease of 85% and 89% respectively, relative to Q0 (Fig. 4). These results suggest that formation of visible polyQ aggregates correlates with neuronal dysfunction.

Unexpectedly, pan-neuronal expression of intermediate polyQ lengths, Q35 and Q40, also affected behavior. In contrast to the clear phenotypic differences observed between transgenic animals that expressed short or long polyQ proteins, individual Q35 and Q40 animals displayed heterogeneity in behavioral phenotypes (Fig. 4). These results did not seem to correlate with the soluble distribution pattern of Q35 and Q40 proteins, visually distinct from the extensive foci formation in Q67 and Q86 animals (Fig. 1E). Behavioral toxicity in intermediate animals may have resulted from aggregation in a sub-set of neurons, or Q35 and Q40 proteins may display subtle differences in biochemical state, both of which could result in neurotoxicity without overt foci formation.

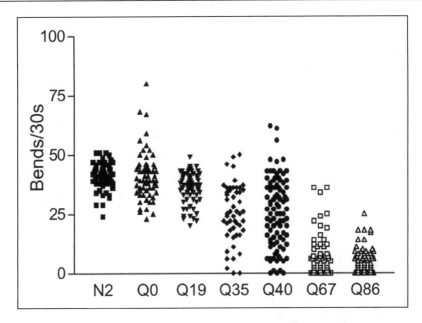

Figure 4. Increases in polyQ length correlate with neuronal dysfunction. Each point represents the average motility of a single young adult animal (4 days) over thirty seconds. Motility was quantified by determining the frequency with which animals thrashed when in liquid.[104]

Dynamic Biophysical Properties of Intermediate polyQ Tracts in the Ventral Nerve Cord

To determine whether there was a link between behavioral phenotypes and biochemical states in Q40 animals we examined neurons associated with motility, where the most significant differences in polyQ length-dependent toxicity were observed. The highest concentration of neurons that regulate motility are found in the VNC. FRAP experiments on multiple neurons distributed throughout the VNC of a single animal reveals that Q40::YFP solubility is heterogeneous. Recovery of Q40 protein ranged from soluble, similar to Q19, or insoluble like Q86 despite the absence of overt visual foci. Areas displaying rapid, but incomplete recovery were also observed, suggesting some neurons may combine both soluble and immobile protein components. While areas of insoluble Q40 protein were observed all along the VNC, only soluble Q40 was observed in neurons in head or tail ganglia.

The heterogeneity of intermediate polyQ protein solubility led us to hypothesize that heterogeneity might also exist in the distribution pattern of Q40 proteins, although it was not detected at the resolution of our earlier microscopy studies. Upon reexamination of the visual distribution of Q40 protein at higher magnification, we found subtle yet consistent changes in distribution, with Q40::YFP proteins localized to areas of increased fluorescent intensity forming oblong shapes with well defined, tapered ends that were not observed in animals expressing either short or long polyQ expansion. Although the distribution pattern of Q40 protein localization at high magnification was distinct from other polyQ containing animals, it could not be detected under a simple dissection microscopes.

To determine whether the intermediate length Q40 protein displays intermolecular interactions distinct from short (Q19) or long (Q86) polyQ proteins we performed FRET analysis. In a single young adult Q40 animal, we tested areas along the entire length of the VNC by FRAP to enable correlation of FRET results with Q40 solubility. Immediately following FRAP, the same region was tested for FRET. In regions of where FRAP results indicate soluble Q40 protein, photobleaching of the acceptor has little effect on donor intensity suggesting that no significant Q40-Q40 protein interactions are occurring. Conversely, in regions where FRAP reveals insoluble Q40 protein, intermolecular interactions of Q40 proteins causes FRET. The efficiency of interaction for Q40 proteins reveals two distinct populations coexist within the neurons of a single Q40 animal whereas Q19 or Q86 animals display only a single type of intermolecular interaction. Together, the complementary techniques of FRET and FRAP present a profile of Q40 proteins in vivo that reveals heterogeneity of biophysical states of an intermediate-length polyQ protein which is distinct from the homogeneous states observed for long or short polyQ proteins in the neurons of a multicellular organism.

These results suggest that Q40 protein can exhibit polymorphic subcellular distribution, solubility, and intermolecular interactions in an individual animal, unlike Q86 that exhibits an invariant immobile state, consistent intermolecular interactions and consistent distribution into foci. Variable biophysical properties of Q40 protein combined with the motility assay in which behavior of individual Q40 animals ranges from wild type to severely impaired, shows that polyQ-mediated neurotoxicity in *C. elegans* can be observed in the absence of overt foci formation.

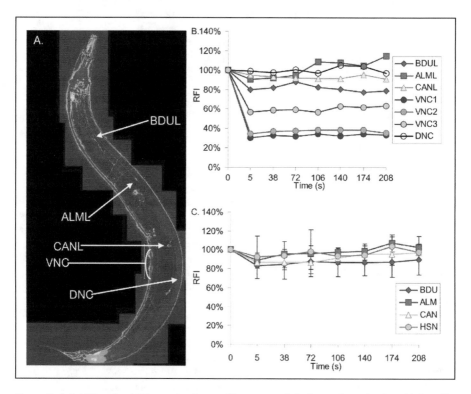

Figure 5. Solubility of polyQ proteins in specific neurons is independent of aging. A) Specific neurons are identified in the flattened z-stack of a representative Q40 animal (8 days old) and (B) FRAP results are shown for each neuron. Q40 protein in the neurons analyzed remains soluble even as animals age. C) FRAP data shown are the average for each specific neuron, with standard deviation, analyzed in animals from 4 to 8 days of age.

Neuron-Specific Responses to polyQ Proteins

Our observation that solubility of Q40 proteins in *C. elegans* can vary between different areas of the nervous system suggests that there are modifiers acting on simple polyQ proteins independent of protein context. The observed variability in polyQ protein solubility could be stochastic rather than a neuron-specific response. To address this, we performed FRAP experiments on several specific neurons in addition to the VNC. We reasoned that if solubility of Q40 protein is stochastic, protein solubility in a specific neuron will vary between animals. However, if polyQ protein solubility is neuron-specific, FRAP results should be consistent in all animals tested. FRAP analysis of specific neurons in multiple animals and show that Q40 protein is consistently soluble whereas the VNC displays polymorphic Q40 solubility in all *C. elegans* examined (Fig. 5A,B). These results suggest that polyQ solubility is consistent in specific classes of neurons. We hypothesize that polyQ solubility is modulated by some property of the neuron in which it is expressed, such as neuronal function, connectivity, activity levels, or expression levels of the polyQ protein.

Another characteristic that could have a major effect on polyQ solubility in neurons is the age of the animal. Aging is well established as a major contributor to the onset of polyQ repeat diseases in both humans and in model systems. We therefore examined specific neurons in animals of different ages show that in specific neurons, Q40 protein remains soluble in animals up to 10 days of age (Fig. 5C). In contrast, the VNC displays insoluble protein as early as three days and polymorphic solubility of Q40 persists as the animals age. These results showed that Q40 can exhibit polymorphic biochemical properties and remained soluble in some neurons while becoming aggregated in the VNC. It also suggested that some neurons displaying soluble Q40 protein are able to maintain solubility as they age. These properties of Q40 are in contrast to the complete aggregation of Q67 or Q86 proteins observed in *C. elegans* nervous system (Fig. 1) and suggest that for intermediate Q lengths, age is not the only modifier of polyQ protein solubility.

C. elegans model provides a unique tool for investigating the basis of neuron-specific toxicity associated with the expression of polyQ proteins. Behavioral and molecular characterization of polyQ expansions in a pan-neuronal system is the first step in establishing a model in which neuron-specific, novel modifiers of polyQ-mediated pathogenesis can be identified. Additionally, establishing this system in neurons provides the baseline for future studies to examine one of the major modifiers of polyQ pathogenesis: tissue type. Development of a neuronal model in *C. elegans* opens the way to comparative studies on how nonneuronal tissue, such as muscle, responds to polyQ proteins.

The *C. elegans* polyQ Series in Muscle Cells

In addition to providing a second tissue for comparative studies, *C. elegans* muscle cells have some advantages for studying the basic properties of misfolded proteins in vivo. Muscle cells are significantly larger than neurons facilitating work under comparatively low magnification. In addition, *C. elegans* muscle cells are sensitive to RNA interference (RNAi) while neurons are refractory. Therefore, a muscle cell model is ideal for rapid, genome wide screening.

Using a similar approach to that described above, fluorescently tagged polyQ proteins were expressed in muscle cells of *C. elegans*. We generated a series of transgenic animals expressing polyQ repeats of different lengths, with greater resolution at the intermediate lengths (Q0, Q19, Q29, Q33, Q35, Q40, Q44, Q64, and Q82). In young adult animals expressing repeats of Q35 or fewer, we observed diffuse fluorescence in all expressing cells while animals expressing muscle Q44, Q64 or Q82 exhibited fluorescence in discreet foci similar to that observed in neurons (Fig. 6A, Fig. 1). Muscle Q40 animals displayed a polymorphic distribution with diffuse fluorescence in some cells and foci in others, similar to neuronal Q40 but more easily observed due to the size of muscle cells. In the *C. elegans* muscle cell model, there is a shift in the cellular distribution of polyQ protein in young adult

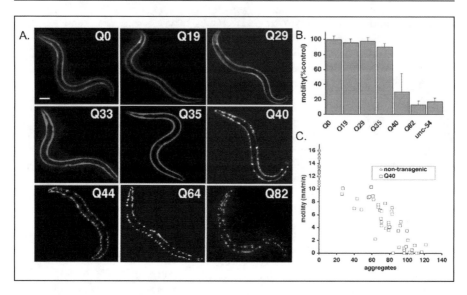

Figure 6. Expression of different polyQ::YFP lengths in *C. elegans* body-wall muscle cells shows a length-dependent aggregation phenotype. Fluorescence images of young adult *C. elegans* expressing different lengths of polyQ::YFP (Q0, Q19, Q29, Q33, Q35, Q40, Q44, Q64, Q82). Animals expressing repeats of ≤Q35::YFP show a diffuse fluorescence in all muscle cells; in contrast, animals expressing ≥Q44 repeats exhibited punctate distribution of fluorescence Animals expressing muscle Q40 showed a polymorphic phenotype with both diffuse and punctate fluorescence distribution. (Bar = 0.1 mm). B) Quantification of motility index for young adult Q0, Q19, Q29, Q35, Q40, Q82, and *unc-54(r293)* animals. Data are mean ± SD for at least 50 animals of each type as percentage of N2 motility. C) Correlation between the number of Q40 foci in the muscle cells of an individual animal and their motility.

animals between 35-40 repeats, reminiscent of the pathogenic threshold, of 30-40 repeats, observed in the human diseases.

To determine whether there is polyQ-mediated dysfunction in muscle, we examined motility and found that Q19, Q29 or Q35 animals behaved similarly to wild type (N2) while Q82 animals displayed severely decreased motility (Fig. 6B). The intermediate Q40 animals, which had aggregates in some cells but not in others, exhibited an intermediate motility defect with a high degree of variation. The polyQ-dependent motility phenotype observed in the muscle cell model was parallel to that of the neuronal model. However, the larger size of body wall muscle cells made it possible to examine how polyQ aggregation correlates with cellular dysfunction. When individual animals were assayed for motility and then scored for the number of Q40 aggregates, it became clear that in muscle cells, aggregation is directly correlation with cellular dysfunction (Fig. 6C).[66]

Aging Influences the Threshold for polyQ Aggregation and Toxicity

Despite the expression of aggregation-prone proteins throughout the lifetime of the organism, pathology associated with diseases such as Huntington's, Alzheimer's and Parkinson's are typically not manifest until late in life. This motivated us to study the behavior of polyQ proteins during aging. We performed experiments in which individual Q0, Q29, Q33, Q35, Q40, and Q82 animals were examined daily for the appearance of protein aggregates and motility. Relative to Q40 and Q82 animals that quickly accumulated aggregates and exhibited a rapid decline in motility, Q33 and Q35 animals exhibited an initial lag prior to the gradual accumulation of aggregates, however to levels much lower than for Q40 or Q82

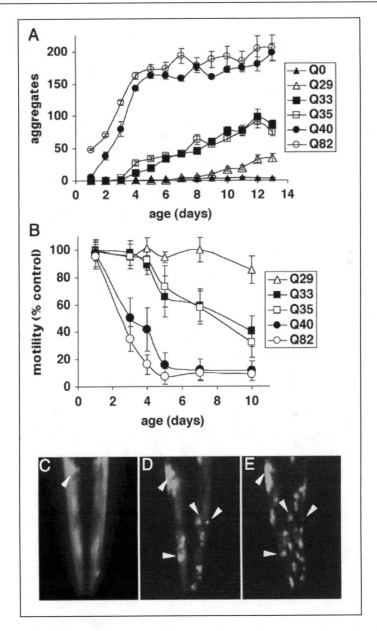

Figure 7. Influence of aging on polyQ aggregation and toxicity. A) Accumulation of aggregates in Q82, Q40, Q35, Q33, Q29, and Q0 during aging. Data are mean ± SEM. Twenty-four animals of each type are represented at day 1. Cohort sizes decreased as animals died during the experiment, but each data point represents at least five animals. B) Motility index as a function of age for the same cohorts of animals described in A. Data are mean ± SD as a percentage of age-matched Q0 animals. C-E) Fluorescence images of the head of an individual Q35 animal at 4 (C), 7 (D), and 10 (E) days of age, illustrating age-dependent accumulation of aggregates. Arrowheads indicate positions of the same aggregates on different days. In E, the animal is rotated slightly relative to its position in D.

(Fig. 7A). For example, aging-dependent aggregate accumulation can be readily observed by comparison of the same Q35 animal at 4, 7, and 10 days (Fig. 7C-E).

In all cases, we observed an aging-dependent loss of motility relative to controls. These results reveal that the threshold for polyQ aggregation and toxicity in *C. elegans'* muscle cells is not static, but rather age-dependent. At 3 days of age or less, only animals expressing Q40 or greater exhibit aggregates (Figs. 6, 7), whereas at 4-5 days the threshold shifts and aggregates appear in Q33 and Q35 animals (Fig. 7A). In Q29 animals, the threshold again shifts (>9-10 days) (Fig. 7). Thus, the threshold for polyQ aggregation is dynamic and likely reflects a balance of different factors including repeat length and changes in the protein folding environment.

Longevity Genes Influence Aging-Dependent Aggregation and Toxicity of polyQ Proteins

Our results reveal that the threshold for polyQ aggregation and cytotoxicity in vivo is dynamic over the lifetime of an animal. Does this dynamic behavior result from the intrinsic properties of a protein motif, or, do changes over time reflect the influence of aging-related alterations in the cell? The idea that the molecular determinants of longevity might influence

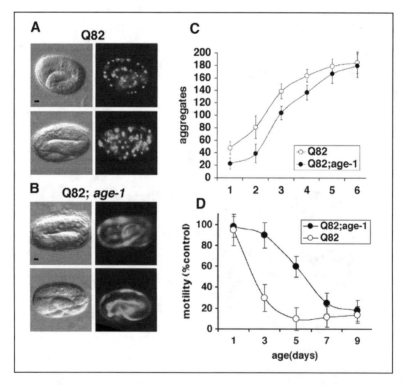

Figure 8. An extended lifespan mutation delays polyQ aggregate accumulation and onset of toxicity. A) Differential interference contrast (Left) and fluorescence (Right) micrographs showing embryos expressing Q82 in wild-type (A), *age-1(hx546)* (B) genetic backgrounds. (Bars = 5 μm.). C) Aggregate accumulation in larval animals expressing Q40 or Q82 in the indicated genetic backgrounds relative to aggregate accumulation in wild-type background. Mean ± SEM. D) Motility index for animals expressing Q40 or Q82 in the indicated genetic backgrounds. Data are mean ± SEM for 30 animals of each type. Motility of nontransgenic wild-type and age-1 animals was similar to that of wildtype (N2).

polyQ-mediated toxicity is supported by observations that the time until pathology develops - days in *C. elegans*, weeks in *Drosophila*, months in mice, and years in humans - correlates approximately with the lifespan of the organism. Here, again, the *C. elegans* model was a tremendous advantage as the availability of mutants with extended lifespans allowed us to test these ideas directly.

To accomplish this, we generated transgenic animals expressing Q82 in the background of the *age-1(hx546)* mutation or *age-1* RNAi. *age-1* encodes a phosphoinositide-3 kinase that functions in an insulin-like signaling (ILS) pathway, and mutations in this gene can extend lifespan by 1.5-2 fold.[67,68] Q82 in the *age-1* background (Q82;age-1) exhibited reduced aggregate formation in embryos relative to Q82 in the wild type background (Fig. 8). Q82 aggregate formation was also reduced 30-50% during larval stages (1-2 days old) in *age-1* animals compared to wild type background and remained significantly lower through 4-5 days of age.[66] Parallel motility assays also demonstrated a delay in onset of the motility defect, consistent with slower aggregate accumulation in Q82;age-1 animals (Fig. 8).

In wild type animals, the kinase activity of AGE-1 is required in a signaling cascade that results in constitutive repression of the fork head transcription factor DAF-16 leading to normal lifespan.[67,69,70] Derepression of DAF-16 in age-1 animals results in extended lifespan, and daf-16 mutations suppress the longevity phenotype. To examine whether age-1 effects on longevity and polyQ aggregation and toxicity are mediated through similar regulatory pathways, we tested whether *age-1* suppression of Q82 phenotypes was affected by inactivation of daf-16 using RNAi. Q82;age-1;daf-16 animals exhibited aggregation and motility phenotypes similar to Q82 expressed in wild type background. Thus, the dual effects of *age-1* on longevity and polyQ-mediated toxicity share a common genetic pathway.

Our demonstration that a mutation conferring longevity also delays polyQ aggregation and toxicity suggests a novel link between the genetic regulation of aging and aging-related disease. In subsequent studies, we and others have demonstrated that the molecular link between these pathways is regulated, in part, by factors that detect and respond to misfolded proteins - namely heat shock transcription factor (HSF) and molecular chaperones/heat shock proteins. For example, it has been shown that inhibition of HSF-1 function leads to decreased lifespan and an accelerated aging phenotype in *C. elegans*.[71-73] Conversely, overexpression of HSF-1 in *C. elegans* extends lifespan.[72,73] Additionally, inactivation of *daf-16*, *hsf-1* or small heat shock proteins in *C. elegans* accelerated the aggregation of polyQ expansion proteins supporting the idea that ILS could coordinately influence aging and protein aggregation through the action of DAF-16, HSF-1 and molecular chaperones.

Genome-Wide RNAi Screening Identifies Novel Regulators of polyQ Aggregation and Toxicity

The results described above identify ILS as a genetic pathway that can influence the course of polyQ-mediated phenotypes. What other pathways might exert similar effects? Numerous over-expression and genetic studies in mammalian and *Drosophila* models have identified various enhancers or suppressors of polyQ-mediated aggregation and toxicity.[74-76] This is well-illustrated by the large number of approaches in which various molecular chaperones either alone or in combination have been shown to influence polyQ-mediated phenotypes.[5,7,74,76-78] While one could argue which of these modifiers is the "key" to determining the fate of aggregation-prone proteins, we interpret these results to suggest that the transition of polyQ proteins from a soluble to an aggregated state is the result of a balance in which multiple pathways are likely to cooperate. To identify the complete protein-folding buffer involved in polyQ transition from a soluble to an aggregated state we used a genome-wide RNAi approach.

RNAi is a commonly used reverse-genetics approach in *C. elegans* allowing the targeted down-regulation of specific genes by introduction of small fragments of cognate double-stranded RNAs.[79,80] This technique has been adapted to genome-wide screens with a library consisting of 16,757 bacterial clones covering 86% of the predicted *C. elegans* genome.[81,82] RNAi has the

Table 1. Genome-wide RNAi screen for modifiers of polyQ aggregation identified 186 genes, in five classes. These genes all function normally to suppress aggregation; therefore knockdown by RNAi resulted in increased aggregation.

↑Production Misfolded Proteins		↓Clearance of Misfolded Proteins		
RNA synthesis & processing (31)	Protein synthesis (59)	Protein folding (11)	Protein transport (18)	Protein degradation (16)
Transcription regulation (5)	Intitiation (6)	Chaperonin (6)	Endocytosis (1)	19S (8)
T27F2.1 (Skip)	Y39G10AR.8 (eIF-2-γ)	cct-5(TCP-1-ε)	Y105E8A.9 (γ-adaptin)	rpt-5 (26S 6A)
Transcription (2)	Elongation (4)	Hsp70 (4)	Nuclear import (4)	20S(10)
ama-1 (RPB1)	eft-3 (eEF1A-2)	hsp-1 (Hsc70)	C53D5.6 (Importin β -3)	pas-4 (α type 7)
Splicing (17)	Ribosomal subunit (43)	DnaJ (1)	Cytoskeleton (6)	Ub ligase (2)
T13H5.4 (SAP61)	rps-26 (S26)	F18C12.2A	K01G5.7 (Tubulin β -2)	C47E12.5 (E1)

additional advantages of allowing detection of lethal positives and the immediate identification of the target genes. Although genome-wide RNAi screens may miss certain genes secondary to variation in mRNA depletion and relative inefficiency of the mechanism in neurons, this approach offers an extremely powerful and rapid tool to identify the set of genes that modify a given phenotype.

To identify genes that prevent polyQ aggregate formation we used *C. elegans* strains expressing polyQ lengths close to the aggregation threshold, Q33 and Q35 strains, in an RNAi genetic screen. To validate this strategy, we first tested the full range of polyQ strains with a group of candidate modifiers, genes containing the TPR domain and therefore likely to act as molecular chaperones. Animals with knock down of candidate modifiers were analyzed for premature appearance of aggregates upon selective RNAi. Of the 72 candidate gene modifiers analyzed only four showed an earlier appearance of foci in Q33 and Q35 animals revealing an unexpected specificity among known modifiers of protein folding.[83] These effects were dependent on the presence of an expanded polyQ motif as no foci were observed for any of the 72 RNAi bacterial clones in either Q0 or Q24 animals. The fluorescent foci induced by RNAi-treatment were biophysically and biochemically identical to aggregates of long polyQ stretches as judged by FRAP analysis and SDS-PAGE.

Having validated the reliability of our approach, we screened the *C. elegans* genome and identified a total of 186 genes that induced earlier onset of the aggregation phenotype in Q35 animals. These modifiers fall into 5 major classes: genes involved in RNA metabolism, protein synthesis, protein folding, protein trafficking and protein degradation (Table 1). Examples of some of these genes are: RNA helicases, splicing factors and transcription factors for RNA metabolism; initiation and elongation factors and ribosomal subunits for protein synthesis; chaperonins and Hsp70 family members for protein folding; nuclear import and cytoskeletal genes for protein trafficking and proteasomal genes for protein degradation. A common feature is an expected imbalance between protein synthesis, folding and degradation triggered by their disruption. In this context, these 5 classes can then be further grouped in two major categories: genes whose disruption leads to an increase in misfolded protein production and genes that when disrupted lead to a decreased clearance of misfolded proteins and proper protein turnover. Our results reveal that the transition between soluble and aggregated states of polyQ proteins is regulated by a much more complex integration of events, extending beyond the immediate involvement of chaperone-mediated folding and proteasomal degradation.

Our findings suggest a model in which each step in the birth, life and death of a protein influences the capacity of the cellular protein folding buffer. For example, perturbation of the RNA-processing machinery caused accelerated aggregation of Q35 perhaps due to an increased burden of abnormal proteins requiring the activity of the protein folding buffer. Uncovering the role of these genes in the disease process suggests that there may be no single molecular mechanism responsible for polyQ-mediated pathology.

Global Disruption of Folding Homeostasis by polyQ Proteins

The widespread effect of polyQ proteins leads to the question of whether polyQ proteins cause cellular dysfunction by a generalized destabilization of the cellular protein folding environment. We utilized our *C elegans* models of polyQ aggregation in neuron or muscle cell models to examine the effect of polyQ protein on cells' ability to maintain protein-folding homeostasis. To test the hypothesis that polyQ proteins disrupt protein homeostasis, we took a genetic approach using diverse *C. elegans* temperature-sensitive (ts) mutations to examine whether their functionality at the permissive condition is affected by expression of aggregation-prone polyQ-expansions. Since many ts mutant proteins are highly dependent on the cellular folding environmentthey represent highly sensitive indicators of a disruption in protein homeostasis. The neuronal and muscle polyQ models described above were crossed into ts mutants to determine the effect of polyQ proteins on cellular homeostasis.[84-86]

Animals expressing ts mutant of the *C. elegans* homolog of a muscle paramyosin (UNC-15) were crossed to *C. elegans* polyQm strains, and phenotypes were examined at both permissive and restrictive temperature conditions. With the paramyosin ts mutation alone, animals do not display overt phenotypes at the permissive temperature, however at the restrictive temperature the ts mutation disrupts thick filament formation and leads to phenotypes including embryonic and larval lethality and slow movement in adults (Fig. 9A). In animals expressing only Q40m, neither lethality nor paralysis was observed.[87] In contrast, more than 40% of embryos coexpressing Q40m and paramyosin ts failed to hatch or move at the permissive temperature (Fig. 9B). This effect was polyQ length-dependent, because coexpression of smaller polyQ expansions with paramyosin(ts) decreased penetrance of ts phenotypes (Fig. 9B). Examination of muscle structure in paramyosin(ts) + Q40m embryos at the permissive temperature revealed a disrupted pattern of actin staining, similar to the pattern in paramyosin(ts) embryos at the restrictive temperature and absent at the permissive temperature or in wild-type animals. Thus, in muscle cells expression of an aggregation-prone polyQ protein is sufficient to cause a paramyosin ts mutation to exhibit its mutant phenotype at the permissive condition.

To determine whether this effect of polyQ expansions is specific to muscle or extends to other tissues, we examined neurons. Animals expressing a ts mutation in the neuronal protein dynamin-1 [dynamin(ts)] become paralyzed at the restrictive temperature but have normal motility at the lower, permissive temperature.[88] With the coexpression of pan-neuronal Q40, dynamin(ts) animals display severe impairment of mobility at the permissive temperature (Fig. 9C). This effect was length dependent since no phenotypes were observed in animals coexpressing the nonaggregating Q19n. Thus, expression of polyQ expansions phenocopies temperature-sensitive mutations in both muscle and neuronal cells at permissive conditions, and this genetic interaction reflects the propensity of polyQ proteins to aggregate. These observations were confirmed by testing a wide range of characterized ts mutations. Strains expressing ts mutant proteins UNC-54, UNC-52, LET-60 (*C. elegans* homologs of myosin, perlecan, and ras-1, respectively), or UNC-45 together with Q24m or Q40m were scored for specific ts phenotypes at the permissive temperature. For all lines generated, the ts mutant phenotype was exposed at permissive conditions in the presence of the aggregation prone Q40m but not by non aggregating Q24m (Table 2). Thus, the chronic expression of an aggregation-prone polyQ protein interferes with the function of multiple structurally and functionally unrelated proteins.

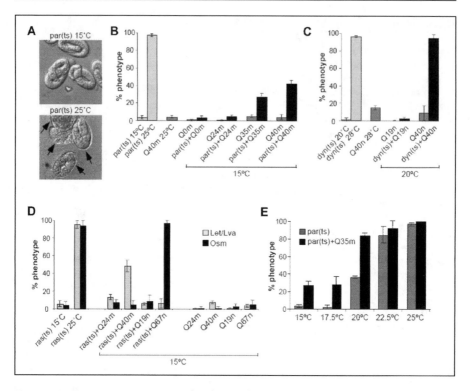

Figure 9. Aggregation-prone proteins alter the conditions required to expose phenotypes dependent on temperature sensitive mutations. Aggregation-prone proteins expose temperature sensitive phenotypes of paramyosin(ts) (B,F), dynamin(ts) (C) and ras(ts) (D) mutants at permissive
temperatures. A) DIC images of age-synchronized 3-fold paramyosin(ts) embryos at indicated
temperatures. Arrows indicate embryos with abnormal body shape. (B) Percentage of unhatched
embryos and paralyzed L1 larvae. Data are the mean ± SD, ≥380 embryos per data point. C)
Percentage of uncoordinated age-synchronized young adult animals. D) Percentage of animals
exhibiting either Osm (black) or the combined Let/Lva (grey) phenotypes. Data are the mean ±
SD, ≥70 synchronized adults for Osm and ≥270 embryos for Let/Lva. ras(ts) + Q40m denotes
ras(ts) animals heterozygous for Q40m. E) Synergistic effect of elevated temperature and polyQ
expansions on paramyosin(ts). Percentage of unhatched embryos and paralyzed L1 larvae for
paramyosin(ts) (grey) and paramyosin(ts)+Q35m (black) at indicated temperatures. Data are the
mean ± SD, ≥300 embryos for each data point. In animals expressing polyQn (neuronal)-YFP
proteins (Q19n, Q40n, Q67n). Data are the mean ± SD, ≥80 animals per data point.

The interaction between polyQ expansions and ts mutant proteins could be cell autonomous or dependent on intercellular interactions. We took advantage of tissue-specific phenotypes caused by expression of ras(ts) protein to examine this question. At the restrictive temperature, animals carrying the ras(ts) mutation display phenotypes of embryonic lethality/larval
development phenotype (Let/Lva), a defect in osmoregulation (Osm) likely reflecting neuronal
dysfunction and a multivulva phenotype (Muv), resulting from dysfunction in the hypodermis.[89] We scored these phenotypes upon expression of polyQ in neuronal or muscle cells.
PolyQ expansions in neurons led to exposure of the Osm phenotype in ras(ts) animals at the
permissive temperature but had no effect on the Let/Lva phenotype (Fig. 9D). Conversely,
expression of Q40m in muscle cells of ras(ts) animals caused increased penetrance of Let/Lva
phenotype but did not expose the neuronal Osm phenotype (Fig. 9D, Table 2). Neither neuronal nor muscle cell expression of polyQ expansions caused the hypodermal Muv phenotype.

Table 2. PolyQ expansions affect the functionality of unrelated ts mutant proteins

Proteins Expressed	TS Allele	Phenotype Scored (N Value)	% Animals Displaying Phenotype		15°C	
			15°C	25°C	Q24	Q40
PolyQm	-	slow movement (n > 300)			6.0 ± 5.3	4.6 ± 4.2
myosin(ts)	e1301		5.6 ± 2.6	84.7 ± 13.5		
myosin(ts) + Qm	e1301				11.2 ± 6.7	51.9 ± 19.3
myosin(ts)	e1157		5 ± 4	98.7 ± 1.4		
myosin(ts) + Qm	e1157				5.9 ± 1.5	55 ± 6
PolyQm	-	abnormal body shape (stiff paralysis) (n > 100)			0	0
perlecan(ts)	su250		1 ± 1.2	97.6 ± 2.2		
perlecan(ts) + Qm	su250				0.8 ± 1.1	48.4 ± 6.5
PolyQm	-	egg laying defect (n > 85)			0	0
UNC-45(ts)	e286		8.4 ± 2.1	93.8 ± 4.9		
UNC-45(ts) + Qm	e286				5.1 ± 7.3	87.7 ± 8.1
PolyQm	-	embryonic lethality + larval development arrest (n > 270)			1 ± 0.15	7.9 ± 1.8
ras(ts)	ga89		5.6 ± 3.4	95.2 ± 4.3		
ras(ts) + Qm	ga89				13.3 ± 3	100*

Specific phenotype of each ts mutation alone or in polyQm background was scored at indicated temperatures. Data are the mean ± SD for at least the indicated number (n) of animals for each phenotype scored. Double homozygous ras(ts) + Q40m animals could not be scored, as they did not reach mature adulthood.

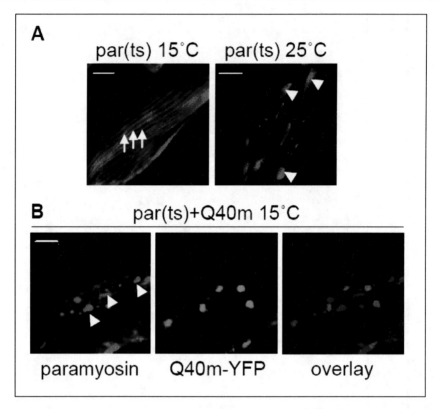

Figure 10. PolyQ expansions affect the folding of a ts mutant of paramyosin. A,B) Confocal images of anti-paramyosin immunostained (red) body wall muscle cells of synchronized young adults expressing indicated proteins. Arrows: normal muscle sarcomeres, arrowheads: abnormal paramyosin(ts) assemblies. Green color is Q40m-YFP fluorescent protein. Scale bar is 10 μm.

A similar control was performed with paramyosin(ts),and was not affected by polyQ expansions in neurons (Q67n). Likewise, polyQ expansions did not affect phenotypes that did not involve ts proteins (caused either by RNAi or gene deletion). Thus, the effect of polyQ expansions on mutant ts proteins reflects specific genetic interactions within the same cell type and does not result from decreased fitness of the organism.

To understand the nature of this interaction, we examined the cellular localization of paramyosin(ts) protein when coexpressed with Q40m. The paramyosin ts mutation affects protein interactions which, at restrictive temperature, results in mislocalization into foci (Fig. 10A).[87] When paramyosin(ts) protein is coexpressed with Q40m, it mislocalizes into foci at the permissive temperature, distinct from Q40m aggregates (Fig. 10B). Thus, expression of Q40m uncovers the protein folding defect in paramyosin(ts) mutant. In view of this, the differential penetrance of ts phenotypes (Table 2) may reflect the sensitivity of each ts mutation to disruption of the folding environment.

Aggregation-prone proteins may exert their destabilizing effects by placing a stress on the folding capacity of the cell. If so, additional stress, such as elevated temperature may act in synergy with polyQ aggregates to further destabilize ts mutants. Indeed, expression of an intermediate (Q35m) expansion shifted the permissive temperature at which paramyosin(ts) was fully inactivated (Fig. 9E). Furthermore, we observed increased penetrance of ts phenotypes in homozygous Q40 animals as compared to heterozygous, consistent with the earlier onset of aggregation in homozygous Q40 animals (Table 2). If the levels of polyQ affect the folding of the ts protein, does the misfolding of the ts protein, in turn, intensify misfolding of polyQ? To answer that question, we quantified mQ40 aggregation in ts backgrounds and found that aggregation was enhanced dramatically. In contrast, mutations that cause a loss of function rather than a ts phenotype did not enhance aggregation. From a genetic perspective, temperature-sensitive mutations in proteins unrelated to cellular folding or clearance pathways behaved as modifiers of polyQ aggregation. This suggests that a positive feedback mechanism exists to enhance the disruption of cellular folding homeostasis.

The appearance of misfolded protein in the cell normally activates a stress response that increases protein refolding and turnover and thus rebalances the folding environment.[90,91] In contrast, our results point to the unexpected sensitivity of cellular folding homeostasis to the chronic expression of misfolded proteins under physiological conditions. It is possible that the low flux of misfolded protein in conformational diseases may alone lack the capacity to activate the homeostatic stress response. This suggests that the stress response fails to compensate for the chronic expression of misfolded proteins in human disease.

One potential interpretation of our results is that the protein folding capacity of the cell, integrated at a systems level, is a reflection of expressed protein polymorphisms and random mutations,[92] which in themselves do not lead to disease because of the balance achieved by folding and clearance mechanisms. However, these proteins may misfold and in turn contribute to the progressive disruption of the folding environment when this balance becomes overwhelmed, e.g., by the expression of an aggregation-prone protein in conformational diseases.

Conclusion

In our *C. elegans* model for the expression of isolated polyQ repeats we focused on the effects of polyQ-containing proteins in neurons or muscle cells. Expression of isolated polyQ motifs rather than full-length disease-related proteins enables us to uncover conserved features underlying a range of neurodegenerative diseases. A genome wide RNAi screen identified multiple molecular chaperones that have been implicated in polyQ diseases, as well as AD and PD. These studies have revealed a common set of factors that link the genetic regulation of protein homeostasis, stress responsiveness and longevity. Thus, longevity and fitness may be, at least in part, a consequence of the efficient detection, capture and resolution of misfolded and aggregation-prone proteins. The presence of marginally stable or folding-defective proteins in the genetic background of conformational diseases reveals potent extrinsic factors that can modify aggregation and toxicity. Given the prevalence of polymorphisms in the human genome,[93] the expression of metastable proteins could contribute to variability of disease onset and progression.[94] This interpretation also provides a mechanistic basis to the notion that the late onset of protein misfolding diseases may be due to gradual accumulation of damaged proteins, resulting in a compromise in folding capacity.[95] Cellular degeneration in diseases of protein conformation is unlikely to be due to a single defect. Thus, the many toxic effects on various cellular processes attributed to misfolded proteins[7,59,96-103] could in fact be an integral part of the global disruption of protein homeostasis.

Acknowledgements

We thank members of the laboratory past and present who contributed to this work both intellectually and technically. Support for this research was provided by grants from the National Institute of General Medical Science, the National Institutes for Aging, the Huntington Disease Society of America Coalition for the Cure, and the Daniel F. and Ada L. Rice Foundation.

References

1. Kakizuka A. Protein precipitation: A common etiology in neurodegenerative disorders? Trends Genet 1998; 14:396-402.
2. Kopito RR, Ron D. Conformational disease. Nat Cell Biol 2000; 2:E207-209.
3. Stefani M, Dobson CM. Protein aggregation and aggregate toxicity: New insights into protein folding, misfolding diseases and biological evolution. J Mol Med 2003; 81:678-699.
4. Dobson CM. Protein folding and its links with human disease. Biochem Soc Symp 2001; 68:1-26.
5. Bonini NM. Chaperoning brain degeneration. Proc Natl Acad Sci USA 2002; 99:16407-16411.
6. Chan HY, Warrick JM, Andriola I et al. Genetic modulation of polyglutamine toxicity by protein conjugation pathways in Drosophila. Hum Mol Genet 2002; 11:2895-2904.
7. Cummings CJ, Mancini MA, Antalffy B et al. Chaperone suppression of aggregation and altered subcellular proteasome localization imply protein misfolding in SCA1. Nat Genet 1998; 19:148-154.
8. Kayed R, Head E, Thompson JL et al. Common structure of soluble amyloid oligomers implies common mechanism of pathogenesis. Science 2003; 300:486-489.
9. O'Nuallain B, Wetzel R. Conformational Abs recognizing a generic amyloid fibril epitope. Proc Natl Acad Sci USA 2002; 99:1485-1490.
10. Kawaguchi Y, Okamoto T, Taniwaki M et al. CAG expansions in a novel gene for Machado-Joseph disease at chromosome 14q32.1. Nat Genet 1994; 8:221-228.
11. Koide R, Ikeuchi T, Onodera O et al. Unstable expansion of CAG repeat in hereditary dentatorubral-pallidoluysian atrophy (DRPLA). Nat Genet 1994; 6:9-13.
12. La Spada AR, Wilson EM, Lubahn DB et al. Androgen receptor gene mutations in X-linked spinal and bulbar muscular atrophy. Nature 1991; 352:77-79.
13. Orr HT, Chung MY, Banfi S et al. Expansion of an unstable trinucleotide CAG repeat in spinocerebellar ataxia type 1. Nat Genet 1993; 4:221-226.
14. Kamino K, Orr HT, Payami H et al. Linkage and mutational analysis of familial Alzheimer disease kindreds for the APP gene region. Am J Hum Genet 1992; 51:998-1014.
15. Laing NG, Siddique T. Cu/Zn superoxide dismutase gene mutations in amyotrophic lateral sclerosis: Correlation between genotype and clinical features. J Neurol Neurosurg Psychiatry 1997; 63:815.
16. Lucking CB, Durr A, Bonifati V et al. Association between early-onset Parkinson's disease and mutations in the parkin gene. French Parkinson's Disease Genetics Study Group. N Engl J Med 2000; 342:1560-1567.
17. Mizuno Y, Hattori N, Kitada T et al. Familial Parkinson's disease. Alpha-synuclein and parkin. Adv Neurol 2001; 86:13-21.
18. Polymeropoulos MH, Lavedan C, Leroy E et al. Mutation in the alpha-synuclein gene identified in families with Parkinson's disease. Science 1997; 276:2045-2047.
19. Rosen DR, Siddique T, Patterson D et al. Mutations in Cu/Zn superoxide dismutase gene are associated with familial amyotrophic lateral sclerosis. Nature 1993; 362:59-62.
20. Driscoll M, Gerstbrein B. Dying for a cause: Invertebrate genetics takes on human neurodegeneration. Nat Rev Genet 2003; 4:181-194.
21. Link CD. Transgenic invertebrate models of age-associated neurodegenerative diseases. Mech Ageing Dev 2001; 122:1639-1649.
22. Thompson LM, Marsh JL. Invertebrate models of neurologic disease: Insights into pathogenesis and therapy. Curr Neurol Neurosci Rep 2003; 3:442-448.
23. Westlund B, Stilwell G, Sluder A. Invertebrate disease models in neurotherapeutic discovery. Curr Opin Drug Discov Devel 2004; 7:169-178.
24. Merry DE. Animal models of Kennedy disease. NeuroRx 2005; 2:471-479.
25. Greene JC, Whitworth AJ, Andrews LA et al. Genetic and genomic studies of Drosophila parkin mutants implicate oxidative stress and innate immune responses in pathogenesis. Hum Mol Genet 2005; 14:799-811.
26. Greene JC, Whitworth AJ, Kuo I et al. Mitochondrial pathology and apoptotic muscle degeneration in Drosophila parkin mutants. Proc Natl Acad Sci USA 2003; 100:4078-4083.
27. Faber PW, Alter JR, MacDonald ME et al. Polyglutamine-mediated dysfunction and apoptotic death of a Caenorhabditis elegans sensory neuron. Proc Natl Acad Sci USA 1999; 96:179-184.
28. Faber PW, Voisine C, King DC et al. Glutamine/proline-rich PQE-1 proteins protect Caenorhabditis elegans neurons from huntingtin polyglutamine neurotoxicity. Proc Natl Acad Sci USA 2002; 99:17131-17136.
29. Parker JA, Connolly JB, Wellington C et al. Expanded polyglutamines in Caenorhabditis elegans cause axonal abnormalities and severe dysfunction of PLM mechanosensory neurons without cell death. Proc Natl Acad Sci USA 2001; 98:13318-13323.
30. Warrick JM, Paulson HL, Gray-Board GL et al. Expanded polyglutamine protein forms nuclear inclusions and causes neural degeneration in Drosophila. Cell 1998; 93:939-949.

31. Feany MB, Bender WW. A Drosophila model of Parkinson's disease. Nature 2000; 404:394-398.
32. Link CD. Expression of human beta-amyloid peptide in transgenic Caenorhabditis elegans. Proc Natl Acad Sci USA 1995; 92:9368-9372.
33. Oeda T, Shimohama S, Kitagawa N et al. Oxidative stress causes abnormal accumulation of familial amyotrophic lateral sclerosis-related mutant SOD1 in transgenic Caenorhabditis elegans. Hum Mol Genet 2001; 10:2013-2023.
34. Lee VM, Kenyon TK, Trojanowski JQ. Transgenic animal models of tauopathies. Biochim Biophys Acta 2005; 1739:251-259.
35. Marsh JL, Thompson LM. Can flies help humans treat neurodegenerative diseases? Bioessays 2004; 26:485-496.
36. Shulman JM, Shulman LM, Weiner WJ et al. From fruit fly to bedside: Translating lessons from Drosophila models of neurodegenerative disease. Curr Opin Neurol 2003; 16:443-449.
37. Orr HT. Beyond the Qs in the polyglutamine diseases. Genes Dev 2001; 15:925-932.
38. Ross CA. Polyglutamine pathogenesis: Emergence of unifying mechanisms for Huntington's disease and related disorders. Neuron 2002; 35:819-822.
39. Trottier Y, Lutz Y, Stevanin G et al. Polyglutamine expansion as a pathological epitope in Huntington's disease and four dominant cerebellar ataxias. Nature 1995; 378:403-406.
40. Zoghbi HY, Orr HT. Glutamine repeats and neurodegeneration. Annu Rev Neurosci 2000; 23:217-247.
41. Ross CA. When more is less: Pathogenesis of glutamine repeat neurodegenerative diseases. Neuron 1995; 15:493-496.
42. Margolis RL, Ross CA. Expansion explosion: New clues to the pathogenesis of repeat expansion neurodegenerative diseases. Trends Mol Med 2001; 7:479-482.
43. Davies SW, Turmaine M, Cozens BA et al. Formation of neuronal intranuclear inclusions underlies the neurological dysfunction in mice transgenic for the HD mutation. Cell 1997; 90:537-548.
44. Mangiarini L, Sathasivam K, Seller M et al. Exon 1 of the HD gene with an expanded CAG repeat is sufficient to cause a progressive neurological phenotype in transgenic mice. Cell 1996; 87:493-506.
45. Ordway JM, Tallaksen-Greene S, Gutekunst CA et al. Ectopically expressed CAG repeats cause intranuclear inclusions and a progressive late onset neurological phenotype in the mouse. Cell 1997; 91:753-763.
46. Gusella JF, MacDonald ME. Molecular genetics: Unmasking polyglutamine triggers in neurodegenerative disease. Nat Rev Neurosci 2000; 1:109-115.
47. Sherman MY, Muchowski PJ. Making yeast tremble: Yeast models as tools to study neurodegenerative disorders. Neuromolecular Med 2003; 4:133-146.
48. Andrew SE, Goldberg YP, Kremer B et al. The relationship between trinucleotide (CAG) repeat length and clinical features of Huntington's disease. Nat Genet 1993; 4:398-403.
49. Brinkman RR, Mezei MM, Theilmann J. The likelihood of being affected with Huntington disease by a particular age, for a specific CAG size. Am J Hum Genet 1997; 60:1202-1210.
50. Bargmann CI. Neurobiology of the Caenorhabditis elegans genome. Science 1998; 282:2028-2033.
51. Bargmann CI, Kaplan JM. Signal transduction in the Caenorhabditis elegans nervous system. Annu Rev Neurosci 1998; 21:279-308.
52. Brownlee DJ, Fairweather I. Exploring the neurotransmitter labyrinth in nematodes. Trends Neurosci 1999; 22:16-24.
53. Lippincott-Schwartz J, Patterson GH. Development and use of fluorescent protein markers in living cells. Science 2003; 300:87-91.
54. White J, Stelzer E. Photobleaching GFP reveals protein dynamics inside live cells. Trends Cell Biol 1999; 9:61-65.
55. Scherzinger E, Sittler A, Schweiger K et al. Self-assembly of polyglutamine-containing huntingtin fragments into amyloid-like fibrils: Implications for Huntington's disease pathology. Proc Natl Acad Sci USA 1999; 96:4604-4609.
56. DiFiglia M, Sapp E, Chase KO et al. Aggregation of huntingtin in neuronal intranuclear inclusions and dystrophic neurites in brain. Science 1997; 277:1990-1993.
57. Perutz MF, Johnson T, Suzuki M et al. Glutamine repeats as polar zippers: Their possible role in inherited neurodegenerative diseases. Proc Natl Acad Sci USA 1994; 91:5355-5358.
58. Ross CA, Poirier MA, Wanker EE et al. Polyglutamine fibrillogenesis: The pathway unfolds. Proc Natl Acad Sci USA 2003; 100:1-3.
59. Kim S, Nollen EA, Kitagawa K et al. Polyglutamine protein aggregates are dynamic. Nat Cell Biol 2002; 4:826-831.
60. Tsien RY. The green fluorescent protein. Annu Rev Biochem 1998; 67:509-544.

61. Berney C, Danuser G. FRET or no FRET: A quantitative comparison. Biophys J 2003; 84:3992-4010.
62. Ross CA, Poirier MA. Protein aggregation and neurodegenerative disease. Nat Med 2004; 10:S10-17.
63. White JG, Southgate E, Thomson JN et al. The Structure of the nervous system of the nematode Caenorhabditis elegans. Phil Trans Royal Soc London Series B Biol Scien 1986; 314:1-340.
64. White JG, Southgate E, Thomson JN et al. The structure of the ventral nerve cord of Caenorhabditis elegans. Philos Trans R Soc Lond B Biol Sci 1976; 275:327-348.
65. Rand JB, Nonet ML. Chapter 22. Synatptic Transmission. In: Riddle DL, Blumenthal T, Meyer BJ, Priess JR, eds. C. Elegans II. Plainview, NY: Cold Spring Harbor Laboratory Press, 1997.
66. Morley JF, Brignull HR, Weyers JJ et al. The threshold for polyglutamine-expansion protein aggregation and cellular toxicity is dynamic and influenced by aging in Caenorhabditis elegans. Proc Natl Acad Sci USA 2002; 99:10417-10422.
67. Guarente L, Kenyon C. Genetic pathways that regulate ageing in model organisms. Nature 2000; 408:255-262.
68. Morris JZ, Tissenbaum HA, Ruvkun G. A phosphatidylinositol-3-OH kinase family member regulating longevity and diapause in Caenorhabditis elegans. Nature 1996; 382:536-539.
69. Lin K, Dorman JB, Rodan A et al. daf-16: An HNF-3/forkhead family member that can function to double the life-span of Caenorhabditis elegans. Science 1997; 278:1319-1322.
70. Ogg S, Paradis S, Gottlieb S et al. The Fork head transcription factor DAF-16 transduces insulin-like metabolic and longevity signals in C. elegans. Nature 1997; 389:994-999.
71. Garigan D, Hsu AL, Fraser AG et al. Genetic analysis of tissue aging in Caenorhabditis elegans: A role for heat-shock factor and bacterial proliferation. Genetics 2002; 161:1101-1112.
72. Hsu AL, Murphy CT, Kenyon C. Regulation of aging and age-related disease by DAF-16 and heat-shock factor. Science 2003; 300:1142-1145.
73. Morley JF, Morimoto RI. Regulation of longevity in Caenorhabditis elegans by heat shock factor and molecular chaperones. Mol Biol Cell 2004; 15:657-664.
74. Cummings CJ, Sun Y, Opal P et al. Over-expression of inducible HSP70 chaperone suppresses neuropathology and improves motor function in SCA1 mice. Hum Mol Genet 2001; 10:1511-1518.
75. Fernandez-Funez P, Nino-Rosales ML, de Gouyon B et al. Identification of genes that modify ataxin-1-induced neurodegeneration. Nature 2000; 408:101-106.
76. Warrick JM, Chan HY, Gray-Board GL et al. Suppression of polyglutamine-mediated neurodegeneration in Drosophila by the molecular chaperone HSP70. Nat Genet 1999; 23:425-428.
77. Carmichael J, Chatellier J, Woolfson A et al. Bacterial and yeast chaperones reduce both aggregate formation and cell death in mammalian cell models of Huntington's disease. Proc Natl Acad Sci USA 2000; 97:9701-9705.
78. Chai Y, Koppenhafer SL, Bonini NM et al. Analysis of the role of heat shock protein (Hsp) molecular chaperones in polyglutamine disease. J Neurosci 1999; 19:10338-10347.
79. Fire A, Xu S, Montgomery MK et al. Potent and specific genetic interference by double-stranded RNA in Caenorhabditis elegans. Nature 1998; 391:806-811.
80. Wang J, Barr MM. RNA interference in Caenorhabditis elegans. Methods Enzymol 2005; 392:36-55.
81. Fraser AG, Kamath RS, Zipperlen P et al. Functional genomic analysis of C. elegans chromosome I by systematic RNA interference. Nature 2000; 408:325-330.
82. Kamath RS, Fraser AG, Dong Y et al. Systematic functional analysis of the Caenorhabditis elegans genome using RNAi. Nature 2003; 421:231-237.
83. Nollen EA, Garcia SM, van Haaften G et al. Genome-wide RNA interference screen identifies previously undescribed regulators of polyglutamine aggregation. Proc Natl Acad Sci USA 2004; 101:6403-6408.
84. Brown CR, Hong-Brown LQ, Welch WJ. Correcting temperature-sensitive protein folding defects. J Clin Invest 1997; 99:1432-1444.
85. Pedersen CB, Bross P, Winter VS et al. Misfolding, degradation, and aggregation of variant proteins. The molecular pathogenesis of short chain acyl-CoA dehydrogenase (SCAD) deficiency. J Biol Chem 2003; 278:47449-47458.
86. Van Dyk TK, Gatenby AA, LaRossa RA. Demonstration by genetic suppression of interaction of GroE products with many proteins. Nature 1989; 342:451-453.
87. Gengyo-Ando K, Kagawa H. Single charge change on the helical surface of the paramyosin rod dramatically disrupts thick filament assembly in Caenorhabditis elegans. J Mol Biol 1991; 219:429-441.
88. Clark SG, Shurland DL, Meyerowitz EM et al. A dynamin GTPase mutation causes a rapid and reversible temperatureinducible locomotion defect in C. elegans. Proc Natl Acad Sci USA 1997; 94:10438-10443.

89. Eisenmann DM, Kim SK. Mechanism of activation of the Caenorhabditis elegans ras homologue let-60 by a novel, temperature-sensitive, gain-of-function mutation. Genetics 1997; 146:553-565.
90. Goff SA, Goldberg AL. Production of abnormal proteins in E. coli stimulates transcription of lon and other heat shock genes. Cell 1985; 41:587-595.
91. Morimoto RI. Regulation of the heat shock transcriptional response: Cross talk between a family of heat shock factors, molecular chaperones, and negative regulators. Genes Dev 1998; 12:3788-3796.
92. Rutherford SL, Lindquist S. Hsp90 as a capacitor for morphological evolution. Nature 1998; 396:336-342.
93. Sachidanandam R, Weissman D, Schmidt SC et al. A map of human genome sequence variation containing 1.42 million single nucleotide polymorphisms. Nature 2001; 409:928-933.
94. Wexler NS, Lorimer J, Porter J et al. Venezuelan kindreds reveal that genetic and environmental factors modulate Huntington's disease age of onset. Proc Natl Acad Sci USA 2004; 101:3498-3503.
95. Oliver CN, Ahn BW, Moerman EJ et al. Age-related changes in oxidized proteins. J Biol Chem 1987; 262:5488-5491.
96. McCampbell A, Taylor JP, Taye AA et al. CREB-binding protein sequestration by expanded polyglutamine. Hum Mol Genet 2000; 9:2197-2202.
97. Muchowski PJ, Schaffar G, Sittler A et al. Hsp70 and hsp40 chaperones can inhibit self-assembly of polyglutamine proteins into amyloid-like fibrils. Proc Natl Acad Sci USA 2000; 97:7841-7846.
98. Perez MK, Paulson HL, Pendse SJ et al. Recruitment and the role of nuclear localization in polyglutamine-mediated aggregation. J Cell Biol 1998; 143:1457-1470.
99. Ii K, Ito H, Tanaka K et al. Immunocytochemical colocalization of the proteasome in ubiquitinated structures in neurodegenerative diseases and the elderly. J Neuropathol Exp Neurol 1997; 56:125-131.
100. Bence NF, Sampat RM, Kopito RR. Impairment of the ubiquitin-proteasome system by protein aggregation. Science 2001; 292:1552-1555.
101. Gabai VL, Meriin AB, Yaglom JA et al. Role of Hsp70 in regulation of stress-kinase JNK: Implications in apoptosis and aging. FEBS Lett 1998; 438:1-4.
102. Gu M, Gash MT, Mann VM et al. Mitochondrial defect in Huntington's disease caudate nucleus. Ann Neurol 1996; 39:385-389.
103. Holmberg CI, Staniszewski KE, Mensah KN et al. Inefficient degradation of truncated polyglutamine proteins by the proteasome. EMBO J 2004; 23:4307-4318.
104. Miller KG, Alfonso A, Nguyen M et al. A genetic selection for Caenorhabditis elegans synaptic transmission mutants. Proc Natl Acad Sci USA 1996; 93:12593-12598.

CHAPTER 16

Hsp90 and Developmental Networks

Suzannah Rutherford,* Jennifer R. Knapp and Peter Csermely

Abstract

The most abundant cytoplasmic chaperone of eukaryotic cells, Hsp90 is a hub in developmental regulatory networks and the first example described of the phenomenon of molecular buffering. As a chaperone for many different signaling proteins, Hsp90 maintains the clarity and strength of communication within and between cells, concealing developmental and stochastic variations that otherwise cause abrupt morphological changes in a large variety of organisms, including *Drosophila* and *Arabidopsis*. The chapter provides a framework for understanding how Hsp90 controls the sudden appearance of novel morphologies. We start with a discussion of the longstanding problem of hidden polygenic variation and then introduce the idea of signal transduction thresholds in mediating the effect of Hsp90 on the expression of phenotypic variation. This leads to a discussion of the role of nonlinearity in creating thresholds for sudden changes in cellular responses to developmental signals. We end with speculation on the potentially pivotal role of Hsp90 in controlling the developmental networks that determine morphological stasis and change in evolution.

Introduction

Hsp90 is a hub in developmental regulatory networks. By maintaining the activity of over 150 signal transduction proteins in many different developmental pathways, the Hsp90 chaperone controls the strength of signaling, not only by its client proteins, but through the pathways in which they reside. Hsp90 target pathways regulate multiple processes including cell cycle and transcriptional control, as well as chromatin remodeling, growth control, apoptosis, stress responses and response to differentiation signals.[1-3] Many Hsp90-dependent pathways are both ancient and conserved. For example, Hsp90 controls the activity of four of eight ancient signaling pathways that arose before the protostome-deuterostome split. These pathways form a "basic evolutionary toolkit"[4] that is thought to have enabled the evolution of most of bilatarian diversity (Table 1). It is not yet understood how the same genes and pathways produce such dramatically different morphologies. One possibility is that evolution occurs via small changes in the structure and connectivity of signal transduction networks, the intricate web of cellular communications and responses that orchestrate development.

Perhaps surprising given their intricacy, biological networks are by nature highly error tolerant and self-correcting. Part of this stability has to do with positive and negative feedback and the complexity of inter-connections. In addition, biological networks are "scale-free". The number of interactions between nodes (genes or proteins) follows a power-law distribution - meaning that a few percent of nodes (such as Hsp90) have huge numbers of interactions, but most

*Corresponding Author: Suzannah Rutherford—Division of Basic Sciences, Fred Hutchinson Cancer Research Center, Mailstop A2-168, 1100 Fairview Avenue North, Seattle, Washington 98109-1024, U.S.A. Email: srutherf@fhcrc.org

Molecular Aspects of the Stress Response: Chaperones, Membranes and Networks, edited by Peter Csermely and László Vígh. ©2007 Landes Bioscience and Springer Science+Business Media.

Table 1. *A conserved group of signaling pathways, believed to have arisen before the protostome-deuterostome split, has been deployed to different ends throughout bilateran development. Shown are the eight major pathways and examples of Hsp90 client proteins imbedded in each.*

Pathway	Hsp90 Targets	References
Hedgehog (Hh)	–	–
Wingless related (Wnt)	–	–
Transforming growth factor B	Tak1*	50
Receptor tyrosine kinase (RTK)	IKK; ERB2; AKT	51-54
Notch	–	–
JAK/STAT	Stat1, Stat3	55
Nuclear hormone pathways	Nuclear hormonre receptors	56
Apoptosis	Caspase 9, Bid	57,58

*Tak1 is a noncanonical effector of TGFB signaling.

nodes having very few interactions (~2).[5] The resulting behavior of scale-free networks is therefore robust to the random loss of a high percent of nodes.[6] However rare, highly-connected nodes such as Hsp90 are control points whose loss is devastating to network function. The "structural stability, dynamic behavior, robustness and error and attack tolerance" depend on the integrity of highly connected nodes (hubs), particularly those positioned high in the hierarchy of network structure (Fig. 1).[7,8] For example, hubs such as Hsp90 provide a connection between many different modules of the network (meaning clusters of nodes more interconnected among themselves than to nodes in other modules).

Several years ago, one of the authors (S. Rutherford) discovered that modest (≤50%) reduction of Hsp90 uncovers a remarkable reserve of previously hidden morphogenic variation in *Drosophila*.[9] Hsp90-buffered changes are highly dependent on genetic background and results from the segregation, in normal populations, of likely hundreds to thousands of alleles with generally small effects on the strength of signaling through Hsp90-dependent pathways. Normally this genetic variation is hidden, and therefore has neutral effects on fitness. However, when the strength of Hsp90 signaling pathways is reduced beyond critical threshold levels, normally stable phenotypes can abruptly change.[10] Waddington called the error tolerance and self-correcting behavior of developmental processes canalization.[11,12] In line with Waddington's original concept of canalization, Hsp90-dependent changes in morphology are not quantitative in nature, but are either highly discrete or qualitative (Rutherford et al, submitted for publication). Indeed, a characteristic and almost unifying feature of life is its highly discrete nature. From species to individuals, organs, cell types and physiological and sub-cellular systems: perturbations change probabilities of alternative types, but rarely result in the generation of intermediate forms.

Here we present a framework for understanding Hsp90 buffering and variation in development. We begin with a discussion of the longstanding problem of hidden genetic variation. We then cover Hsp90 and signal transduction thresholds, the role of nonlinearity in creating thresholds in development, and end with speculation on Hsp90 potentially pivotal role in controlling the networks that control the balance of developmental stasis and change.

Hidden Genetic Variation

"We are largely ignorant of the answer to a question of fundamental importance: *How much [and which] variation has direct phenotypic or functional effects that influence the survival and reproduction of individuals?*" Quote from special committee charged with evaluating the Human Genome Diversity Project.[13]

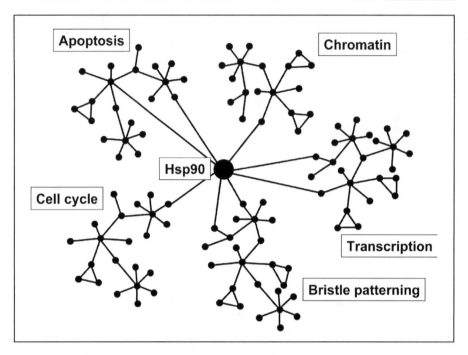

Figure 1. Meta-network of cellular and developmental processes (modules or sub-networks) controlled by Hsp90. A hierarchical network model predicts many properties of Hsp90 buffering, such as control of the modularity, robustness and balance between the functions making up the meta-network of development. Within each module Hsp90 has targets in both activating and inhibitory pathways, suggesting that balance within each module is also maintained.

Estimates of nucleotide polymorphism in *Drosophila* or human populations suggest that there are hundreds of thousands to millions of nucleotide differences between the genomes of unrelated individuals.[14,15] However, extensive surveys of wild populations of flies reveal very little variation in phenotype.[16] Indeed, the Neutral Theory asserts that a large amount of the genetic variation in populations has little or no effect on phenotype.[17] Yet a substantial, if unknown, amount of silent variation is 'conditionally cryptic' and can contribute to phenotypic variance or disease in the appropriate context.[18-20] Many factors allow the accumulation and maintenance of cryptic variation. It has long been known that under differing environmental or genetic conditions, latent genetic variation is expressed.[20] The hidden genetic variance attributable to conditionally cryptic alleles has traditionally been the province of evolutionary and quantitative genetics. For example, dominance interactions between alleles at a given locus, epistatic interaction between alleles at different loci, and genotype-by-environment interactions all buffer the expression of genetic variation.[18] However, the molecular basis of natural quantitative variation, whether conditionally cryptic or not, remains an open question.[21]

Early *Drosophila* geneticists attempted to catalog second-site modifier mutations and environmental effects on mutant phenotypes. However, both expressivity (severity) and penetrance (fraction affected) of mutant phenotypes are altered by so numerous factors it seems likely that their sheer number and the difficulty of their isolation eventually simply overwhelmed the early efforts. The gradual reduction in the phenotypic expression of deleterious mutations by ongoing fitness selection in the mutant stocks has been extensively documented. Perhaps most dramatic is the polygenic suppression of *Drosophila eyeless* (*ey*) mutations. When first isolated, *ey* mutations cause the complete loss of visible eye structures. The *ey* mutations are never-the-less extremely sensitive to environmental effects, and are easily suppressed by the rapid accumulation of ubiquitous and abundant polygenic modifiers that restore a normal eye structure and vision.[22] Outcrossing these stocks again restores the *ey* phenotype and shows that the mutatation was still present, but "hidden" by modifier alleles. As eye structure is normally invariant, these modifier alleles normally constitute cryptic genetic variation with hidden potential to affect eye development. Perhaps most astounding, ectopic expression of the *ey*/Pax-6 transcription factor drives the formation of *Drosophila* eye structures in multiple tissues of the adult fly (for example on wings, legs, antennae).[23] Because Pax-6 function is both necessary and sufficient for eye development, and is highly conserved throughout evolution, it has been called the "master control gene" for eye development.[24] While Pax-6 may be the one of the best examples of a master control gene, its singularity and importance is questionable as its leading role in this process can be by-passed by polygenic variants and environmental effects.[10] These studies demonstrate the abundance of developmental variation and alternate routes to the development of indistinguishable morphological endpoints.

Conversely, the increased variation in phenotype revealed by major developmental mutations is common and well-documented, and is a form of gene-interaction or epistasis.[10,25] Reducing Hsp90 either genetically, or with an inhibitor, has pleiotropic effects on many morphogenic processes, uncovering variation depending on both Hsp90 reduction and its interaction with trait-specific genetic backgrounds. The diversion of Hsp90 to proteins damaged by heat or other stress provides environmental control of Hsp90-dependent genetic and genotype-by-environment variation. Recent work has shown that other chaperones involved in protein folding, such as Hsp70 and Hsp60 family members, also buffer genetic variation, likely through a direct molecular effect on sequence variants that cannot fold normally without their assistance.[26,27] By contrast, we believe Hsp90 effects are largely indirect.[28] Rather than buffering variant client proteins directly, allowing them to fold, accumulating evidence indicates that Hsp90 acts through enhancing and suppressing effects of genetic variation and the environment on the strength and fidelity of developmental signaling pathways.

Hsp90 and Signal Transduction Thresholds

Our thinking about Hsp90-buffered variation centers on the idea of thresholds for the expression of phenotypes in response to continuously varying strengths of signaling through Hsp90 target pathways. The genetic interactions of Hsp90 with client proteins in signal transduction pathways demonstrate the existence of thresholds, and are easily understood in light of the large body of previous work on the evolutionarily conserved biochemical and genetic requirement for Hsp90 by specific client proteins.[1-3] When Hsp90 levels are decreased by mutation or by inhibitors, signal transduction clients begin to loose activity, and the strength of target pathways becomes severely reduced.[29-31] Threshold trait models were developed in the context of quantitative and population genetics and are often used to describe the genetics of human disease.[32] In development, thresholds are transitions between continuous inputs and discrete outputs. Many binary cell-fate decisions are controlled by thresholds. Specific genetic interactions between Hsp90 and signaling pathways dominantly reveal phenotypic thresholds. Loss of Hsp90 function is lethal to most cells in most organisms, in *Drosophila* heterozygous (recessive) Hsp90 mutations dominantly enhance and suppress sensitizing (cryptic) mutations in many Hsp90 client pathways. Hsp90 heterozygotes fail to complement Cdc37 heterozygous mutations,[33] suppress an over-activated phenotype of the Torso receptor tyrosine kinase,[34] enhance Cdc2 mutant heterozygotes[35] and enhance or suppress under- and over-activated components of the *sevenless* (*sev*) mitogen activated protein kinase (MAPK) pathway.

The well-characterized MAPK signaling pathways initiated at the Drosophila *sev* receptor is a model for Hsp90 interactions with signaling near trait thresholds. High-level activation of the *sev* pathway promotes the normal photoreceptor cell fate, while lower-level activation of the pathway results in a default nonneuronal fate and loss of a photoreceptor. This decision is an all-or-none switch; no intermediate cell-types are found. If signaling is too high or in the wrong place, extra photoreceptors are produced, giving flies a gain-of-function rough eye phenotype. A series of molecular genetic screens were instrumental in identifying the components of receptor tyrosine kinase signaling pathways. Endogenous members of the *sev* cascade were replaced with marginally over- or under-activated analogs and the temperature adjusted so that flies have normal or gain-of-function phenotypes.[35-37] Heterozygous mutations in other components of the pathway (*m*/+), which are normally 'cryptic' recessive mutations, behave like dominant enhancer or suppressor mutations in the appropriate genetic background. Tellingly, when *sev* signaling is reduced toward a lower threshold, Hsp90 mutations also reduce signaling further, resulting in the loss-of-function phenotype seen with heterozygous mutations in other pathway members.[35,36] Similarly, when the *sev* pathway is over-activated beyond an upper threshold (gain of function phenotype), heterozygous Hsp90 mutations reduce activity and restore a normal, smooth eye phenotype.[38,39] The genetic interaction of Hsp90 in many client pathways shows that it can reduce signaling near thresholds for the expression of mutant phenotypes. If target pathways are sensitive to reduction of Hsp90 client functions, the same result is produced simply reducing Hsp90 activity.

The level of extrinsic or intrinsic noise in signaling and the cell sensitivity could also modulate the expression of phenotypes near signal transduction thresholds. This phenomenon is called either stochastic resonance or stochastic focusing, depending on the whether the noise is extrinsic or intrinsic, respectively. If noise is low, a low level signal may not pass the signal transduction threshold. On the contrary, if the noise is high, the same low level signal will pass the same threshold a higher fraction of the time (Fig. 2).[40,41] By decreasing the "noisiness" of cellular signals, molecular chaperones would also modulate the sensitivities of signal transduction near trait thresholds. When Hsp90 is inhibited, increased developmental and/or purely environmental noise may allow normally-lower signals to pass the same sensitivity thresholds, increasing phenotypic diversity at the whole-cell, or whole-organism level.[42]

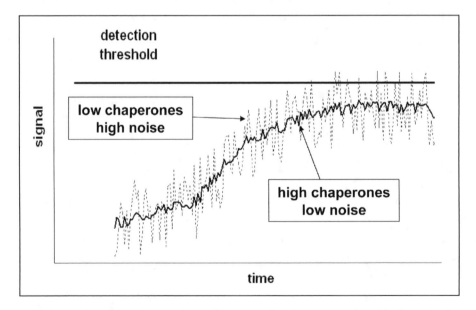

Figure 2. Stochastic resonance as a possible reason for chaperone-induced buffering in the diversity of developmental changes. Stochastic resonance is the phenomenon, where intrinsic or extrinsic noise helps a low level signal to surpass a signal transduction threshold. If molecular chaperones decrease the noise of cellular processes, this may also contribute to their buffering effects of developmental changes by not allowing low-level signals to act.

Nonlinearity in Developmental Responses to Signal Transduction

The ability to produce all-or-none or switch like behavior in response to continuous variation in the strength of the underlying developmental signal may be a common theme in biology.[40,43] For example, the addition of upstream kinases in MAPK signaling creates a sigmoidal (threshold) response over a narrow range.[44] The addition of positive feedback can sharply amplify a graded input to produce a steep and sudden ("switch-like") change in output.[45] We suggest that such nonlinearity in the response of phenotype to underlying developmental signals is a common feature of development, and produces much of the observed discrete behavior of biological systems. Given nonlinearity in Hsp90 target pathways, when Hsp90 is reduced, signaling is reduced. We believe that the sudden changes in morphology, or changes in previously invariant quantitative traits reflects the fact that thresholds have been crossed.

A Pivotal Role for Hsp90 in Network Evolvability?

The activity of signal transduction clients and therefore client pathways decrease sharply with modest decreases in Hsp90 function. Through these target proteins, Hsp90 simultaneously controls output of multiple target pathways imbedded in many different developmental processes such as regulation of cell-cycle, apoptosis or differentiation responses. Each of these processes is controlled by a complex sub-network of signaling pathways that integrate developmental cues to generate appropriate responses. Hsp90 may be uniquely positioned to balance the stability and modularity of the 'meta-network' of processes that make up development, linking network sensitivity and behavior to the environment.

If the meta-network of development were a mobile with groups of hanging objects (nodes), we suggest that Hsp90 would be at the center of the mobile, supporting several groups of objects ('sub-networks' or modules). Linkages between each sub network and Hsp90 are represented by the strings that support the mobile. The effective concentration of Hsp90 available for signal transduction determines the overall height (activity) of the meta-network, and the connectivity or separation of the modules—complete loss of Hsp90 world fragment the modules into isolated sub-networks. By reducing or increasing several target protein functions simultaneously, we suggest that Hsp90 maintains the balance of the system. The overall height (activity) between the subnetworks is shifted ensemble with the degree of environmental stress.

Pushing the metaphor a little further, we suggest that Hsp90 also maintains balance within each sub-network. For example, both cdc28 (a key activator of the cell cycle) and wee1 (an inhibitory kinase that acts on cdc28) are Hsp90 targets. Decreases in both activating and inhibiting pathways would balance out the effect of varying Hsp90 on the rate of the cell cycle. It seems reasonable to expect that in most cases selection has sculpted the set of targets in activating and inhibitory pathways such that the balance of their function is maintained. Normal changes in Hsp90 availability would have no net effect. However, if natural variation or the environment has already destabilized the network balance, bringing an activating or inhibitory pathway close to its threshold, further reduction of signaling by loss of Hsp90 function could result in thresholds being breached and previously hidden phenotypes revealed.

Extending our thoughts beyond the role of Hsp90 to a general molecular buffering effect of chaperones, we have to note that chaperones provide low affinity, transient contacts with other proteins. Weak links are known to help system stability in a large variety of networks from macromolecules to social networks and ecosystems, which may be a general network-level phenomenon explaining many of the genetic buffering effects.[42,46] Besides molecular buffering by Hsp90 and other chaperones, a large number of other mechanisms may also control the diversity of the phenotype.[25,47-49] A single common mechanism, such as involvement in signaling or epigenetic modifications of histones and DNA structure, seems unlikely to explain all of the effects observed. If a general explanation is sought, we believe it is more likely to be related to the overall structure of developmental networks and their emergent properties.

Acknowledgements

Work in the authors' laboratory was supported by research grants from NIH R01 GM068873 the EU (FP6506850, FP6-016003), Hungarian Science Foundation (OTKA-T37357 and OTKA-F47281), Hungarian Ministry of Social Welfare (ETT-32/03), Hungarian National Research Initiative (1A/056/2004 and KKK-0015/3.0).

References

1. Csermely P, Schnaider T, Soti C et al. The 90-kDa molecular chaperone family: Structure, function, and clinical applications. A comprehensive review. Pharmacol Ther 1998; 79:129-168.
2. Pratt WB, Toft DO. Regulation of signaling protein function and trafficking by the hsp90/hsp70-based chaperone machinery. Exp Biol Med (Maywood) 2003; 228:111-133.
3. Zhao R, Davey M, Hsu YC et al. Navigating the chaperone network: An integrative map of physical and genetic interactions mediated by the hsp90 chaperone. Cell 2005; 120:715-727.
4. Knoll AH, Carroll SB. Early animal evolution: Emerging views from comparative biology and geology. Science 1999; 284:2129-2137.
5. Jeong H, Tombor B, Albert R et al. The large-scale organization of metabolic networks. Nature 2000; 407:651-654.
6. Albert R, Jeong H, Barabasi AL. Error and attack tolerance of complex networks. Nature 2000; 406:378-382.
7. Ravasz E, Somera AL, Mongru DA et al. Hierarchical organization of modularity in metabolic networks. Science 2002; 297:1551-1555.
8. Barabási AL. Linked: The New Science of Networks. Cambridge: Perseus Pub, 2002:280.
9. Rutherford SL, Lindquist S. Hsp90 as a capacitor for morphological evolution. Nature 1998; 396:336-342.
10. Rutherford SL. From genotype to phenotype: Buffering mechanisms and the storage of genetic information. Bioessays 2000; 22:1095-1105.
11. Waddington CH. The Strategy of the Genes; a Discussion of Some Aspects of Theroetical Biology. Vol. ix. New York: Macmillan, 1957:262.
12. Waddington C. Evolutionary systems - Animal and human. Nature 1950; 183:1634-1638.
13. Evaluating Human Genetic Diversity. Commission on Life Sciences, National Research Council. Washington, DC: National Academy Press, 1997.
14. Begun DJ, Aquadro CF. Levels of naturally occurring DNA polymorphism correlate with recombination rates in D. melanogaster. Nature 1992; 365:519-520.
15. Moriyama EN, Powell JR. Intraspecific nuclear DNA variation in Drosophila. Mol Biol Evol 1996; 13:261-277.
16. Powell JR. Progress and Prospects in Evolutionary Biology: The Drosophila Model. Vol. xiv. New York: Oxford University Press, 1997:562.
17. Kimura M. The Neutral Theory of Molecular Evolution. Vol. xv. Cambridge [Cambridgeshire], New York: Cambridge University Press, 1983:367.
18. Lynch M, Walsh B. Genetics and Analysis of Quantitative Traits. Vol. xvi. Sunderland: Sinauer, 1998:980.
19. Hartman JL, Garvik B, Hartwell L. Principles for the buffering of genetic variation. Science 2001; 291:1001-1004.
20. Perera FP. Environment and cancer: Who are susceptible? Science 1997; 278:1068-1073.
21. Mackay TF. Quantitative trait loci in Drosophila. Nat Rev Genet 2001; 2:11-20.
22. Morgan TH. Variability of eyeless. Publs Carnegie Instn 1929; 399:139-168.
23. Halder G, Callaerts P, Gehring WJ. Induction of ectopic eyes by targeted expression of the eyeless gene in Drosophila. Science 1995; 267:1788-1792.
24. Gehring WJ. The master control gene for morphogenesis and evolution of the eye. Genes Cells 1996; 1:11-15.
25. Bergman A, Siegal ML. Evolutionary capacitance as a general feature of complex gene networks. Nature 2003; 424:549-552.
26. Fares MA, Ruiz-Gonzalez MX, Moya A et al. Endosymbiotic bacteria: groEL buffers against deleterious mutations. Nature 2002; 417:398.
27. Chow KC. Hsp70 (DnaK) — An evolution facilitator? Trends Genet 2000; 16:484-485.
28. Rutherford SL. Between genotype and phenotype: Protein chaperones and evolvability. Nat Rev Genet 2003; 4:263-274.
29. Sreedhar AS, Soti C, Csermely P. Inhibition of Hsp90: A new strategy for inhibiting protein kinases. Biochim Biophys Acta 2004; 1697:233-242.
30. Whitesell L, Mimnaugh EG, De Costa B et al. Inhibition of heat shock protein HSP90-pp60v-src heteroprotein complex formation by benzoquinone ansamycins: Essential role for stress proteins in oncogenic transformation. Proc Natl Acad Sci USA 1994; 91:8324-8328.

31. Zhang H, Burrows F. Targeting multiple signal transduction pathways through inhibition of Hsp90. J Mol Med 2004; 82:488-499.

32. Falconer DS, Mackay TFC. Introduction to Quantitative Genetics. Vol. xiii. Essex, England: Longman, 1996:464.

33. Kimura Y, Rutherford SL, Miyata Y et al. Cdc37 is a molecular chaperone with specific functions in signal transduction. Genes Dev 1997; 11:1775-1785.

34. Doyle H, Bishop J. Torso, a receptor tyrosine kinase required for embryonic pattern formation, shares substrates with the sevenless and EGF-R pathways in Drosophila. Genes Dev 1993; 7:633-646.

35. Cutforth T, Rubin GM. Mutations in Hsp83 and cdc37 impair signaling by the sevenless receptor tyrosine kinase in Drosophila. Cell 1994; 77:1027-1036.

36. Daga A, Banerjee U. Resolving the sevenless pathway using sensitized genetic backgrounds. Cell Mol Biol Res 1994; 40:245-251.

37. Simon M, Bowtell D, Dodson G et al. Ras1 and a putative guanine nucleotide exchange factor perform crucial steps in signaling by the sevenless protein tyrosine kinase. Cell 1991; 67:701-716.

38. Dickson BJ, van der Straten A, Dominguez M et al. Mutations Modulating Raf signaling in Drosophila eye development. Genetics 1996; 142:163-171.

39. van der Straten A, Rommel C, Dickson B et al. The heat shock protein 83 (Hsp83) is required for Raf-mediated signalling in Drosophila. EMBO J 1997; 16:1961-1969.

40. Hasty J, Pradines J, Dolnik M et al. Noise-based switches and amplifiers for gene expression. Proc Natl Acad Sci USA 2000; 97:2075-2080.

41. Paulsson J, Berg OG, Ehrenberg M. Stochastic focusing: Fluctuation-enhanced sensitivity of intracellular regulation. Proc Natl Acad Sci USA 2000; 97:7148-7153.

42. Csermely P. Weak Links: Stabilizers of Complex Systems from Proteins to Social Networks. Heidelberg: Springer Verlag, 2006.

43. Gardner TS, Cantor CR, Collins JJ. Construction of a genetic toggle switch in Escherichia coli. Nature 2000; 403:339-342.

44. Goldbeter A, Koshland Jr DE. Ultrasensitivity in biochemical systems controlled by covalent modification. Interplay between zero-order and multistep effects. J Biol Chem 1984; 259:14441-14447.

45. Ferrell Jr JE, Machleder EM. The biochemical basis of an all-or-none cell fate switch in Xenopus oocytes. Science 1998; 280:895-898.

46. Csermely P. Strong links are important, but weak links stabilize them. Trends Biochem Sci 2004; 29:331-334.

47. True HL, Berlin I, Lindquist SL. Epigenetic regulation of translation reveals hidden genetic variation to produce complex traits. Nature 2004; 431:184-187.

48. Wade MJ, Johnson NA, Jones R et al. Genetic variation segregating in natural populations of Tribolium castaneum affecting traits observed in hybrids with T. freemani. Genetics 1997; 147:1235-1247.

49. Sangster TA, Lindquist S, Queitsch C. Under cover: Causes, effects and implications of Hsp90-mediated genetic capacitance. Bioessays 2004; 26:348-362.

50. Bouwmeester T, Bauch A, Ruffner H et al. A physical and functional map of the human TNF-alpha/NF-kappa B signal transduction pathway. Nat Cell Biol 2004; 6:97-105.

51. Broemer M, Krappmann D, Scheidereit C. Requirement of Hsp90 activity for IkappaB kinase (IKK) biosynthesis and for constitutive and inducible IKK and NF-kappaB activation. Oncogene 2004; 23:5378-5386.

52. Xu W, Mimnaugh E, Rosser MF et al. Sensitivity of mature Erbb2 to geldanamycin is conferred by its kinase domain and is mediated by the chaperone protein Hsp90. J Biol Chem 2001; 276:3702-3708.

53. Chiosis G, Timaul MN, Lucas B et al. A small molecule designed to bind to the adenine nucleotide pocket of Hsp90 causes Her2 degradation and the growth arrest and differentiation of breast cancer cells. Chem Biol 2001; 8:289-299.

54. Fujita N, Sato S, Ishida A et al. Involvement of Hsp90 in signaling and stability of 3-phosphoinositide-dependent kinase-1. J Biol Chem 2002; 277:10346-10353.

55. Sehgal PB. Plasma membrane rafts and chaperones in cytokine/STAT signaling. Acta Biochim Pol 2003; 50:583-594.

56. Pratt WB, Welsh MJ. Chaperone functions of the heat shock proteins associated with steroid receptors. Semin Cell Biol 1994; 5:83-93.

57. Pandey P, Saleh A, Nakazawa A et al. Negative regulation of cytochrome c-mediated oligomerization of Apaf-1 and activation of procaspase-9 by heat shock protein 90. EMBO J 2000; 19:4310-4322.

58. Zhao C, Wang E. Heat shock protein 90 suppresses tumor necrosis factor alpha induced apoptosis by preventing the cleavage of Bid in NIH3T3 fibroblasts. Cell Signal 2004; 16:313-321.

Index